OCTOPUS

Physiology and Behaviour of an Advanced Invertebrate

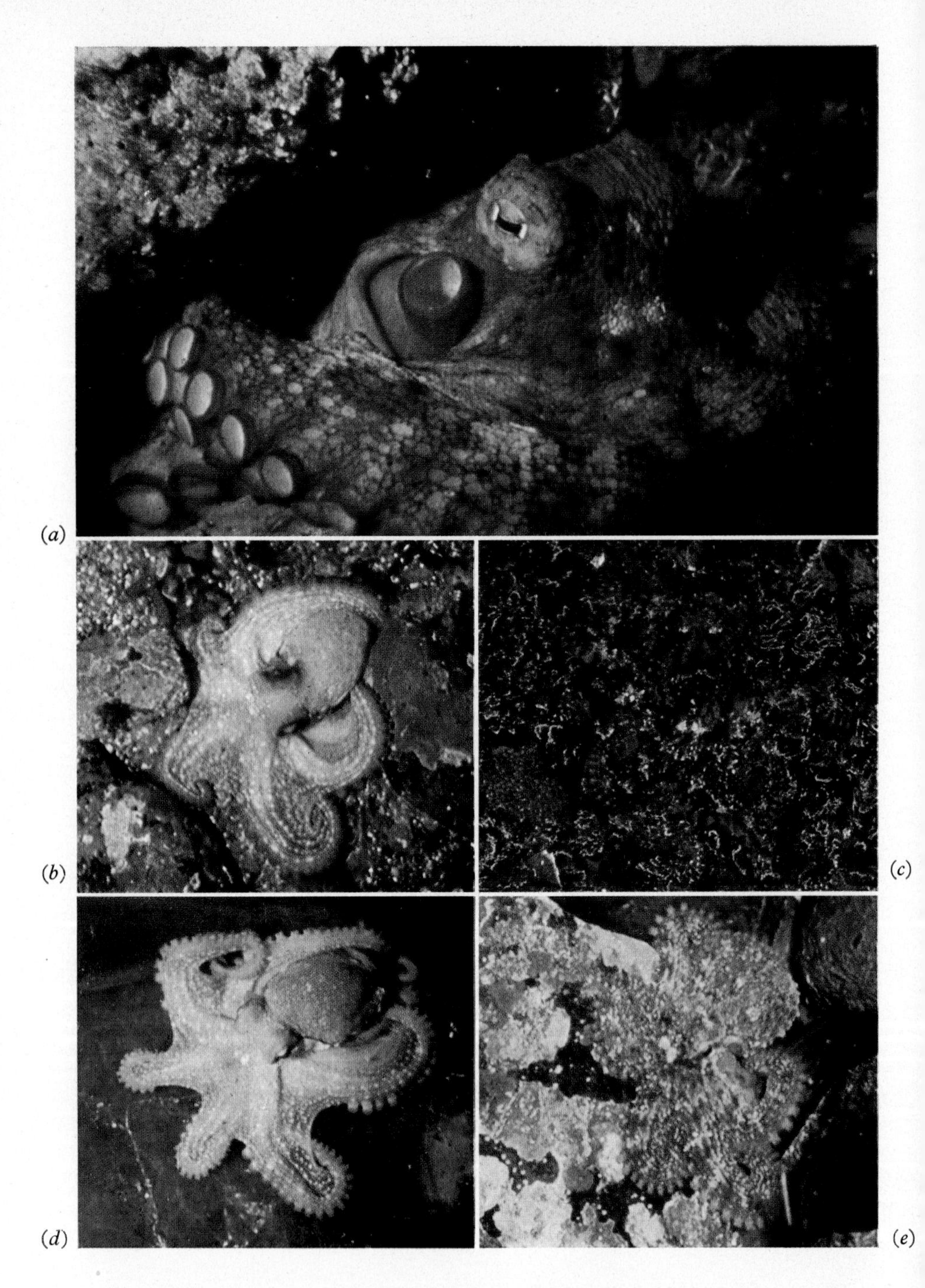

Frontispiece (*a*) A female *Octopus* of about 1 kg watching from her home in the sea. (*b*)–(*e*) Smaller individuals in aquaria, all save (*c*) are photographs of the same small (20 g) animal. (*b*) shows the dymantic response to a sudden movement overhead. (*c*) Animal among coralline algae, cryptic. (*d*) Chromatophores largely contracted, leaving the web translucent. (*e*) Typical mottled appearance of an octopus, cryptic against a broken background. (Photograph (*c*) from Packard and Sanders, 1969; rest by M.J.W.).

OCTOPUS

Physiology and Behaviour of an Advanced Invertebrate

M.J. WELLS
M.A., Sc.D.
Department of Zoology
University of Cambridge

LONDON
CHAPMAN AND HALL

A Halsted Press Book
John Wiley & Sons, New York

First published 1978
by Chapman and Hall Ltd
11 New Fetter Lane, London EC4P 4EE

Typeset by Hope Services
and printed in Great Britain by
University Printing House, Cambridge

ISBN 0 412 13260 6

Distributed in the U.S.A. by Halsted Press,
a Division of John Wiley and Sons, Inc., New York

Library of Congress Cataloging in Publication Data
Wells, Martin John.
Octopus.

Bibliography: p.
Includes index.
1. Octopus. I. Title.
QL430.3.02W44 594'.56 77-2795
ISBN 0-470-99197-6

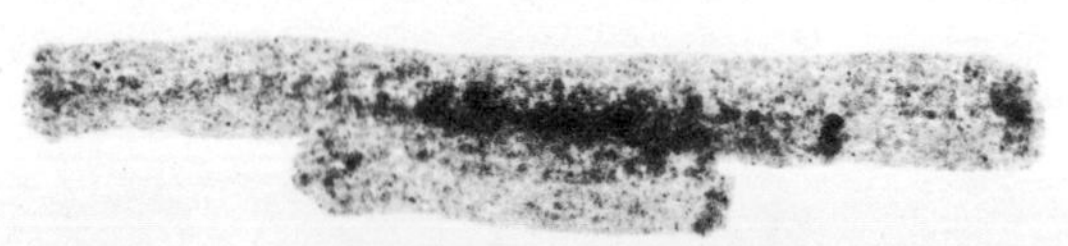

For J.W.

Contents

Acknowledgements

It is easy enough to start. Professor J.Z. Young and Dr Marion Nixon read and commented upon drafts of most of the chapters in this book; if mistakes occur they almost certainly lie in sections that I failed to show them; Dr Nixon has also helped me by compiling the index. John Rodford, of the Zoology Laboratory, Cambridge, has drawn the more complicated anatomical pictures, often from my own very rough drafts and Mrs Joy Schreiber did almost all of the typing, no small task in view of my execrable handwriting. I am grateful too to the following for permission to reproduce copyright material: Academic press; American Psychological Association; American Society of Zoologists; Ballière Tindall and Cassel; Cambridge University Press; Centre National de la Recherche Scientifique; Company of Biologists; Experimental Psychology Society; Gordon and Breach; John Wiley & Sons; Macmillan Journals; North-Holland Biomedical Press; Oxford University Press; Pergamon Press; Plenum Press; Royal Society of London; Society for the Experimental Analysis of Behaviour; Society for Experimental Biology; Springer-Verlag; Swedish National Research Council; Zoological Station, Naples and to a large number of individual authors; their names appear at the end of the captions to figures and tables taken or derived from their works. But the list of people to whom I am indebted does not really end there, either. The persons truly responsible for this book are the teachers whose enthusiasm for their subject convinced me that biology really was a worthwhile career; Carl Pantin and Mark Pryor at Cambridge, and I.T. Hamilton and Bill Coulson, who taught me as a schoolboy. It is thanks to them that I am here now, writing about things that fascinate me and still, at rising fifty, astonished that the world is somehow prepared to pay me to play with living things for a living.

November 1976

M.J.W.
Zoology Department, Cambridge

CHAPTER ONE
Introduction

1.1 Background to a brain

1.1.1 *Behaviour and physiology*

Textbooks of comparative physiology tend to begin with biochemistry and stop short of a discussion of animal behaviour. Their authors are obliged to handle things the way they do because they are hunting for statements that can be made about animals generally, and the further one proceeds towards the molecular basis of life, the easier it seems to be to do this. Their problem is how to hold things together after the first few chapters. Krebs cycle and the genetic code seem to apply to all of us, but the attempt to compile a universal story begins to go to pieces and the account becomes little more than a catalogue as soon as contact is made with the great range of solutions that have arisen at an organ rather than at a cellular level. The eyes, brains, guts and motor systems of animals always seem to include features unique to the creature in question which one could hardly expect to predict simply from a knowledge of its cell biology. Why was the basic molluscan design extended in the particular ways that we find in an octopus? Chemistry cannot tell us. But perhaps behaviour can. This is why this book, unfettered by the need to be comparative, includes much about behaviour, and rather little about biochemistry. If we examine the behaviour of cephalopods now, and the probable behaviour of cephalopods in the past, we can come a long way towards understanding their present structure and physiology.

Ethology points the way of escape from what at first sight may appear to be a Lamarckian heresy. The natural selection of behaviour patterns as signals shows repeatedly how particular movements, recognized by other members of the species, are rendered more conspicuous and effective by colouration or structure. Where this confers some advantage in communication – signalling sexual condition or bad temper – any mutant exaggeration of colour or structure will tend to be retained. Behaviour comes first; the peacocks tail is there because the success or failure of the peacock's behaviour provided a basis on which natural selection could act.

So too, it is arguable, with any other aspect of physiology. If a species is pressing up the estuaries, or onto the shore, there is selection for the capacity to osmoregulate. If the animal includes vegetable matter in its diet, there is a basis for the selection of enzyme systems that can digest plants, or the extension of parts of the gut to form cesspits for the accommodation of friendly bacteria. If a predator is already straining its sense organs and brain to the limit to trap a wary prey it is almost inevitable that the structure and physiology of these parts will change. The directions of the changes are, in broad terms, predictable. Given sufficient information about the starting conditions, the changes may be as predictable as those brought about by human selection of a breeding stock. Organ physiology, in short, must be considered and indeed perhaps can only be fully understood in terms of what the individuals of an animal species were trying to do with their bodies in the recent and remote past.

If, as seems inescapable, behaviour is determining the direction of evolution in this manner, it is also inevitable that the pressure on change will increase with central nervous elaboration. In particular the rate of change is liable to be stepped up enormously as soon as an animal type evolves the capacity to learn. The learning animal can discover from its own experience how to recognize the warning signals that show some physiological mechanism is being unacceptably stressed, and it can learn by experience how to react in the circumstances; the ache in the stomach teaches it what not to eat. A creature that can learn can nibble at the fringes of disaster, close to the edge of its physiological range without much likelihood that it will blunder into extinction. The closer the population as a whole can live to the limits set by its current physiology, the more rapidly can natural selection proceed. Behaviourally, the species is leaning against one edge of the bell-shaped curve of its own variability, determining the direction of its further evolution.

Octopus vulgaris, the subject of this monograph, is, whatever else, a learning animal and any account of its physiology must be related to this. It would be absurd to discuss its digestion or circulation – let alone the operations of its sense organs – without reference to the creature's life style since it is the life style of *Octopus* and its squid ancestors that has forced the evolution of the enormous differences between the organ systems of cephalopods and those of less ambitious molluscs. *Octopus* does, as we would predict, live close to the limits set by its own physiology.

The circulation, to take one example, is barely adequate for such an active animal, mainly because of the absence of any system for packaging the blood pigment; haemocyanin in solution is a poor oxygen carrier. Cephalopod blood can transport less than 5 millilitres of oxygen per 100 ml of blood (compared with about 15 vol% in fish) and the whole supercharged system of triple hearts, high blood pressure and pulsating blood vessels succeeds only in returning blood that retains less than 30% of its dissolved oxygen by the time it reaches the gills. This at rest; the effect of exercise is immediate and surprisingly long-lasting even in octopuses as small as 300 g, which must very swiftly run into oxygen debt when they flee from predators or pursue their prey (Sections 3.2.2, 3.2.4). Digestion, too would seem to be limiting. As with other molluscs, digestion in *Octopus* is based on secretion-absorption cycles by a massive diverticulum of the gut, an adequate system in a less hectic past, but scarcely appropriate in a predator that must be an opportunist in the matter of feeding. *Octopus* feeds mainly at night, and spends a great deal of every day sitting at home. Even so a digestive process that requires between eighteen and twenty-four hours to dispose of a crab must be sub-optimal. In some squids, where there is an even greater behavioural pressure on swift digestion, the absorptive function of the hepatopancreas has been abandoned altogether and the process takes only 4–6 hours (Section 4.3.7).

Further evidence that the octopus is a rapidly evolving creature comes from the systematics of the group. The genus *Octopus* includes more than 100 known species (Voss, 1977). Many of these are difficult to distinguish one from another. Sometimes the only obvious differences lie in such taxonomically awkward features as colour or the size of the eggs (Pickford and McConnaughy, 1949). Most descriptions are based on museum specimens. While this can sometimes lead to a multiplication of apparent species (*O. rugosus*, for example, turns out to be *O. vulgaris*, fixed with the skin papillae raised; Pickford, 1955) it is far more likely to lead to an underestimate in the case of a genus like *Octopus*, where so many of the important features are obliterated in the course of preservation. We know, for example, that transient patterns of chromatophore display are important in sex recognition and mating and that different species are active at different times of day. When behavioural isolation of sympatric forms and the differences in behaviour of the same morphological species from different habitats are taken into account, even the 100 or so species that systematists recognize today may have to be increased considerably.

Octopus vulgaris itself is both widespread and variable (Robson, 1929). Animals attributable to this species are found in temperate and tropical seas from the Mediterranean to Japan. Its food and habits and details of its antomy differ in ways that may or may not be significant in subdividing the population into groups now unlikely to interbreed. Again, we cannot be sure whether we are dealing with one good species, or several; it becomes a matter of definition. What is plain is that the shallow water octopods – a recent group, geologically speaking, with the first finned octopus in cretaceous rocks and the truly benthic littoral forms presumably much later than that – are still unequivocally 'on the make', speciating as they specialize and extend their range into the whole variety of ecological niches open to animals with such truly remarkable behaviour.

1.1.2 *The history of the cephalopods*

To understand how *Octopus* has come to be as it is it is necessary not only to consider what the animal is now trying to do with its body and what its recent ancestors were trying to do with theirs, but also to think back to the origins of the cephalopods and try to reconstruct the selective pressures acting upon the group since its appearance in late Cambrian times.

Packard (1972) has argued persuasively that much about the evolution of cephalopods can be understood once one realises that they have been competing directly with fish ever since these vertebrates first came down to the sea from freshwater.

In Ordovician and Silurian times, there was no vertebrate opposition (Fig. 1.1). The early cephalopods had chambered shells with a siphuncle and they presumably achieved neutral buoyancy by abstracting salts and allowing gas to replace fluid in the manner that Denton and Gilpin-Brown (1961, 1966) have described for *Sepia* and *Nautilus.* They were probably the first free-swimming animals of any considerable size. For millions of years before the fish evolved swim-bladders from lungs the cephalopods were enjoying an unrivalled freedom of manœuvre coupled with what was very possibly a virtual immunity from attack by creatures other than their own kind.

The situation began to deteriorate for them in the Devonian and

Carboniferous periods, with a rapid and still more significant stiffening of the competition at the beginning of the Mesozoic, as bony fish and

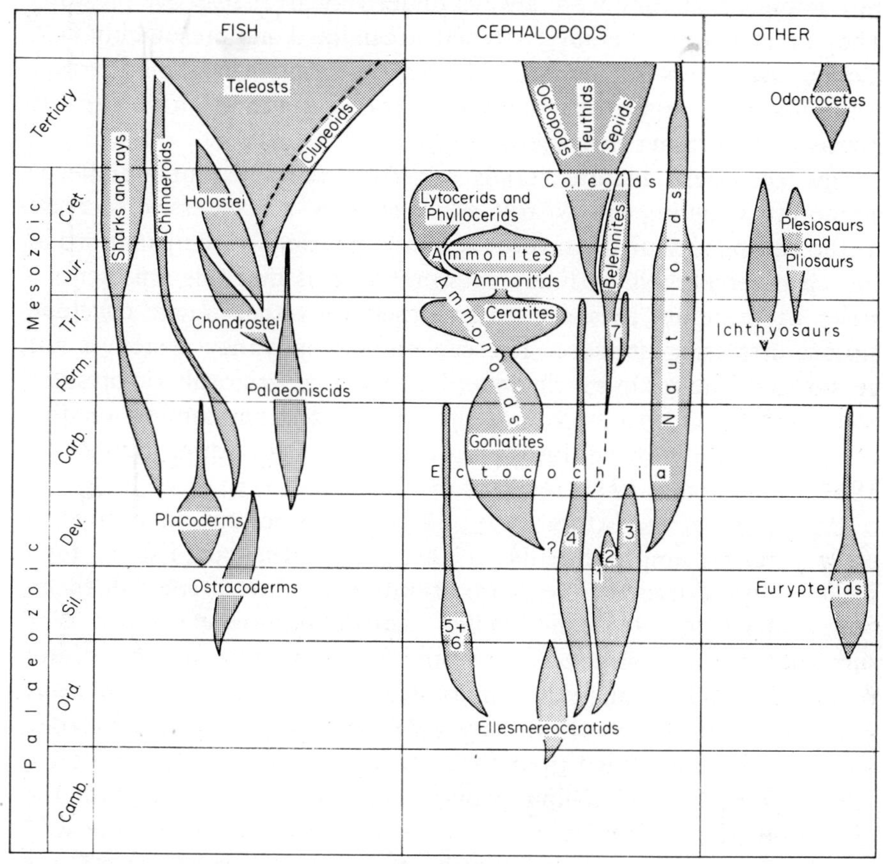

Fig. 1.1 Diagram of the evolution and distribution in time of the main cephalopod orders, fish, certain other vertebrates and one invertebrate group, the Eurypterids. Relative abundance is shown by horizontal extent. Early radiations from the Ellesmereoceratida are: 1. Discosorida, 2. Tarphycerida, 3. Oncocerida, 4. Orthocerida, Aulacocerida and Bactitida, 5. Actinocerida, 6. Endocerida. 4, 5 and 6 were straight-shelled (? fast swimming) forms. Derivation of the coleoids is obscure; teuthoids and sepioids may or may not have arisen independently of belemnitids; their origin could be considerably earlier than shown here (from Packard, 1972).

later reptiles spread into the seas. The cephalopod response was diversification and retreat into deeper water. Nautiloids persisted as heavily armoured forms, while the goniatites gave rise to an array of relatively light-shelled and presumably more manœuvrable animals, the ammonites. The ammonites enjoyed considerable success, for a considerable while, though their explosive radiations and high extinction rate implies that they were always under very great selective pressure. They ranged from laterally flattened streamlined and presumably fast-moving swimmers to vermetid-like sessile forms. The shell aperture was sometimes very restricted and at least some must have become microphagous rather than predatory.

One consistent feature stands out from the apparent muddle of ammonite experiments. Nearly all lines show a progressive increase in the complexity of the sutures defining the edges of the partitions in the chambered shell. A likely interpretation is that this represents a series of attempts to shore up an otherwise rather fragile envelope against increasing pressure. The cephalopod buoyancy system is not pressurized, the gas in the chamber is at less than atmospheric pressure and there is a constant risk of implosion. The ammonites, on this evidence, were moving into progressively deeper water (Donovan, 1964; Westermann, 1971; see Packard, 1972).

Driven out of the littoral into mid and deep oceanic environments by progressive improvements in the performance of fish (a fate, incidentally, shared by the more primitive families of the fish themselves; Marshall, 1965) the shelled cephalopods were trapped. The siphuncle cannot extract water from the chambers at depths greater than 240 m and although the ammonites, like the present-day *Nautilus* and *Spirula,* may sometimes have penetrated to greater depths, they could not have gas-filled their shells there (Denton and Gilpin-Bown, 1961, 1966; Denton, Gilpin-Brown and Howarth, 1967). With the exception of *Nautilus* itself, which persists successfully as a heavily armoured scavenging animal on the slopes of indopacific reefs, the tetrabranchiate coiled-shell cephalopods all died out by the end of the Mesozoic.

The eventual failure of the tetrabranchiates did not, however, mean the end of the cephalopods. In late Palaeozoic-early Mesozoic times a new range of fish-like forms had begun to appear, apparently derived from the straight-shelled orthocerids which were themselves weighted to adopt a horizontal fish-like posture as they swam. The new animals had considerably reduced the shell, which first became internal, and finally lost its chambered form altogether to remain as a simple stiffening rod in the roof of the elongate mantle. The belemnites were unequivocally predatory, equipped with an impressive array of hooks on their grasping arms, finned and very probably possessed of the fully manoeuvrable funnel of modern squids. They had, presumably, good eyes on the modern coleoid pattern, since they could hardly have competed with fish had their performance been limited by the pin-hole vision of *Nautilus*.

The belemnites reinvaded shallow water. They are found, for example, in the gut-contents of ichthyosaurs which were air-breathing, visual hunters; it seems most unlikely that these mesozic 'dolphins' had the echo-sounding capability of the Cetacea.

The belemnites, in their turn, died out, replaced by a range of yet more lightly-built squids, by the sepioids and by the octopuses, all groups with representatives that have managed to remain competitive in the difficult but productive shallow-water environments of reefs and sandbays.

These 'new' cephalopods are characterized by great manoeuvrability, by a capacity to change colour and by a quite exceptional development of the eyes and central nervous system. The animals are unarmoured, usually non-poisonous, and withal predatory, obliged to make decisions as to when to attack and when to retreat and well able to learn from individual experience. The brain-to-body weight ratio of modern

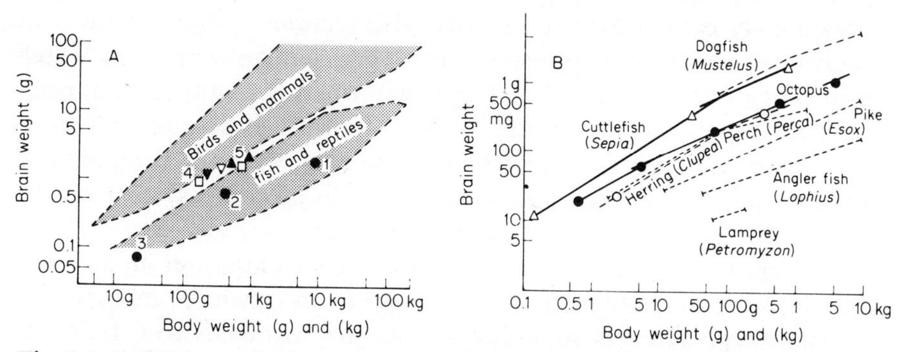

Fig. 1.2 Brain size and relative growth. Log-log plots of brain and body weights. (A) Adult cephalopod brains compared with higher and lower vertebrate brains. ▲, *Sepia*; ▽, *Loligo*; 1. *Octopus vulgaris*; 2. *O. salutii*; 3. *O. defillipi*; 4 and 5. oesgopsid squids *Illex* and *Todarodes.* (B) Growth curves of the brain of the cuttlefish (△), octopus (●) and various fish (from Packard, 1972).

cephalopods exceeds that of most fish and reptiles (Fig. 1.2). When one further remembers that the brain weight of an *Octopus* represents only the more specialized sensory integrative, higher movement control and learning parts of a rather diffuse nervous system – the ganglionated cords of the arms alone contain almost three times as many neurones – it becomes clear that one is dealing with an animal that might well be expected to possess a central nervous capability approaching or exceeding that of many birds and mammals.

1.1.3 *Man and the octopus*

There is, as Packard (1972) points out, a haunting similarity in the history of *Octopus* and ourselves. Both of us have evolved from groups obliged to spend a period in the wilderness, ekeing out a peripheral existence while other animals, temporarily better adapted, dominated the more desirable habitats. The teleost fish and the great reptiles respectively forced upon us a way of life that depended upon adaptability rather than armour, a capacity to know when to run and how to detect trouble in the making that has eventually placed the coleoids and the mammals in a unique position with respect to their invertebrate and vertebrate competitors. Educated in dangerous schools, the octopods and the primates in particular have graduated as unspecialized but nevertheless highly successful exploiters of some of the most complex environments in the whole Earths' ecosystem.

The result is a mollusc that a primate can recognize as a fellow creature. It is very easy to identify with *Octopus vulgaris*, even with individuals, because they respond in a very 'human' way. They watch you. They come to be fed and they will run away with every appearance of fear if you are beastly to them. Individuals develop individual and sometimes very irritating habits, squirting water or climbing out of their tanks when you approach – and it is all too easy to come to treat the animal as a sort of aquatic dog or cat.

Therein lies danger. It is always dangerous to interpret an animal's reactions in human terms, but with dogs or cats or white rats there is a certain reasonableness in doing so. We are mammals too. It is not absurd to suppose that the other warm-blooded inhabitants of our labs and homes can feel cold or hungry or thirsty and that they will respond to stimuli in much the same way as we would ourselves. Our physiologies are similar. Our behavioural priorities must be more or less similar, rat motivation and human motivation are not unlikely to be comparable.

The octopus is an alien. It is a poikilotherm, never had a dependent childhood, has little or no social life. It may never know what it is to be hungry. Depriving an octopus of food makes it less, not more, liable to emerge from its home and attack things that it sees – motivation is *not* adjustable in the same way as it would be in a mammal (Section 12.1.11). It knows sex, but it does not get excited about it. The heartbeat of a male octopus in copulation is as steady as in a resting animal (Section 3.2.5) while all the evidence suggests that the function of the sexual displays of males is limited to establishing the sex of other individuals rather than stimulation of the female (Section 5.1.5). The animal, it is true, learns under conditions that would lead to learning in a mammal but the facts that it learns about its visual and tactile environment are sometimes very different from those that a mammal would learn in similar circumstances (Chapters 8 and 9). Simply because it is evidently intelligent and possessed of eyes that look back at us, we should not fall into the trap of supposing that we can interpret its behaviour in terms of concepts derived from birds or mammals. This animal lives in a very different world from our own.

1.1.4 *The contents of the present review*

In this book I have concentrated upon the physiology of a single species, *Octopus vulgaris,* picking up information from other cephalopods only where direct evidence of the state of affairs in *O. vulgaris* is lacking. The main reason for limiting the account in this way was the feeling that in this species there was an almost unique opportunity to produce a comprehensive physiology of a single invertebrate type. Here it should be possible, for once, to get a feel for the workings of an animal as a whole, rather than as a series of isolated organ systems. It is also the cephalopod species we know most about and the animal that I have worked upon myself.

From what has already been said it is apparent, I hope, that the uniqueness of *Octopus* lies in its complex behaviour, and in the exceptional development of some of its organ systems. I have concentrated on these aspects rather than on the lower levels of its physiology, where *Octopus* has much more in common with other animals. To that justifiable bias I should confess that I have never, personally, been able to find much excitement in chemistry and I have cravenly evaded discussion at levels where I am frankly incompetent and reluctant to do the necessary homework. A monograph is, inevitably, a rather personal thing and the account is bound to be skewed by too much print squandered on things that interest the author, and too little on the things that may interest others. I am sorry about this. But not very.

In an attempt to keep the reference list down to a handleable size,

I have tended to cite only the more recent work, wherever this includes adequate coverage of the older literature. This is in no way an attempt to play down the value of the pioneering studies of men like Bert or von Uexküll. It is simply that their experiments have been repeated and extended since by people with the advantage of better apparatus and a wider range of techniques. The reference lists in the later reports show on whose shoulders they stand.

Finally, the animal we are discussing is known correctly as *Octopus vulgaris* Cuvier. It has often been referred to as *O. vulgaris* Lamarck. Lamarck did describe it, later in the same year (1797) as the baron but Cuvier takes precedence. 'Although Cuvier's description gives only two characters to distinguish *O. vulgaris* (which he gives in error as *O. vulgare*) there is no doubt as to the species intended since he says this is the common octopus of our shores and there is a figure. He also mentions its large size' (Opinion 233, 1954 *International Commission on Zoological Nomenclature,* 4, 277–295).

All the statements and figures in this book are derived from observations of this species unless otherwise specified.

CHAPTER TWO
An outline of the anatomy

2.1 External features and layout of the main internal organs

Rather surprisingly for such a common and well-known animal, there is no readily available dissection guide for *Octopus vulgaris.* Some parts have been described in great detail; the brain, notably, has had a very thorough treatment in Young's (1971) *Anatomy of the nervous system of Octopus vulgaris* (690 pp and 645 figures!) but there is no obvious source for a general picture of the anatomy of the animal as a whole. To find an illustration covering the major vessels of the blood system, for example, one must go to Isgrove's (1909) monograph on *Eledone,* while the morphology and function of the excretory system is best described in the work of Harrison and Martin (1965) for *O. dofleini.* In this chapter some rather scattered information about anatomy is gathered together and given in just sufficient detail to define terms and to show how the systems described in the chapters that follow fit in to the overall layout of the animal.

2.1.1 *Orientation and the naming of the arms*

The octopus is a mollusc, and the Mollusca as a group are characterized on the one hand by great conservatism in the matter of their basic internal anatomy and on the other by extreme adaptability in bodily form. The cephalopods have a mantle cavity, ctenidia draining into a systemic heart, kidneys and gonads in coelomic spaces that connect with the pericardium and open into the mantle, like any other class of mollusc. They have shells, albeit often very much reduced shells; the vestiges, in *Octopus,* are limited to two small stiffening rods in the mantle above the gills. They are soft on the outside. The structural modifications that are *not* typical of molluscs as a whole, notably the addition of accessory gill hearts and the almost total enclosure of the haemocoele to form a capillary blood system, the changed rôle of the digestive gland (no longer absorptive in some of the squids, and only intermittently so in octopuses, Section 4.3.5), the very great development of the funnel and the use of the mantle as a locomotor organ, are all changes associated with an increase in size and/or the adoption of an active predatory life-style.

In contrast with this essentially conservative internal structure, the external form of *Octopus* has little or nothing in common with that of the bivalves or gastropods, chitons or scaphopods. Quite apart from the matter of a shell, the animal at first sight lacks any equivalent of the creeping or swimming molluscan foot while the head seems to be in the middle rather than at the front of the body. Comparative anatomists and embryologists have shown that the arms are formed from elements that would otherwise form the foot so that the 'proper' orientation of the animal is arms = ventral, abdomen = dorsal, with the head between the eyes anterior and the mantle posterior, as it would be in a pretorsional gastropod (Fig. 2.1a).

There is little point in sticking to this orientation for present purposes. The animal does not sit like that. In a live *Octopus* the suckered surface of the arms, the funnel and the mantle cavity are all ventral, the eyes latero-dorsal while the supra- and suboesophageal parts of the brain are respectively dorsal and ventral rather than anterior and posterior as the classically 'correct' orientation would imply. The orientation adopted for descriptive purposes throughout this book is that shown in Fig. 2.1b, with the arms anterior to the head and mantle. Looked at from above, they can be labelled in terms of side and position in the sequence L1 R1, L2 R2 etc. working from front to back (Fig. 2.1c). The second and third pairs of arms are longer than the first and fourth, and while the animal is liable to use any or all of its arms to examine objects, L2 and R2 are perhaps most frequently extended so that in life these give the impression of being a little longer than L3 and R3 which Robson (1929) has measured as generally the longest in fixed specimens. R3 is the 'hectocotylus' in males, specialized for the transfer of spermatophores and generally a little shorter than L3.

2.1.2 *The hectocotylus and sexual dimorphism*

The shape of the tip of the hectocotylus has been much used in octopod systematics. In many species it is considerably elaborated, and in some it actually breaks off and is left in the mantle of the female

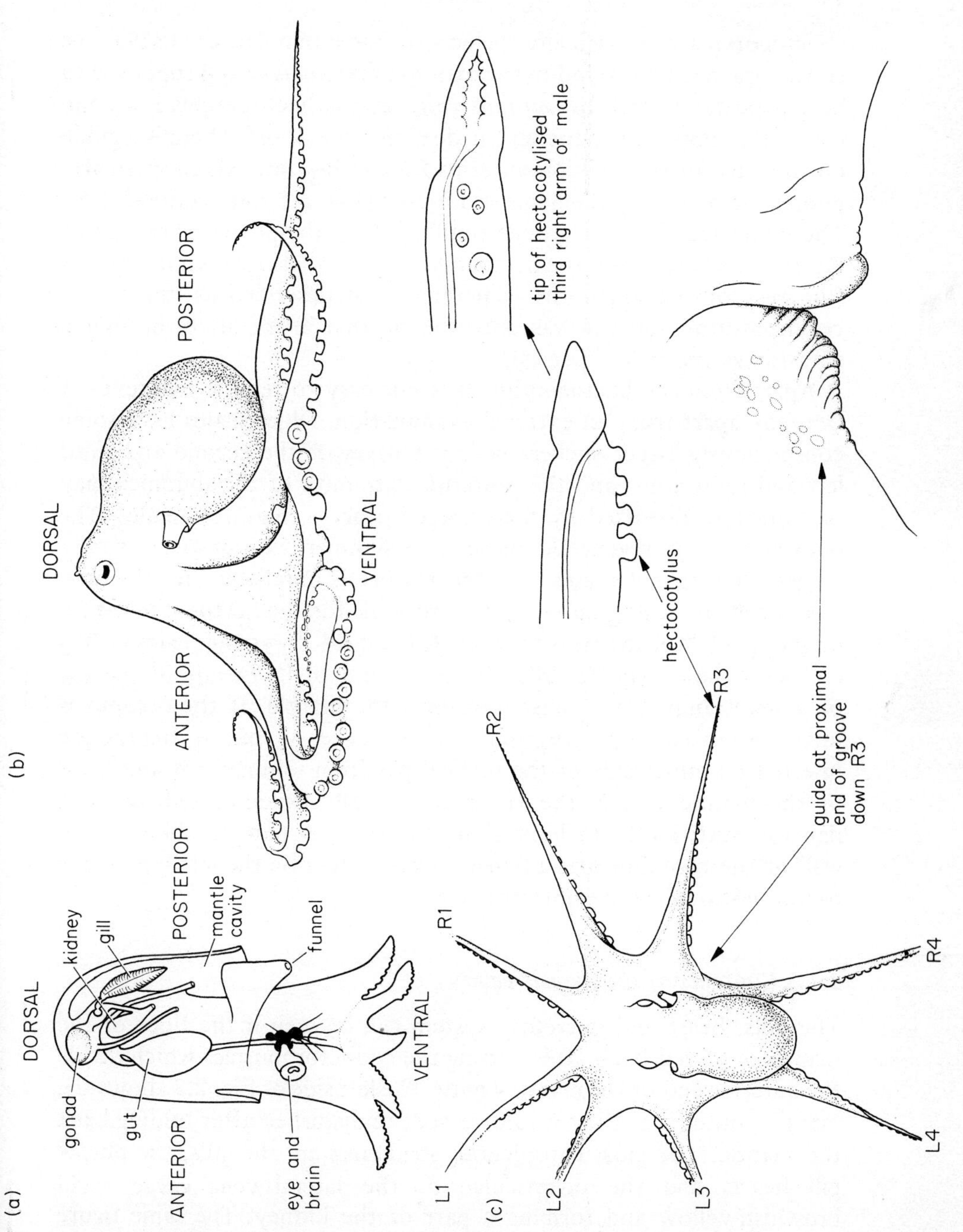

Fig. 2.1 (a) 'Classical' orientation of a cephalopod and (b) the natural orientation adopted in descriptions throughout this book. (c) shows the animal from above, with the arms identified and details of the hectocotylized third right arm of males.

(*Hectocotylus* was originally the generic name that Cuvier (1829) gave to the organism he found in the mantle of *Argonauta* and supposed to be a parasitic worm). In *Octopus vulgaris,* the hectocotylus is a comparatively simple structure and it does not break off. There is a guide on the edge of the web joining arms R3 and R4, into which spermatophores are placed by the penis and funnel (Fig. 2.1 and Section 5.1.5). The guide leads into a groove running along the posterior margin of R3 to a suckerless tip, the hectocotylus proper, also shown in Fig. 2.1 (the rest of the apparatus is perhaps best described as 'the hectocotylized arm' since it was only the tip that broke off to become a parasitic worm in any species).

Apart from the hectocotylus, it is not easy to tell the two sexes of *Octopus* apart from an external examination. Older males have some conspicuously larger suckers near the bases of the second and third left and right arms, and the posterior extremity of the abdomen may be noticably distorted by the enlarged ovary in mature females. The bodies of males, in general, appear to be slimmer, with proportionately longer arms than females, but the feature is unreliable and the only sure means of sexing an octopus is to look for the hectocotylized tip or the 'guide' on the web between R3 and R4. In young males (100 g and less) these may be difficult to see and to be certain of the sex of a small animal one must look into the mantle. If the octopus is unanaesthetized and struggling this is not easy; but if fingers are inserted on either side of the vertical partition joining roof and floor of the mantle cavity, the latter can usually be depressed, without damage, sufficiently to look along the roof. A pair of ducts in the wall of the roof indicates a female, a single duct on the left side of the partition shows the specimen to be a male.

2.1.3 *Contents of the mantle cavity*

The gills, heart and excretory system can be seen if the floor of the mantle is folded back after cutting the adductor muscles which divide the anterior end of the mantle cavity. The left side of Fig. 2.2 shows the mantle contents as they would be seen undisturbed after folding back the mantle. The most conspicuous structures are the gills, the purple gill hearts and the diverticulae of the lateral vena cavae, vivid brownish-yellow and forming a part of the kidney. The same figure (RHS of the figure, LHS of the animal) shows the mantle contents at a later stage in dissection, following removal of the left lateral vena cava

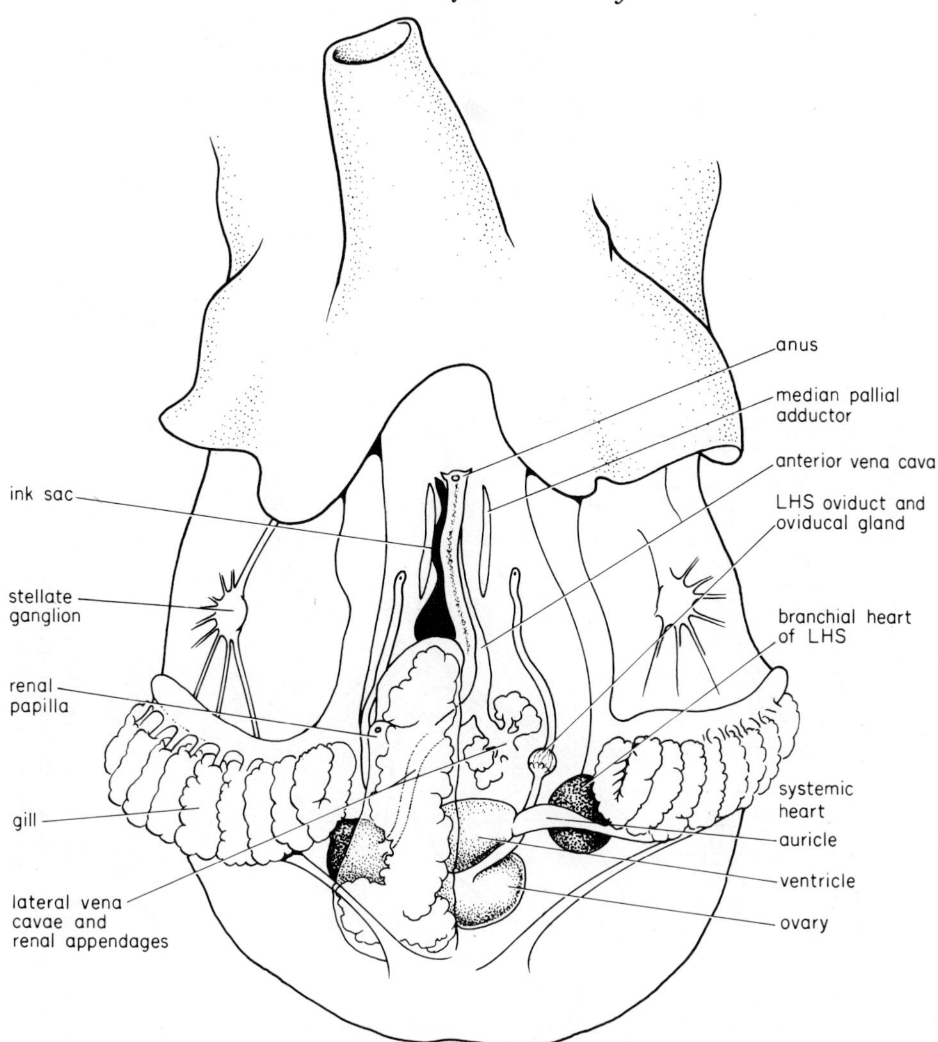

Fig. 2.2 Contents of the mantle cavity. The median pallial adductor muscles, on either side of the rectum, have been cut and the mantle turned inside out. On the left hand side most of the lateral vena cava has been removed together with its renal diverticulae and the kidney sac.

and its associated kidney tissues. The efferent branchial vessel can be seen leading from the gill to the left auricle of the systemic heart, and the origin of the oviduct is exposed. Fig. 3.11, p. 41, shows some further details in a male.

An alternative means of displaying the mantle contents is shown in Fig. 2.3, where the dissection has been made from the side, by splitting

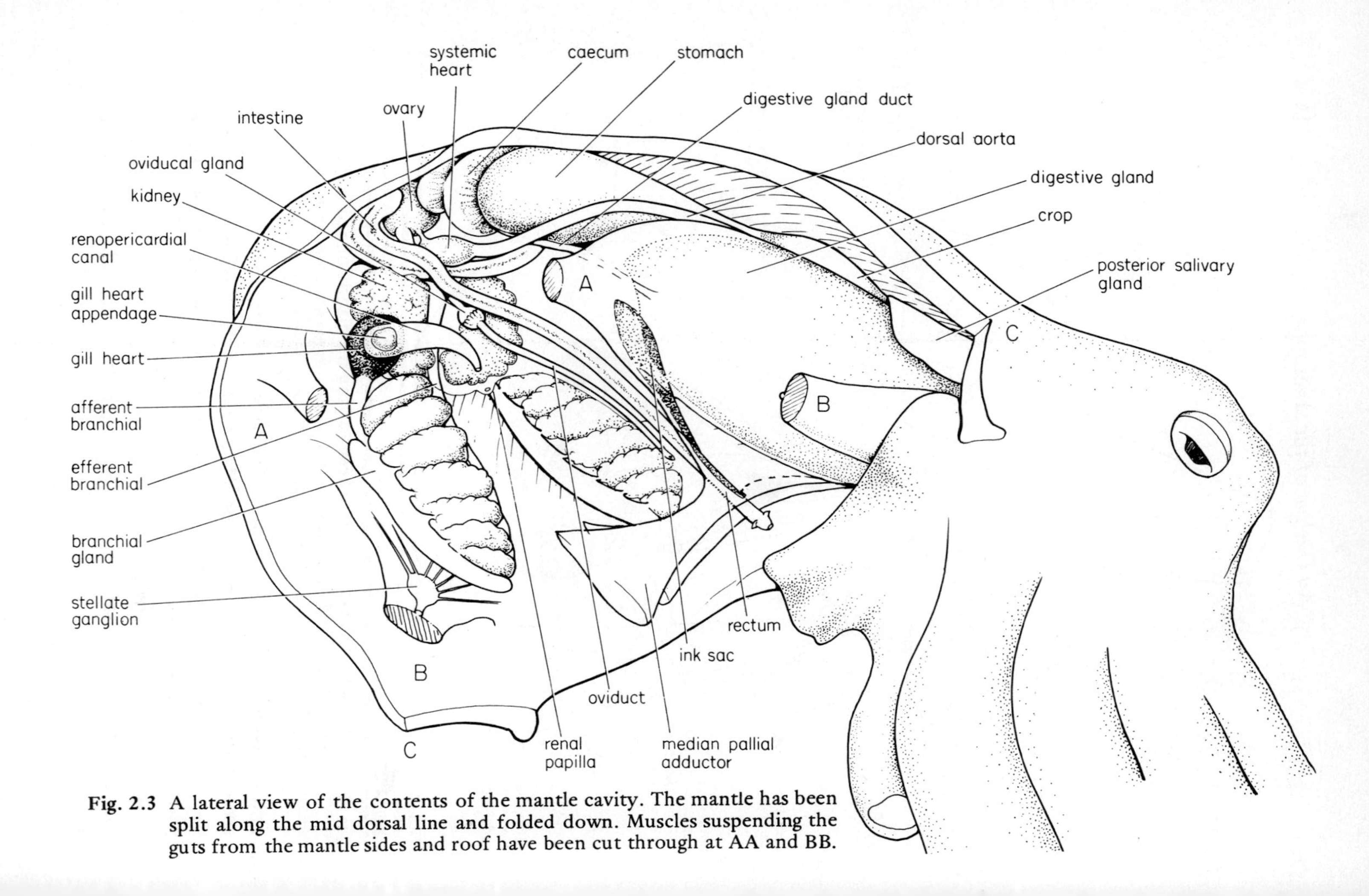

Fig. 2.3 A lateral view of the contents of the mantle cavity. The mantle has been split along the mid dorsal line and folded down. Muscles suspending the guts from the mantle sides and roof have been cut through at AA and BB.

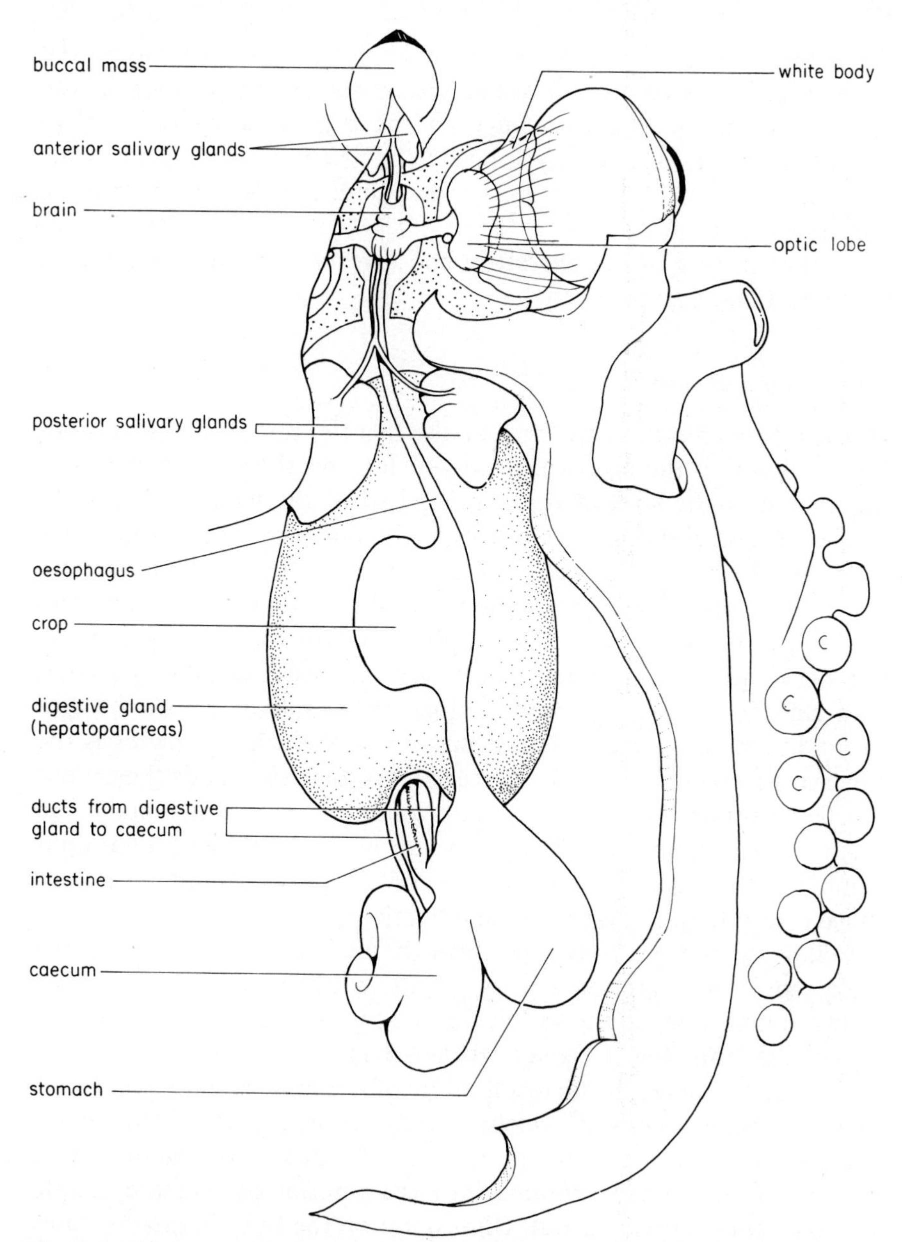

Fig. 2.4 The gut. Dissection from the dorsal surface. The mantle roof has been cut away and the connective tissue sheath surrounding the gut has been removed, along with the dorsal aorta that would otherwise lie beside the oesophagus. Extending the dissection forward, the cartilage around the brain has been pared down to the level of the optic tracts. Fig. 4.4, p. 81, shows details of the innervation of the gut.

the mantle along its dorsal surface, and rolling it back to expose the systems inside. This gives a better impression of the position of the internal organs in life, suspended in the mantle cavity rather than stuck to its roof, as they may appear to be when dissected in the usual way, from below. The gut is best shown in dissections made from the dorsal surface (Fig. 2.4).

Further details of the excretory and digestive systems are given in Chapters 3 and 4 respectively.

2.1.4 *Head and funnel*

Working forwards from the rear of the mantle, the most conspicuous large structure is the digestive gland which forms the main thickness of the animal at the level of the anterior half of the mantle. Ventral to this lie the anterior vena cava (returning blood from the head and arms to the gill hearts via the kidneys), the ink sac, the gonadial ducts and the rectum. Dorsally there is the forward-running anterior aorta and the backward-leading crop and oesophagus of the foregut (Fig. 2.4). In the 'neck' between head and mantle are the posterior salivary glands, with fine ducts running forward along either side of the oesophagus to the buccal mass. Immediately anterior to the salivary glands is the concave posterior face of the cartilage box that surrounds the central ganglia, cradling the eyes, the optic lobes and the 'white body', an extensive structure, which appears to be the origin of enormous numbers of wandering amoebocytes and which may in addition have an excretory function (Fig. 2.4, and Section 3.3.9).

Following the gut forwards through the brain one comes to the beak, radula and anterior salivary glands, which together form a near-spherical mass lying in the socket formed by the head cartilage and the base of the arms (Fig. 4.2 and Section 4.3.1).

The funnel, beneath the head, is one of the most remarkable organs found in cephalopods. Developmentally, it is a part of the 'foot'. Functionally, it is a part of the mantle. The two work together, with the rim of the funnel jamming into the anterior edge of the mantle when the animal breathes out. On inspiration the flexible rim collapses as water is drawn in. The funnel is quite extraordinarily mobile. It can be directed to either side of the head, forwards or lateroposteriorly. The animal uses it to propel itself when swimming, to direct jets of water at over-curious fish, to carry spermatophores from mantle to 'guide' and to aerate the eggs, besides its more obvious function in blowing to a distance the used water and refuse from the animal's respiratory, excretory and digestive systems. The funnel has an abundant nerve supply from the suboesophageal lobes of the brain.

2.1.5 *Hearts and circulation*

Octopus has an enclosed circulatory system, with a capillary bed linking arteries and veins. This produces a pressure drop which necessitates further pumps if the returning blood is to pass through the gills.

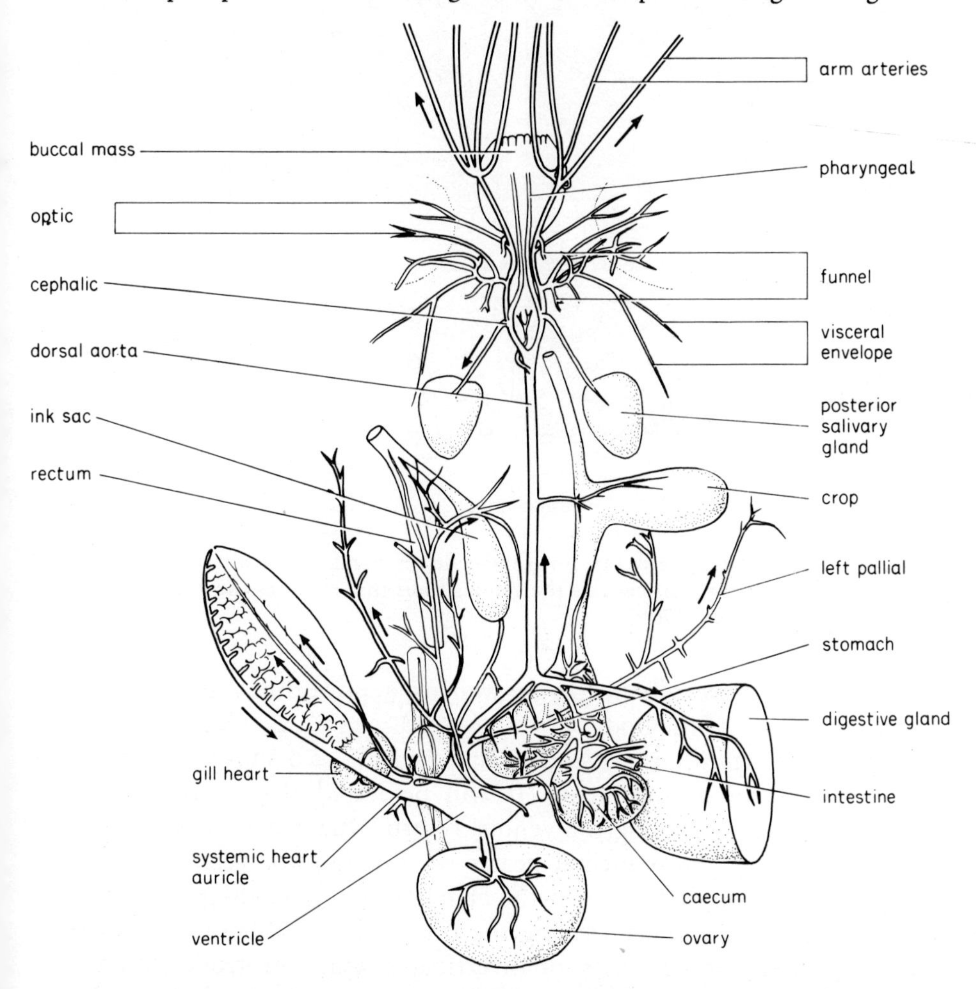

Fig. 2.5 The arterial system. The left gill and gill heart have been removed, and the digestive gland dissected out and reflected back to the left, exposing the dorsal aorta (*Eledone,* after Isgrove, 1909).

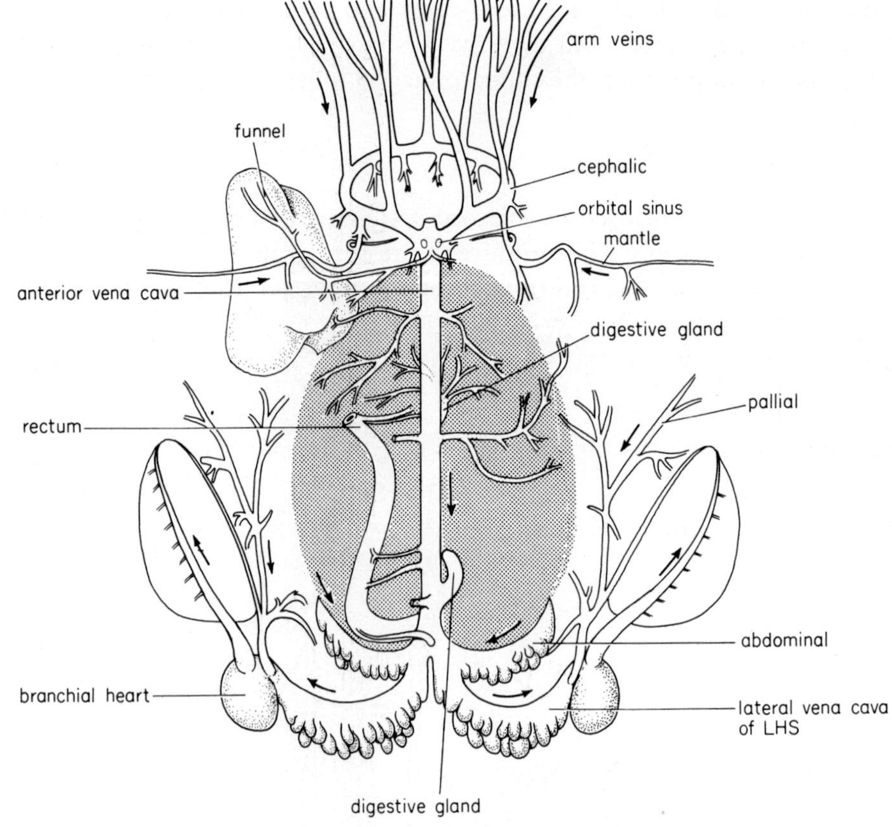

Fig. 2.6 The venous system. (a) shows the main veins and (b) the alternative return route to the vena cava via the venous sinus that encloses the gut (*Eledone*, after Isgrove, 1909).

Cephalopods have gill hearts as well as a powerful systemic heart serving the rest of the body.

The main arteries are shown in Fig. 2.5; Fig. 2.6 shows the corresponding venous arrangements. It will be noted that there are two distinct systems handling the venous return. One, mainly draining the arms, arrives at the branchial hearts via the anterior vena cava. The other is a system of sinuses in the membranes surrounding the gut. In both instances the veins forming the final link to the branchial hearts are expanded into a series of diverticulae which protrude into the kidney sacs, covered by the yellow-brown secretory and absorptive cells of the sac walls.

The functioning of the circulatory system is discussed in Chapter 3.

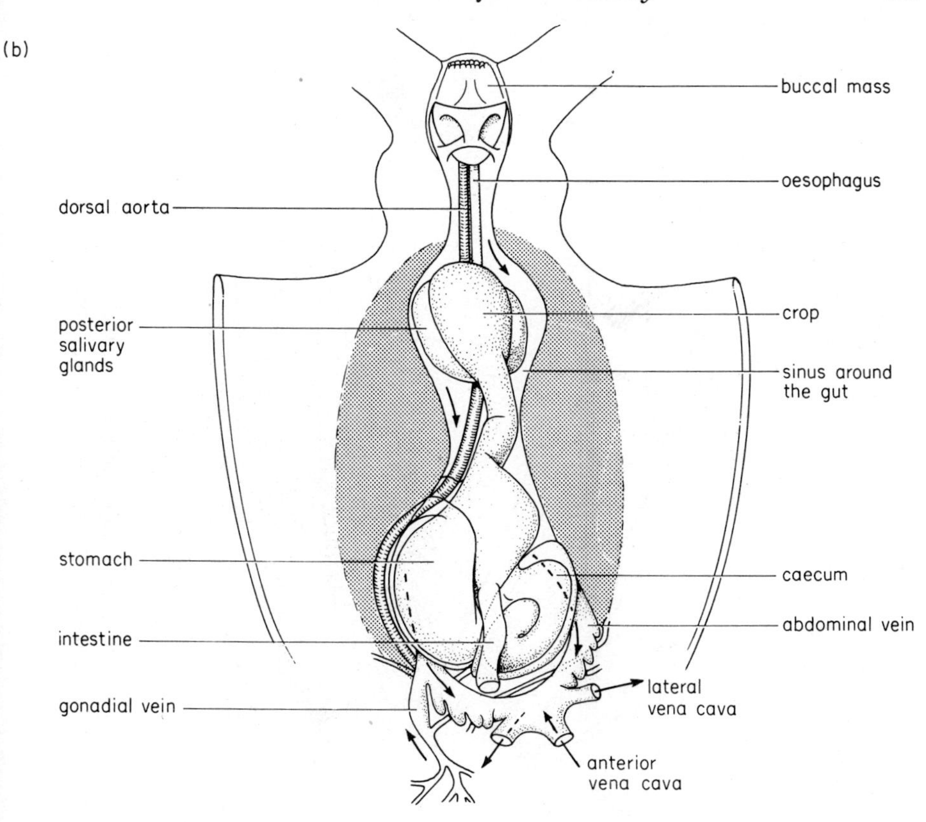

2.1.6 *Nervous system*

Most of the chapters in this book are concerned, in one way or another, with the performance of the nervous system which is quite remarkably developed in cephalopods. The brain (Fig. 2.7) is large, by invertebrate or vertebrate standards, yet it contains only about one third as many neurones as the arm nerve cords. In addition there is an extensive plexus of nerves around the gut with major peripheral ganglia at the level of the buccal mass and stomach, and a considerable system of nerves and ganglia serving the hearts. The blood vessels, veins as well as arteries, have an elaborate nerve supply and there would seem to be little doubt that the detailed operation of the blood vessels is under direct nervous control in a matter quite unlike anything to be found in vertebrates. The same extraordinary degree of detail is seen in the control of skin texture and chromatophore change, both operations served by very large numbers of neurones with cell bodies located within the central nervous system.

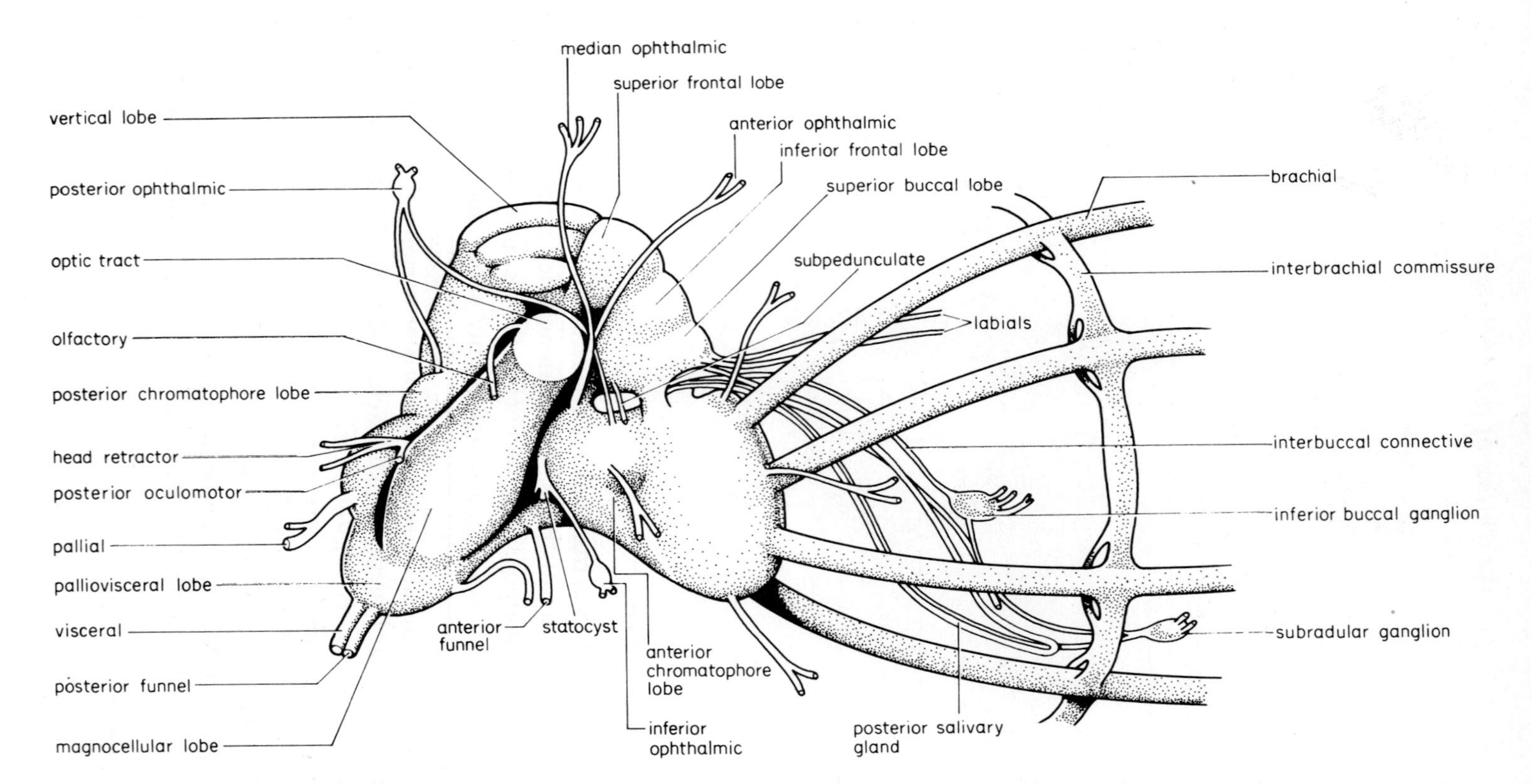

Fig. 2.7 Lobes and nerves of the brain of *Octopus* seen from the right side, after removal of the optic lobe by section of the optic tract on that side. A corresponding picture of the brain from above is given in Fig. 6.1 (after Young, 1971).

Beyond Fig. 2.7, which shows the brain from one side, with the principal nerves radiating from it, no attempt will be made to describe the nervous system here. The structure of the major sense organs, the nervous system in the arms, gut and circulatory system, brain structure, and the innervation of specific organs are described as they become relevant in the chapters that follow.

CHAPTER
THREE

Respiration, circulation and excretion

This chapter deals with three interdependent systems. Pressurized filtration by the kidneys is affected by the affairs of the three hearts and these in turn by the movements of the mantle. Each is under nervous control and each must take into account the activities of the other two if it is to perform correctly. Research on the performance of the three systems has, however, usually been concerned with one system at a time and generally with one aspect of performance at a time. Sooner or later we shall have to try and monitor nerve impulses, pressure changes, oxygen and metabolite concentrations simultaneously, since these observations will undoubtedly reveal some very elegant control mechanisms. In the meantime, one is obliged to consider the workings of the three in the classical manner, one at a time; remembering that, eventually, the bits will patch together to yield a much more beautiful story.

The section that follows immediately deals with respiratory movements and their control. The main purpose of the movements is to oxygenate the blood and assist the escape of metabolites shed into the mantle cavity. The respiratory functions of the blood are dealt with below in Section 3.2.13; excretion, apart from the elimination of carbon dioxide which is considered in discussing the blood pigment, is reviewed in Section 3.3.

3.1 Respiration

3.1.1 *Mantle movements*

Like other molluscs, cephalopods have gills in a mantle cavity. Unlike the rest, the walls of the cavity are contractile. This immediately presents a problem because there is no hard skeleton for the muscles to work against in expanding the respiratory chamber. The dibranchiate cephalopod solution to this difficulty has been the development of massive connective tissue lattices in the walls of the mantle, coupled with radial muscles in the thickness of the mantle wall.

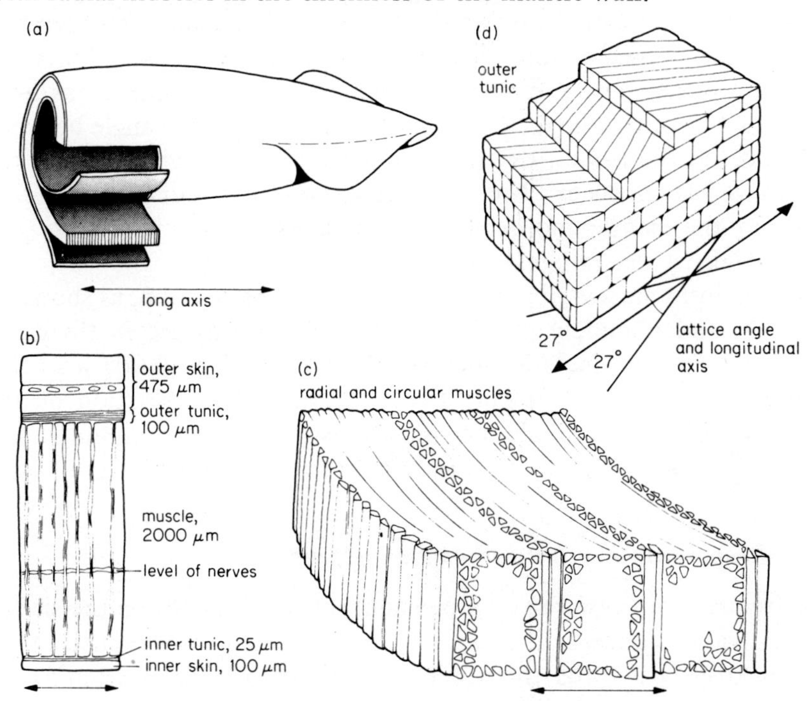

Fig. 3.1 Diagrams showing structure of the squid mantle. (a) Diagram of *Lolliguncula* mantle with skin and pen removed showing the thick muscle layer between outer and inner tunics. (b) Diagram of a longitudinal radial section of *Lolliguncula* mantle to show relative thickness of the components in an animal with a mantle length of 120 mm. (c) Diagram showing the arrangement of muscle cells in the mantle of *Lolliguncula.* Radial muscles, shown oriented vertically, attach to outer and inner tunics. (d) Diagram of the arrangement of large collagen fibres in the outer tunic of *Lolliguncula.* Each of the seven layers is one fibre thick. Close packing of fibres in this array gives them rectangular cross-sections (from Ward and Wainwright, 1972).

The system has been worked out most thoroughly in squids. The main structural elements are shown in Fig. 3.1. There are circular and radial muscles and longitudinal lattices of collagenous fibres running helically around the mantle on either side of the muscle layer. In addition, and not shown in Fig. 3.1, there is a further longitudinal lattice in the plane of the radial muscles, between the inner and outer connective tissue tunics. The fibres in all three lattices lie at an angle of about 27° to the long axis of the body.

The presence of the lattices ensures that muscular contractions will only lead to extension at right angles to the length of the mantle. As either set of muscles contracts the lattice in that plane is compressed, so that the corresponding cross-sectional area is decreased. Since the volume of the muscle remains unchanged and the system has little capacity for longitudinal extension because of the small angle between the fibres in the tunics, extension in a direction at right angles to the long axis has to occur. The circulars expand the radials, the radials thin out the mantle and expand the circulars (Ward and Wainwright, 1972; Packard and Trueman, 1974).

The mantle muscle fibres are triangular in cross-section, as shown in Fig. 3.1, and obliquely striated. Sarcomere lengths are in the range 1.6 ± 0.3 μm and include thick (0.02 μm) and thin (0.005 μm) filaments (Ward and Wainwright, 1972; for further information on the fine structure of cephalopod muscle see Gonzales-Santander and Garcia-Blanco, 1972; Kawaguti and Ikemoto, 1957; Hanson and Lowy, 1960).

3.1.2 *Respiratory rhythm and oxygen uptake*

The frequency of respiratory movements at rest depends on the size of the individual concerned. Table 3.1, from Polimanti (1913) lists observations from *O. vulgaris*; the respiratory rate at 15 °C drops from 51 per minute at 2.5 g to 12 per minute at 8 kg. Johansen and Martin (1962)

Table 3.1 The respiration rate of *O. vulgaris* (observations made at Naples in March, T 15 °C; Polimanti 1913).

Body wt, g (range)	*Inspirations per minute (mean)*	*Number of animals observed*
2.5–3	51	15
8	45	10
9–23	40	7
40–50	36	6
100–150	30	12
200–400	27	18
500–800	26	20
900	25	4
1000–2000	24	13
2500–3000	23	3
3500	22	2
4000	18	3
8000	12	1

and Smith (1962) show figures of 8 per minute for *O. dofleini* twice this size at 7–9 °C, a useful indication that this very large animal is physiologically comparable with the smaller species.

The efficiency of the respiratory cycle is similar to that of bottom-living fish and gastropods; Hazelhof (1939) gives figures ranging from 50–80% of the available oxygen removed from each mantle-full of water.

Ventilation rate is increased by oxygen lack or carbon dioxide excess, mainly by an increase in the volume taken in at each inspiration, which can rise by a factor of ten from the resting level; there is a simultaneous though less marked increase in frequency (Winterstein, 1925).

Borer and Lane (1971) have recorded the effect of temperature on the respiratory frequency of *O. briareus* at the different oxygen tensions. At 20 °C, the frequency remained constant at about 21 inspirations per minute (animals of 250–450 g) until the oxygen content fell to about 4 parts per million (mg O_2/l). After this it rose systematically at 4 additional breathing cycles per minute for every 1 ppm drop in

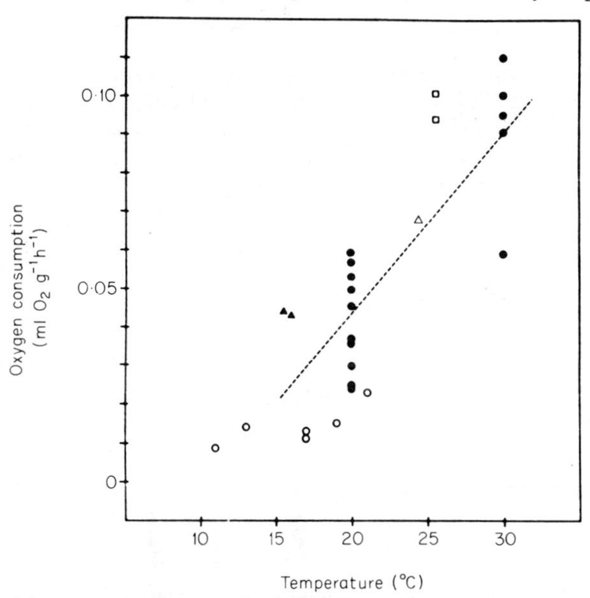

Fig. 3.2 Rates of oxygen consumption at different temperatures; ● 250–442 g *Octopus briareus* (Borer and Lane, 1971); ▲ 2300 g *O. vulgaris* (Jolyet and Regnard, 1877); △ 280 g *O. vulgaris* (Montuori, 1913); ○ 7.1 g *O. vulgaris* (Vernon, 1896); and □ 240–412 g *O. cyanea* (Maginnis and Wells, 1969); the dashed line connects mean values for the O_2 uptake of *O. briareus* at 20 and 30°C (from Borer and Lane, 1971).

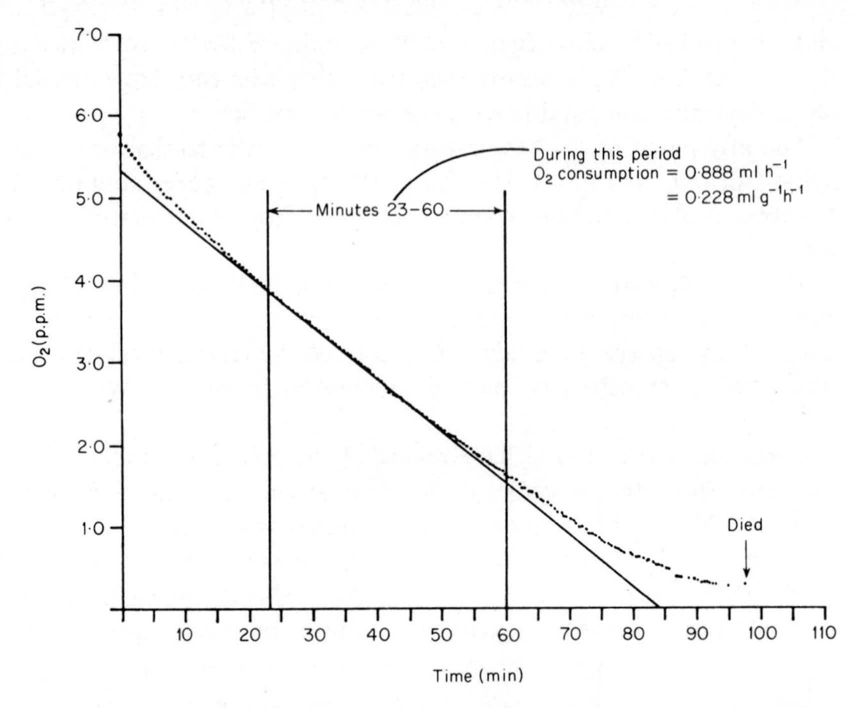

Fig. 3.3 Oxygen consumption by an octopus of 3.9 g kept in a closed container until it died. Once the animal had settled down in its jar, it consumed oxygen at a steady rate down to 2 ppm. The animal first showed signs of distress at 0.6 ppm and died 20 min later at 0.25 ppm (from Maginnis and Wells, 1969).

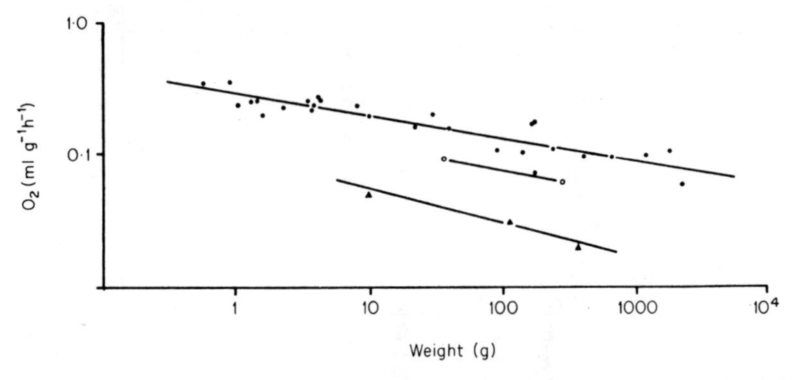

Fig. 3.4 Weight-specific oxygen consumption of *Octopus cyanea*, plotted ●. Oxygen consumption measured as ml $g^{-1} h^{-1}$ falls off with increasing weight. ○ and ▲ show values for *O. vulgaris* quoted in Montuori (1913). The figures of *O. cyanea* and those for *O. vulgaris* plotted ○ were obtained at 25±3°C. The three results plotted ▲ are quoted from Buytendijk by Montuori without stating the temperature; they were, presumably, run in colder water (from Maginnis and Wells, 1969).

oxygen content, to the point at which the animals began to show respiratory distress. At 30 °C the rate and the total oxygen consumption was about doubled; calcuation gives a Q_{10} of 2.18 (Borer and Lane, 1971; Fig. 3.2).

Octopus is able to regulate its oxygen uptake over the whole of the range that it is likely to encounter in nature. Experimentally, *O. cyanea* at rest in its aquarium showed no signs of distress until the oxygen content fell to 0.6 ppm., although there was a slight progressive decline in oxygen uptake from 2.0 ppm downwards (Fig. 3.3). Total oxygen consumption by the resting animal rose as weight$^{0.833}$, over the range 0.5 to 2300 g (Fig. 3.4). It is, predictably, somewhat greater than for most molluscs, being comparable with the rates found in poikilothermic vertebrates (Hemmingsen, 1960). Consumption did not appear to be different in males and females (Maginnis and Wells, 1969).

3.1.3 *Control of mantle contractions*

The nervous control of mantle contraction has been investigated by a number of workers and most recently by Gray (1960) and Wilson (1960), both working with *O. bimaculatus* and/or *O. bimaculoides* (sibling species, see Pickford and McConnaughy, 1949).

The mantle musculature is controlled through the pallial nerves from the suboesophageal part of the brain. In *O. vulgaris* some 4000 fibres run to the stellate ganglion on each side and synapse there with neurones that give rise to axons in the stellate nerves; there is an approximately twenty-fold increase in numbers at this level (Lund, 1971; Young, 1972). A second group of some 2300 neurones in the pallial nerve passes directly through the ganglion to the chromatophores in the skin (Sereni and Young, 1932). The chromatophore fibres are all quite small (up to 4 μm in diameter) in contrast to those controlling the mantle musculature, which run up to 18 μm, with a conduction velocity of 3.8 m s^{-1} in the largest fibres (Burrows, Campbell, Howe and Young, 1965). The stellar nerve efferents are smaller and slower, up to 15 μm and 1.75 m s^{-1} (Wilson, 1960); unlike squids *Octopus* has no third order giant fibres.

Running in the opposite direction, up the stellar nerves and into the stellate ganglion are an estimated 80 000 afferents, some 10 000 of which continue through to the pallial nerve and the brain (Lund, 1971, in Young, 1971).

The usual response to stimulation of one of the stellar nerves is contraction of the circular muscles, reducing the area of a part of the mantle. Less commonly, the same area is thinned and expands as the radial muscles contract.

There is evidence for a double innervation, with 'fast' and 'slow' nerves; in *Octopus* both produce mechanical responses that summate (in *Loligo* the 'fast' nerves produce an 'all or nothing' contraction); graded contractions are probably brought about by local potentials rather than spike potentials and recruitment of units (Wilson, 1960).

Recording from the pallial nerve of *O. bimaculatus/bimaculoides*, Gray (1960) found bursts of impulses that corresponded to inspiratory and expiratory movements of the mantle. Regular discharges continued when the pallial nerve was cut distal to the point of recording, and evidently originated in the posterior half of the suboesophageal brain. The rhythm was not dependent upon sensory feedback since it continued when both pallial nerves and both visceral nerves had been cut; the latter operation has sometimes (incorrectly, see Section 3.2.9 below) been said to stop the respiratory rhythm in *O. vulgaris* (Ransom, 1884; Mislin, 1955).

The apparent irrelevance of sensory feedback was rather surprising and Gray followed the matter further by cutting windows through the mantle musculature to record from individual stellar nerves in the nearly intact animal. 'No impulses were eyer seen during normal breathing' (Gray, 1960). This is odd, because afferents are present in plenty and the animal is known to have stretch receptors in the mantle musculature (Alexandrowicz, 1960b; Section 7.4). If the mantle is pinched or prodded, there is reflex local contraction of that part of the mantle, operating through the stellate ganglion. If the skin is stripped from an acute preparation and local pressure or stretch applied to the mantle musculature, phasic afferents fire in the corresponding stellar nerve; the input produces efferent discharges in neighbouring stellar nerves. Simulation of the mantle, moreover, facilitates motor discharges through the stellate ganglia when the pallial nerve is excited (Gray, 1960). All these observations would suggest a mechanism designed to ensure uniform contraction of the mantle. The absence of feedback in respiration could mean that mantle contraction is normally so well organized centrally that no corrective feedback through the stellate ganglion is required. Only the more variable movements made in locomotion or posturing need adjustment at this level and these, perhaps, are organized through a quite separate set of motor nerves, synapsing on the inner face of the stellate ganglion which alone receives the stellar afferents (Young, 1972). Whichever case pertains, it is evident that more experiments are needed before we can claim to understand even the working of this comparatively simple part of the *Octopus* nervous system.

3.2 Circulation

3.2.1 *The anatomy of the circulatory system*

Cephalopods have a closed circulation with arteries and veins linked by a capillary network. The gross anatomy of this system has already been described in Chapter 2. All save the branchial arteries, served by the branchial hearts, arise from a single powerful ventricle, which is fed from two auricles draining the gills. The venous system includes large sinuses. The size of these in life is a little difficult to assess because they become inflated in injected specimens but it must be considerable, since the blood volume is estimated to form about 6% of the body volume (Martin, Harrison, Huston and Stewart, 1958; see below).

The distribution of the main arteries and veins has been described many times. Figs. 2.5 and 2.6 in Chapter 2 are taken from Isgrove (1909); Grimpe (1913) gives a comparative account. Alexandrowicz (1928b) has described the innervation of the blood vessels, and Boletzky (1968) has followed the growth of the blood system in the developing embryo. Young (1971) provides details of the circulation to the central nervous system and Fig. 3.5, which is taken from his account, shows how very elaborate the system of vessels can be.

The arteries are thick-walled. A lining of epithelial cells is surrounded by a basement membrane wrapped with longitudinal and circular smooth muscle layers. Both the latter are innervated. In the longitudinal muscles, typical 'chemical' synapses are found with clefts of the order of 25 nm and many vesicles; in the circular muscles the cleft is much narrower, 10 nm or so, which perhaps indicates electrical transmission. In the finer arterioles there seem to be fewer synapses than muscle cells, and it has been suggested that transmission between pericytes may replace innervation of all the contractile elements (Barber and Graziadei, 1967b). In the finest vessels the epithelial lining is often incomplete (Fig. 3.6, Barber and Graziadei, 1965).

Veins resemble arteries, with a single layer of muscle cells and relatively few fibrils. Capillaries are structurally similar to veins with

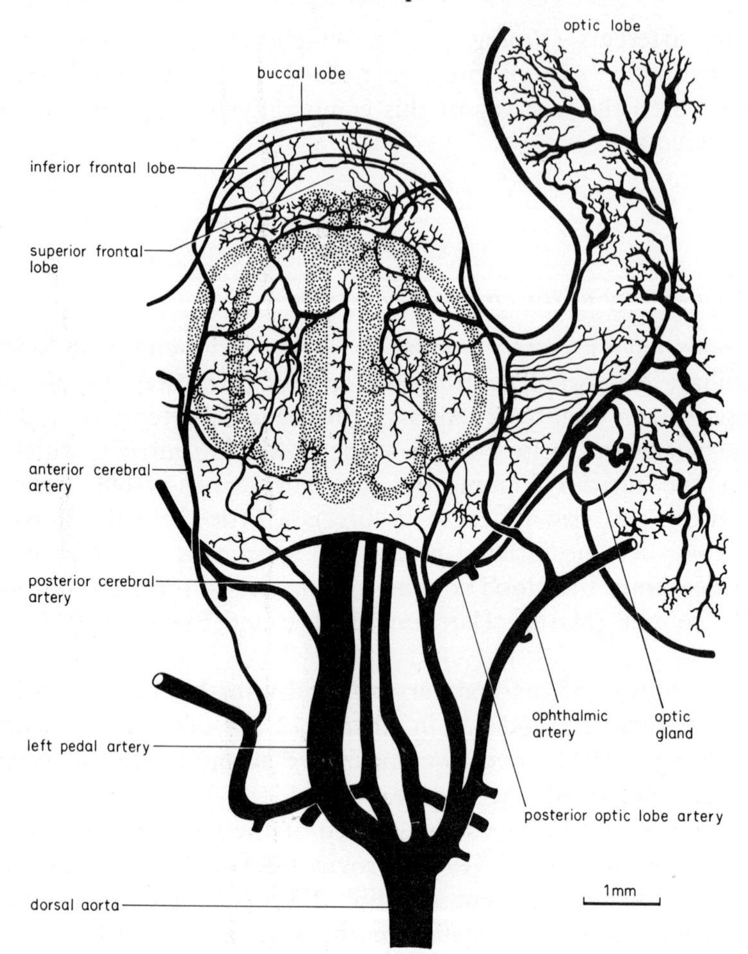

Fig. 3.5 The arterial supply to the supraoesophageal part of the brain and to the optic lobe on the RHS. The outlines of the vertical and subfrontal lobes are shown dotted. Tracing from a photograph of a cleared specimen seen from above after perfusion with indian ink. The left pedal artery is regularly the larger of the two in *Octopus* (after Young, 1971).

contractile fibres in the pericytes of even the finest vessels (decapods; Kawaguti, 1970).

In addition to the blood vessels, there is a system of extracellular spaces, that may or may not be confluent with the vessels themselves. Fig. 3.6 summarizes what is currently believed to be the anatomical situation in the brain. Injected materials, such as indian ink, and haemocyanin, can be found in this 'glio-vascular' system. It is just

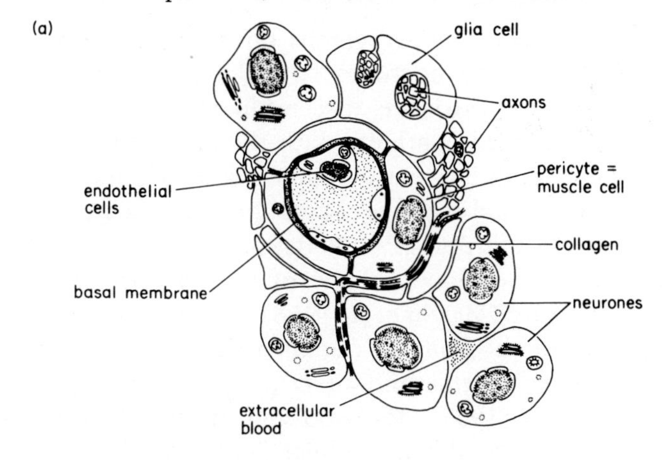

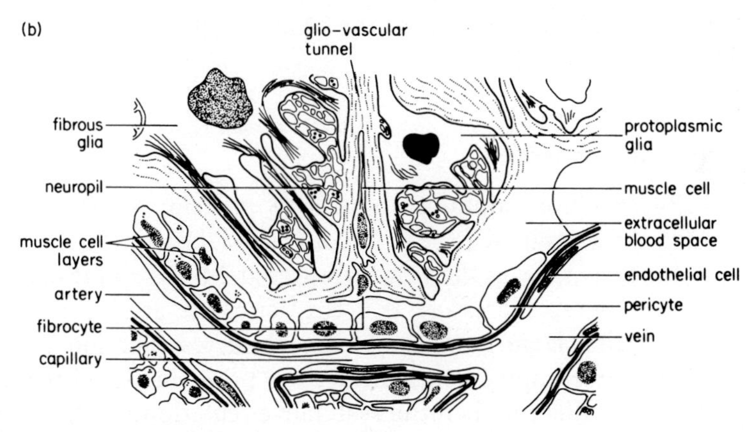

Fig. 3.6 The relation between blood vessels and extracellular spaces in the brain. (a) Shows the range of relations found in EM studies; neurones may be separated from blood vessels by pericyte cells, a collagen layer and glia, or bathed directly in extracellular blood. (b) Is a similar summary in L.S. In addition to a capillary net there is an extensive network of gliovascular tunnels. (a) Redrawn after Barber and Graziadei, 1967a and (b) after Gray, 1969.

possible that their occurrence there is a result of inadequate fixation techniques, or of the pressure associated with perfusion (Young, 1971). The latter explanation, at least, seems improbable; pressures of a few tens of centimetres of water are quite sufficient to allow perfusion, and local pressures considerably larger than this must be generated whenever an octopus squeezes itself between rocks, or grips and contracts its body musculature. Barber and Graziadei (1967a), responsible for the electronmicrographs on which Fig. 3.6a is based, state specifically that the neurones bathed directly in blood 'look quite well-preserved', On balance, it seems likely that the glio-vascular 'lymph' spaces are indeed confluent with the blood. If this is so, and there is any comparable system of open channels outside the CNS, then estimates of blood volume based on dyes which attach to haemocyanin, or the dilution of colloidal mercuric sulphide, could be considerably awry. The blood volume of *Octopus dofleini* has been measured in these ways (Martin *et al.* 1958) and the figure of 5.8%±1% of the wet weight of the body so obtained has formed the basis of a number of estimates in this and in *O. vulgaris* (see for examples, Chapman and Martin (1957), Harrison and Martin (1965) or O'Dor and Wells (1973)). The total extra-cellular fluid space, including blood, estimated from inulin dilution, is of the order of 28±7.3% of the body weight (Martin *et al.* 1958).

3.2.2 *Pumps and pressures*

Octopus cell-free blood contains about 10% protein, nearly all of it haemocyanin. As a result it is very viscous (flow through capillaries is about four times as slow as the flow of water at 24 °C) and it must be pumped at considerable pressure if it is to circulate at all rapidly. It sometimes contains large numbers of amoebocytes (Section 3.2.16, below). Add to these impediments the need to press the blood through two capillary beds (gills and then the systemic circulation) in the course of each circuit, and the complexity of the octopus contractile system ceases to be much of a surprise.

In addition to the systemic and the two branchial hearts many, and perhaps all, of the major arteries and veins are contractile. This, together with the observation that the isolated systemic heart will beat regularly for hours provided only that it is perfused under pressure (Fredericq and Bacq, 1939, and many earlier experiments), gave rise to the view that the whole system should be regarded as serially contractile - each section of blood vessel expanding the next, which contracts in its turn (Skramlik, 1941). An alternative interpretation is that the hearts are the source of the beat and that the principal veins, far from being themselves contractile, merely reflect the filling of the branchial hearts and/ or the respiratory movements of the gills and mantle (Johansen and Martin, 1962).

To choose between these two alternatives (both of which, of course, turn out to be too simple) one must know about the pressures concerned. The problem is that octopuses manipulate objects with their suckers. They are liable to rip out cannuli, so that Fredericq (1878), Fuchs (1895) and other workers using *O. vulgaris* felt obliged to nail the arms of their animals to a board or otherwise tie them down in order to make their observations. The value of measurements of blood pressure or heartbeat rate taken from an animal nailed to a board is very questionable, and most of the early work is perhaps best forgotten now that recordings have been made under more 'natural' conditions.

The breakthrough, made during the early 1960's arose from the discovery that the very large *Octopus dofleini* is a quite unusually phlegmatic and gentle creature. It makes little or no attempt to remove cannuli and recordings of blood pressures can be made from several places at once in unrestrained animals.

3.2.3 *Arterial and venous pressures in the resting animal*

The systolic aortic pressure of *O. dofleini* at rest is 45–70 cm H_2O. In the afferent branchials corresponding pressures are 25–50 in systole and about 15 cm in diastole. Similar aortic pressures are found in *O. vulgaris*.

The pressure changes in the anterior vena cava (0–17 cm H_2O) tend to coincide with the respiratory movements of the mantle, complicated by local fluctuations arising from muscular contractions of the renal appendages, which are in almost continuous movement (Fig. 3.7; Johansen and Martin, 1962).

The venous return to the vena cava must be pumped. Smith (1962) measured the pressure in the principal arm veins and found it to fluctuate from 0 to about 3 cm of water. This is too low to feed the 0–17 cm of the vena cava or the 15 cm diastolic pressure of the branchial hearts. Peristaltic contractions sufficiently powerful to flatten the vessels can be seen travelling along the arm veins at about 10 cm s^{-1} and pressure measurements indicate that they continue around the cephalic veins towards the vena cava (in *O. dofleini* the cephalic veins are deeply buried in connective tissue and difficult to observe; in young *O. vulgaris* peristaltic contractions are obvious). When simultaneous recordings are made from arm and cephalic veins, and from the vena cava, a succession of pressure changes can be seen (Fig. 3.8). In the first part

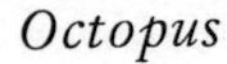

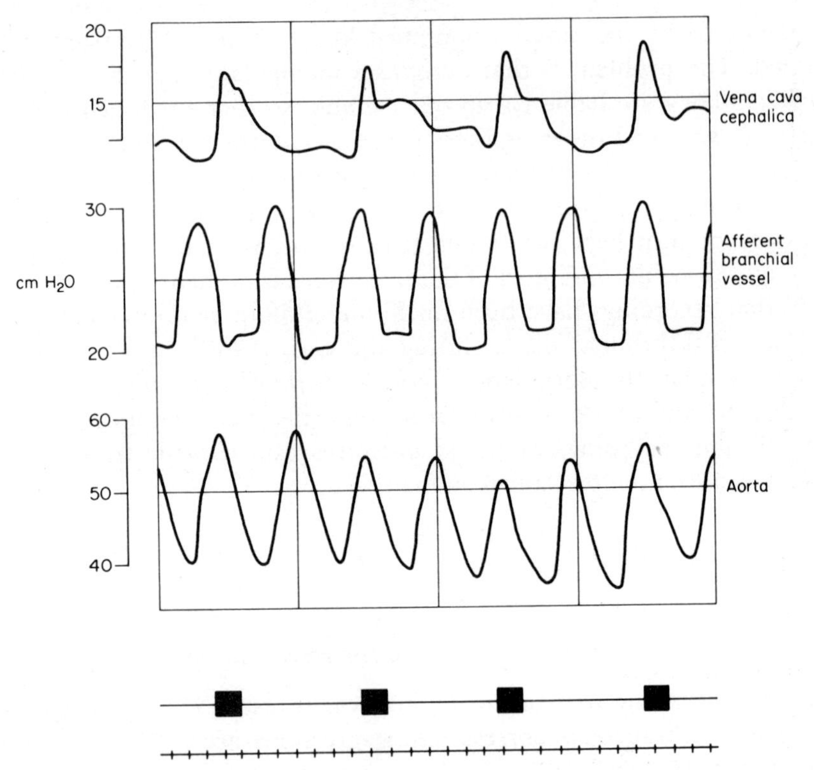

Fig. 3.7 Simultaneous pressure records from vena cava cephalica, one afferent branchial vessel and the dorsal aorta. Time marks: 1 s. Solid black bars indicate time of expirations. The record reads from left to right (from Johansen and Martin, 1962).

of a venous pumping cycle the arm vein pressure maximum occurs during a vena cava minimum, filling the cephalic vein, which in turn empties into the vena cava when this is at minimal pressure during the next inspiration; there is a stepwise increase in mean venous pressure as the returning blood approaches the branchial hearts (Smith, 1962).

Fluctuations in pressure become more and more irregular towards the periphery. Those in the vena cava are normally in time with respiratory movements of the mantle (Fig. 3.7). But the arm veins contract sporadically and often out of phase with one another; one section of a vessel may be contracting while another is quiescent and it is clear that a number of factors, quite apart from pressure further along the line, are contributing to the activity of each section of the venous return at any one time.

There has been some investigation of these factors. One is nervous

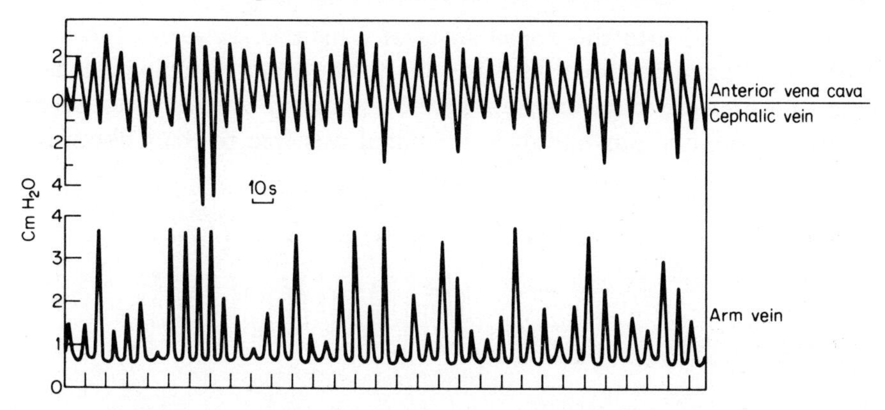

Fig. 3.8 Differential pressures between two catheters in arm and head veins. In the upper record, a rising line indicates a rise in pressure in the anterior vena cava while a falling line indicates a rise in pressure in the cephalic vein; it does not indicate a negative pressure in the vena cava. The recording of the arm vein pressure gives a direct record of pressure. Read results from left to right (from Smith, 1962).

control. Mislin (1950) working with *O. vulgaris* found that the frequency of contraction in the arm veins doubled to twice that of the respiratory rate when the nerves from the visceral lobe to the arms were cut. Mislin and Kauffmann (1948) showed that isolated lengths of vein pulsate faster when their perfusion pressure is increased.

3.2.4 *Blood pressure and heartbeat in exercise*

As *O. dofleini* begins to move slowly about its tank, the systolic and diastolic pressures of the systemic heart rise (from 50 to 70 and from 35 to 50 cm water in a typical series of measurements). Presumably, these rises are at least partly attributable to an increase in peripheral resistance as the animal's muscles contract. At the same time the heartbeat rate rises from 6 to 8 beats per minute. A surprising long time of 10 to 15 minutes is needed to return to resting values after as little as 3 minutes of gentle exercise (Johansen and Martin, 1962).

O. dofleini is an unusually large octopus and by all accounts rather inactive even in the sea (Cousteau and Diolé, 1973). One might expect *O. vulgaris,* which is a smaller, more active creature, to be less extended by this sort of stress. However, it too shows prolonged effects even of gentle exercise. Fig. 3.9 shows pressure recordings made through a

T-piece inserted into the dorsal aorta of a male *O. vulgaris* of 700 g. The animal itself is shown in Plate 5f. Apart from a fine pipeline running to a pressure transducer above the aquarium, which did not seem to disturb the octopus unduly, the animal was free to roam about its

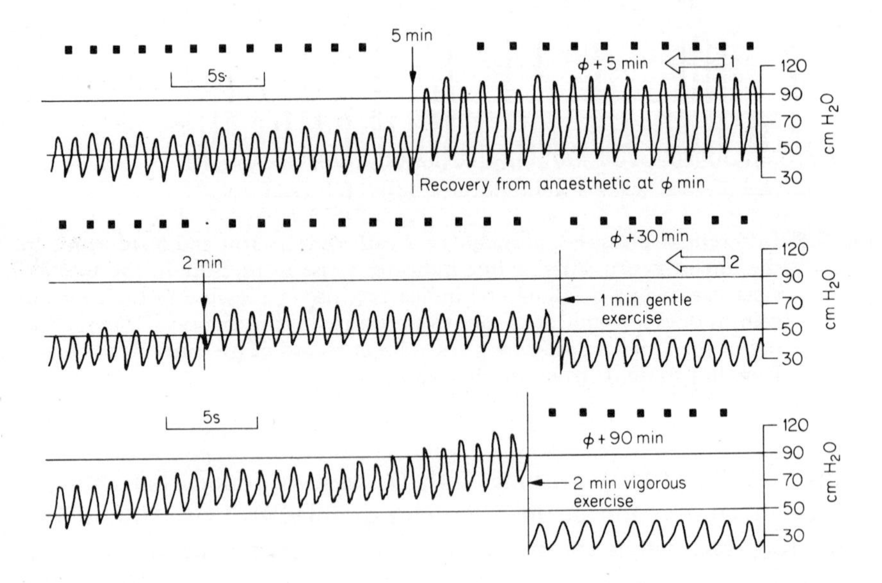

Fig. 3.9 The aortic pressure of *O. vulgaris* and the effect of exercise. Records read from right to left, top to bottom. The record is timed from the moment when respiration began (time ϕ). Black squares show the respiratory rhythm (inspiration). Male of 700 g; the animal itself is shown in Plate 5f. Traced from a cardiograph record (Wells, unpublished results).

tank. As Fig. 3.9 shows, even walking once up and down its aquarium, a distance of 2 m, nearly doubled its blood pressure, while more vigorous and prolonged attempts to escape from a hand in its tank trebled the animals resting systolic pressure from about 40 to about 120 cm of water. The effect would presumably have been even greater had it been possible to record during rather than only after the period of exercise. In this instance, the heartbeat had still not returned to its pre-exercise 'resting' level 10 minutes later.

3.2.5 *Other forms of stress; sudden stimuli and sex*

If the octopus is surprised by a sudden movement on the part of an observer, or a thump on the side of its tank, the systemic heart may miss one or more beats. This is sometimes associated with a 'dymantic' display (Section 5.1.5) and could be due to a sudden tensing of the muscles. Cardiac arrest in response to shocks of these sorts rapidly habituates when the stimuli are repeated.

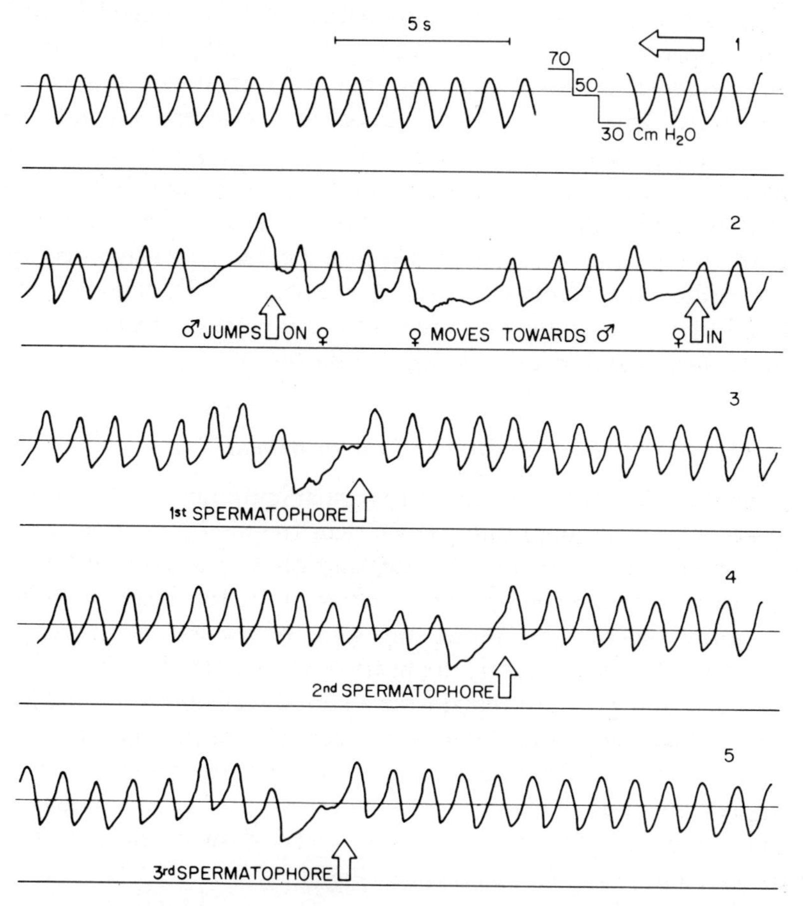

Fig. 3.10 Copulation and heartbeat in *O. vulgaris.* Mating rapidly followed the introduction of the female into the male's tank. Cardiac arrest occurred at 'first sighting', as the female moved towards the male, as he moved over to pin her down and subsequently at each ejaculation. Runs 1–5 start at intervals of about 2 min. Male of 555 g, Naples 26.8.75 (Wells, unpublished).

Fig. 3.10 shows the effect of introducing a female octopus into a tank with a male. The male's heart missed a beat when the female was dumped into his tank, but he lost no time in grabbing her and inserting his hectocotylus. The subsequent records show a beat missed at each spermatophore ejaculation (Section 5.1.5), but there is no indication of a change in beat frequency or amplitude associated with copulation. Octopuses take these things very calmly.

Together with changes due to exercise and arrests attributable to sudden tensing of the body musculature, any prolonged recording reveals apparently spontaneous changes in amplitude that are not obviously correlated with anything that the animal can be seen to be doing. From time to time the heart stops for several seconds (as it does, for longer periods, in *O. dofleini*). The mean pressure, moreover, varies very considerably in the same inactive animal at different times, so that it is probably incorrect to think in terms of a 'resting' level with any fixed values. Control of the heartbeat is evidently as complex as one would expect from the anatomy (Section 3.2.7), and profoundly affected by what is going on in the brain and parts of the body remote from the hearts and mantle (Wells, unpublished).

3.2.6 *Oxygen (carbon dioxide) and the heartbeat*

In general the heartbeat remains very regular in timing, and the response to stress is to alter amplitude rather than frequency. If, however, the animal is placed in a small container and obliged to recycle the same water, the heartbeat slows progressively to only a few beats per minute, long before the octopus shows any signs of respiratory distress. Returned to well-aerated water, the heart beat is restored within seconds, while pulse and pressure remain for a period well above normal levels. The response is apparently due to oxygen lack rather than to carbon dioxide excess, or to the accumulation of some other metabolite since it can be evoked by placing the octopus in fresh seawater that has been boiled. Animals recovering from anaesthesia show a similar elevation of stroke volume and mean pressure (Fig. 3.9).

3.2.7 *Control of the heartbeat; nervous anatomy*

The three hearts all receive branches from the visceral nerves. Fig. 3.11 shows the gross anatomy (in *Octopus*, from Young, 1967), and Fig. 3.12 some of the details, here worked out for *Sepia* by Alexandrowicz (1960a). In octopods, a part of the cardiac ganglion appears to have moved forward, so that an extra relay point, the fusiform ganglion, is found, higher up along the visceral nerve. The auricles are innervated from the fusiform ganglia in octopods, and the commissure connect-

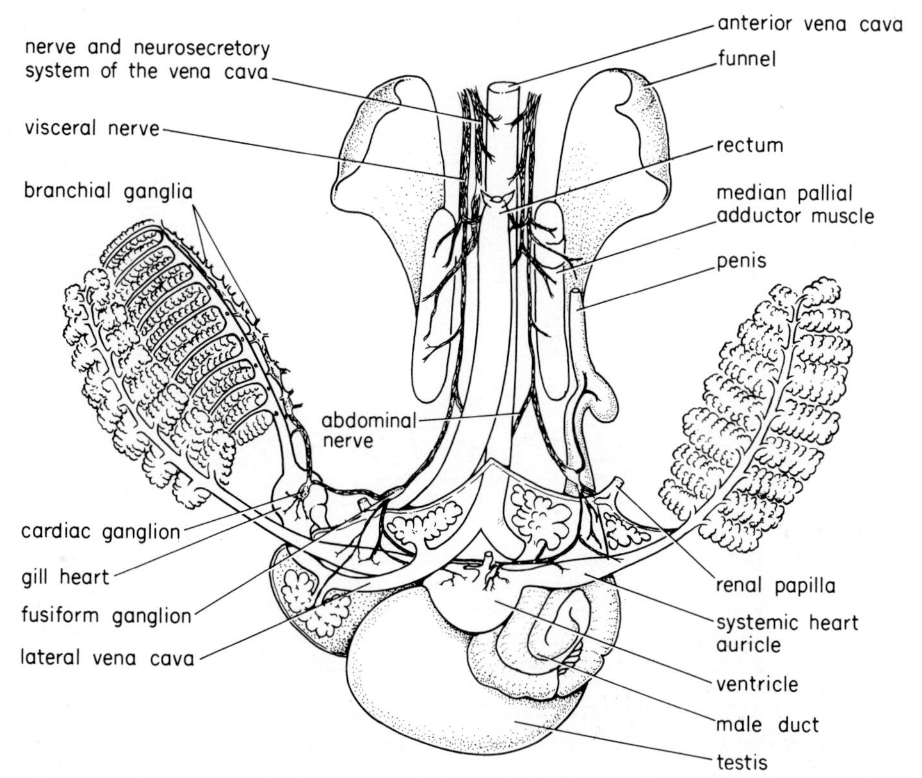

Fig. 3.11 The nerve supply to the hearts. Dissection from the ventral surface similar to that shown in Fig. 2.2. Semi-diagrammatic, with the branchial and cardaic ganglia shown on the right side only (after Young, 1967).

ing the two sides also originates there, rather than from the cardiac ganglia. In octopods (though not in decapods) each cardiac ganglion has a core of muscle fibres, and itself pulsates (Alexandrowicz 1963). In both groups the hearts are innervated both directly from the suboesophageal brain, and indirectly via relays in the fusiform and/or cardiac ganglia. In *Octopus* each of these ganglia contains some 4000 neurones and gets sensory feedback from the hearts and gills as well as inputs from elements in the excretory and reproductive systems. From the anatomy one would expect to find local reflexes superimposed on an overall central nervous control.

3.2.8 *The rhythm of the hearts*

The normal sequence of the heartbeats is a contraction of the vena cava followed by contraction of the branchial hearts. Then there is a

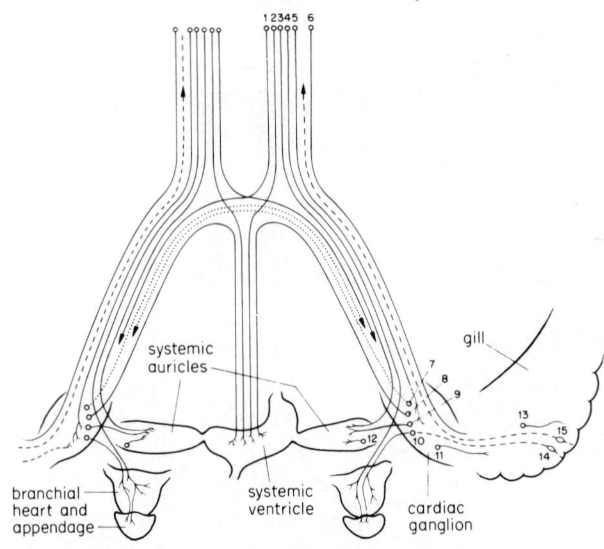

Fig. 3.12 Diagram showing the innervation of the hearts and the gills in *Sepia*. 1–6, neurones with cell bodies in the central nervous system; 7–11, neurones with cell bodies in the cardiac ganglion; 12, neurones in the auricle; 13, neurones with cell bodies in the branchial ganglia; 14, 15, sensory neurones in the gills (after Alexandrowicz, 1960a).

pause. Then the efferent branchials contract and a single peristaltic wave passes down the auricle on each side, filling the systemic ventricle, which contracts in its turn. One-way valves prevent any return flow from the aorta to the ventricle, or from the ventricle to the auricles (Fredericq, 1914). In *O. vulgaris*, ventricular systole occurs about once every second (Figs. 3.9 and 3.10). In the much larger *O. dofleini*, at rest in its aquarium, rates of 6–18 per minute are found (Johansen and Martin, 1962). The heartbeat often coincides with the respiratory rhythm, but is not tied to it; it is commonly more frequent than the movements of the mantle (Figs. 3.7 and 3.9 above) and it may continue when respiratory movements have temporarily ceased. Occasionally, the systemic heart will stop altogether (for a minute or two in *O. vulgaris* (Wells, unpublished results), an hour or more in *O. dofleini* (Johansen and Martin, 1962)). Cardiac arrest does not necessarily mean that circulation stops altogether, because the arteries and veins may continue to move the blood, albeit more slowly than usual.

3.2.9 *Myogenic and neurogenic contractions*

In isolation, every component of the heart's system will contract rhythmically, but some seem to be more dependent upon passive extension than others. Thus the branchial hearts will beat feebly in isolation if laid in a dish under seawater (with or without the cardiac ganglia attached), while the systemic ventricle will not, unless it is fed continuously under slight pressure. Short sections of vena cava will contract regularly, while similar sections of the auricles will not unless they are passively distended. This might suggest a pacemaker based on the branchial hearts or the vena cava, but for the apparent dependence of the whole rhythm on the integrity of the connections between fusiform and cardiac ganglia; if these connections are cut (in the intact animal, pegged out on its back with the mantle split), the branchial hearts contract irregularly, the gills distend with blood and the ventricle ceases to contract because it is no longer filled. A similar condition arises if the visceral nerves are cut above the level of the fusiform ganglia. Ransom (1884), followed by Bottazzi and Enriques (1901, *O. macropus*) and Fredericq (1914) carried out a series of experiments in which the visceral nerve was cut on one side only. Electrical stimulation of the cut end of a visceral nerve stops the ventricle and the auricle on the same side in diastole. The corresponding branchial heart contracts and stays contracted; the branchial heart on the other side relaxes. These results are consistent with the findings from isolated nerve-heart preparations, discussed below, which also show exitatory and inhibitory effects of stimulating the visceral nerve. Stimulation of the central end of the cut nerve produces further relaxation of the branchial heart on the unstimulated side. Ransom (1884) and subsequent reviewers (see Hill and Welsh, 1966) have interpreted these results as showing that the branchial hearts are normally driven reflexly from the ventricular rhythm, itself mainly dependent upon a direct CNS input. Preliminary results with largely intact animals moving freely in their tanks after implantation of electrodes into the branchial hearts and a cannula into the dorsal aorta suggest, however, that this classical story may apply only to acute preparations. In free-moving animals that have fully recovered from anaesthesia the three hearts continue to pulsate in an essentially normal manner, with the two branchials in phase and preceeding the systemic, after section of both visceral nerves. Removal of the fusiform ganglia allows the branchial hearts to beat out of phase, but they generally fall into step after a while and the beat of the systemic is then quite normal. Removal of the cardiac ganglia results in a fluttering beat by the branchial hearts and a drop in blood pressure; the systemic continues to beat normally when it is filled. These results suggest that the pacemakers for the system lie in the cardiac ganglia rather than in the systemic heart. The whole system can operate with both the cardiac and the fusiform ganglia removed, the branchial and systemic hearts continuing to beat in a well-integrated manner (though typically at a rather low pressure) for many hours while the animal is at rest. As soon as the animal is handled or forced to move about, however, the beat is disrupted and blood pressure falls. The hearts may stop altogether, presumably because an increase in peripheral resistance cuts off the venous return in the absence of a preemptive rise in pulse amplitude and mean blood pressure (Wells, in preparation).

3.2.10 *The isolated heart preparation*

The isolated cephalopod systemic ventricle will continue to beat for many hours if perfused with a pressure head of blood or seawater. It is then easy to test the effect of pharmacological reagents. In a slightly more elaborate preparation, the innervation can be retained and the effect of stimulating the visceral nerve can be studied. A number of experiments has been made using these preparations; Krijgsman and Divaris (1955) is a useful review.

All authors agree that stretching or internal pressure greatly increases the regularity, frequency and amplitude of the heartbeat *in vitro*. Co-ordination of the systemic auricles and ventricle is lost in the un-stretched heart, so there is no muscular conduction of the rhythm from auricle to ventricle as in vertebrates. Stimulation of the visceral nerve to the ventricle has complex effects that continue after the end of stimulation. Thus Fredericq and Bacq (1939; working with *Octopus* and *Eledone*) found that stimulation (20 V, 10 μF condenser discharges at 300 min^{-1}) typically inhibited contraction; when stimulation ceased, there was (1) a latent period of 0.5–2 s before the return of the beat; and (2) a subsequent, temporary increase in frequency and amplitude. Both were dependent upon the number of stimuli given; the longer the period of stimulation, the shorter the latent period and the greater the increases in frequency and amplitude. These results are most readily explained by supposing that the ventricle gets a mixed excitatory/inhibitory nerve supply. Simultaneous excitation of both sets of endings releases both sorts of chemical transmitter. The inhibitory transmitter reaches threshold levels first and is destroyed more rapidly than the excitatory.

Fredericq and Bacq (1939) made no guess as to the nature of the transmitters they were postulating, but a variety of pharmacological experiments (references see Krijgsman and Divaris, 1955) suggest that the inhibitory transmitter is acetylcholine. The effect of injecting acetylcholine into the fluid perfusing the heart *in vitro* is cancelled by curare. Eserine and atropine are ineffective, but this is not inconsistent with the results from other molluscs. Acetylcholine has been isolated from the CNS of *Octopus* (Bacq and Mazza, 1935).

The identity of the excitatory transmitter is less well established. *In vitro,* adrenalin and noradrenalin excite, increasing frequency and amplitude (Fänge and Østlund 1954). Noradrenalin, dopamine, octopamine and 5-hydroxytryptamine (5HT) have all been identified from the octopod nervous system (Florey and Florey, 1954; Juorio, 1971; Juorio and Molinoff, 1974; Matus, 1973) and 5HT has been shown to cause increased ventricular contraction in intact, non-anaesthetized *O. dofleini* (Johansen and Huston, 1962).

Hormonal control of the heart beat *in vitro* is discussed in Sections 6.2.1 and 6.2.2 below.

3.2.11 *Drugs and the circulation in the intact animal*

The effect of drugs on the heart of an intact animal will be complicated by effects on the rest of the vascular system. So far only one report has appeared of an attempt to show the effects of heart accelerators and inhibitors in the intact, unanaesthetized octopus. As usual, the collaborative *O. dofleini* has formed the testbed, with cannuli used for injection and pressure recordings inserted into the aorta, the vena cava and the afferent and efferent branchials. Acetylcholine, as expected, slowed the heartbeat and caused vasodilation with a consequent drop in blood pressure. Adrenalin and noradrenalin might have been expected to cause an increase in the heartbeat rate and/or an increase in mean pressure, but do neither; diastolic pressure and heartbeat both fall again, seemingly due to vasodilation. Only 5HT excites as expected (Johansen and Huston, 1962). It is not very easy to relate these experiments to the probable course of events in the normal animal. Pouring neurotransmitters into the bloodstream is not much more physiological than testing them *in vitro*; in life they presumably appear in minute quantities at the nerve endings in the heart muscle and are scarcely likely to escape to the peripheral circulation. These results perhaps mimic effects on the more distant arteries and veins that would normally be produced by the CNS but not necessarily simultaneously with changes mediated via the visceral nerve.

Perhaps the most interesting feature of the 'intact animal' experiments is the speed with which quite distant elements of the heart system react to local injection of drugs. Thus the fall in pressure induced by infusion of noradrenalin into the efferent branchial is reflected almost simultaneously in the corresponding afferent. Infusion into the afferent branchial on one side slows the gill hearts on both sides simultaneously. These must be nervous effects, being far too quick for the blood to circulate (Johansen and Huston, 1962).

3.2.12 *An attempt to summarize*

The three hearts of an octopod are linked by blood vessels that are themselves contractile. All parts of the system seem to be stimulated by stretch, so that a self-regenerating wave of contraction can be transmitted by fluid pressure. In the absence of fluid pressure, there is no transmission of a contractile wave from one element of the system to another (unlike vertebrates). Normal co-ordinated activity of the hearts seems to depend upon connexions, with pacemakers in the cardiac ganglia linked by a commissure through the fusiform ganglia. There are bilateral and unilateral reflexes affecting the branchial hearts which may not have to pass through the CNS, and an over-riding CNS control that can accelerate the hearts in anticipation of exercise, or stop them altogether. Moreover there are bound to be pressure controls associated with mantle movements, flow to the branchial heart appendages (see Section 3.3.4) and the movements of the gills, themselves served by a chain of ganglia connected to the cardiac centres. Every one of these will affect the pressure imposed on some other part of this pressure-sensitive system.

3.2.13 *Blood; oxygen and carbon dioxide transport*

The oxygen carrier in cephalopod blood is haemocyanin, in an approximately 9% solution. The molecular weight is about 25 000, with a minimal oxygen carrying unit of twice that size, since 1 molecule of oxygen combines with two of copper. In fact, the circulating haemocyanin exists as giant stable aggregates with a molecular weight (from sedimentation) of about 2 785 000 in *Octopus vulgaris*; some 98% of the total blood protein is in this form. The physical and chemical

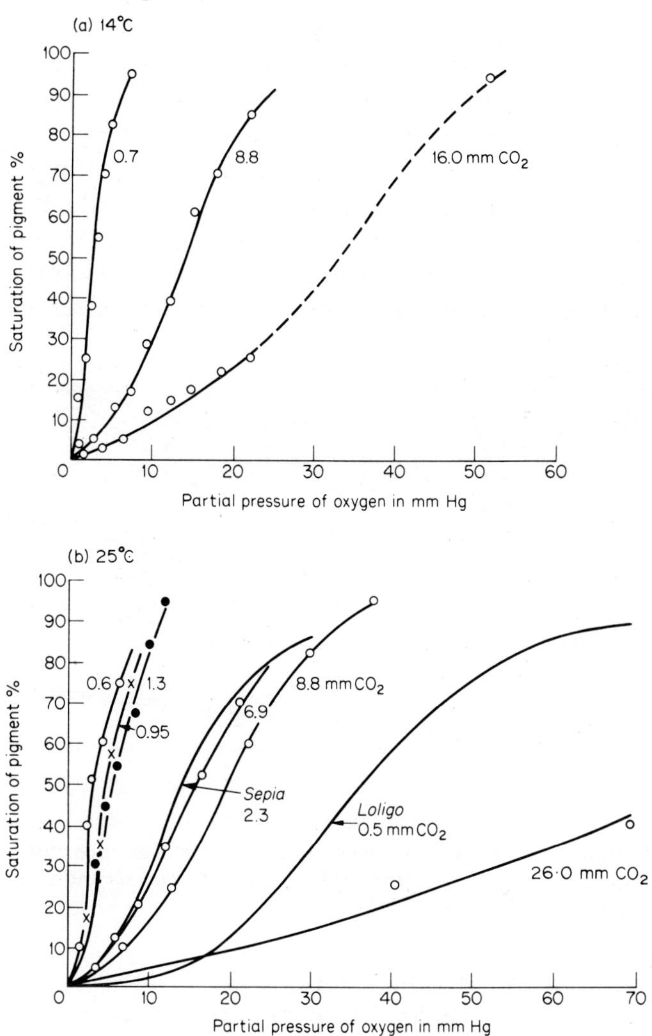

Fig. 3.13 Oxygen dissociation curves for *O. vulgaris* at 14°C and 25°C. There is a marked Bohr effect. (b) includes similar curves for *Sepia* and *Loligo* at 25°C; *Octopus* remains saturated at a much lower partial pressure (after Wolvekamp, 1938).

composition of molluscan haemocyanins has been reviewed by Ghiretti (1966).

As an oxygen carrier cephalopod blood compares poorly with the bloods of vertebrates, having a maximum content of about 4.5 vols % in *Octopus*, where the haemocyanin solution is probably as concentrated as it can reasonably be if the viscous fluid is to pass along the capillary system. There is a marked Bohr effect (Wolvekamp, 1938, Fig. 3.13). 70–80% of the oxygen is removed from the pigment as it passes through the tissues (Hazelhof, 1939; Winterstein, 1909). Haemocyanin, like haemoglobin, is poisoned by carbon monoxide, though its affinity for the gas is not very high and it is readily displaced by oxygen when exposed to the air (Rocca and Ghiretti, 1963).

Octopus blood can carry far more carbon dioxide than seawater. In this it resembles crustacean bloods rather than the blood of less active molluscs; Fig. 3.14 shows the CO_2 dissociation of *O. macropus* blood,

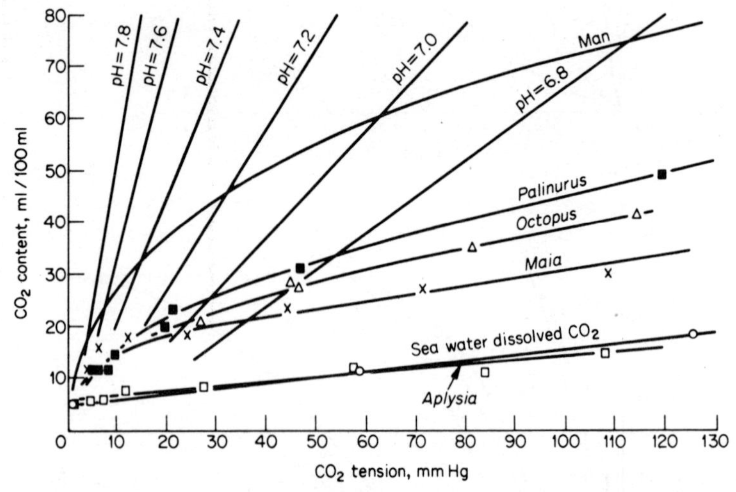

Fig. 3.14 Carbon dioxide-carrying capacity of *Octopus macropus* blood compared with that of man, two decapod crustaceans, *Aplysia* and seawater (after Parsons and Parsons, 1923).

compared with that of man, two decapod crustaceans, and *Aplysia* (Parsons and Parsons, 1923). No quantitative analysis of the causes determining the shape of the CO_2 curve has been made for a cephalopod; but it is known that deproteinized haemocyanin cannot act as a carrier, and it is speculated that the mechanism must be essentially similar to that of CO_2 transport by haemoglobin-pigmented bloods (see Hoar, 1966).

The relation between CO_2 tension and O_2 dissociation shown in Fig. 3.13 must be considered in relation to the amounts of carbon dioxide actually present in the living animal. *Octopus vulgaris* arterial blood contains about 4% CO_2 and even samples from the vena cava carry less than 8% (Parsons and Parsons, 1923; Winterstein, 1909). 4–8% CO_2 is found at partial pressures of around 5 mm Hg, so the physiologically relevant dissociation curves shown in Fig. 3.13 lie well over on the left of the series. Compared with *Sepia* or *Loligo*, the O_2 dissociation curves for *Octopus* show very low unloading tensions. It is not immediately obvious why this should be so, since seawater will always be saturated with oxygen in the open sea (Wolvekamp, Baerends, Kok and Mommaerts, 1942). But the animals choose to live in holes, and may withdraw deeply into these if threatened; the difference between octopus and squid bloods could be an adaptation to the very local oxygen deficiencies that octopuses may generate by their own habits.

3.2.14 *Haemocyanin synthesis in the branchial glands*

The origin of haemocyanin in the body has been something of a puzzle until quite recently. The copper is all derived from the animals crustacean and molluscan food and it is stored, in the first instance, in the hepatopancreas (Ghiretti and Violante, 1964) but the haemocyanin is not, it seems, synthesized there; instead, it is almost certainly formed in the branchial glands, underlying the gills. It is clear from the structure of the branchial glands at EM level that they are synthesizing something in large quantitites; nucleoli are prominent and the cells are packed with rough endoplasmic reticulum. Between the channels in the reticulum, in vacuoles, and elsewhere in the branchial gland cells there are large numbers of spherical particles that appear to be identical with the haemocyanin found in blood vessels (Dilly and Messenger, 1972; Schipp, Höhn and Geinkel, 1973). A second line of evidence arises from work on the fate of ^{14}C-leucine injected into the bloodstream. Much of this material appears later as a labelled protein in the blood; immunological techniques show that the labelled protein is haemocyanin. If the branchial glands are removed surgically, or destroyed by freezing, no labelled haemocyanin is made, and the animals die within a few days (Messenger, Muzii, Nardi and Steinberg, 1974). Removal of one gland permits survival and growth and is invariably followed by hypertrophy of the remaining gland (Taki, 1964).

3.2.15 *Terminal respiration*

Within the cells of the body, iron-based pigments take over from copper. The oxygen uptake of slices from a variety of tissues is very sensitive to oxygen tension (it doubles in a Warburg apparatus when oxygen is substituted for air) and is inhibited almost totally by 10^{-3} M KCN. The tissue system is sensitive to carbon monoxide, and the inhibition is reversed by light. Difference spectra show a complete set of cytochromes a, a_3, b and c. These and the flavins present in octopus tissues appear to be very similar to those found in mammals (Ghiretti, Giuditta and Ghiretti, 1958).

3.2.16 *Blood cells, phagocytosis and blood clotting*

Cephalopods, like other molluscs, have amoebocytes in their blood. The typical size range in *O. vulgaris* is 6–7 μm, with a lobulated or crescent-shaped nucleus filling much of the cell. At EM level, the cytoplasm is found to be packed with mitochondria, endoplasmic reticulum and granules (Barber and Graziadei, 1965). These amoeboid cells can be found throughout the blood system but do not appear to escape from the body; they are not, for example found in the gut, as in most other molluscs (Bidder, personal communication, but see Section 3.3.12). The number of amoebocytes in the blood can vary enormously. A few seem always to be present. But in maturing females, for example, there appears always to be a gross increase in numbers; in one small animal, maturing naturally at the exceptionally small size of 200 g, over 100 mg (wet weight) of wandering cells were recovered from 4 ml of blood. A similar increase has sometimes been observed in octopuses that have become infected with bacteria after surgery.

It seems probable that the wandering cells found in the blood are part of a population that includes the large number of more or less sessile phagocytic cells found among the connective tissue and in the intercellular spaces of certain organs. Fine carbon particles (12 nm ±) injected into the bloodstream of *Eledone* are taken up by phagocytes in the posterior salivary glands and in the gills. Both organs become blackened, while the blood itself is cleared of carbon within a few hours. The accumulation appears to be permanent. Carbon is also found in the white body, where phagocytosis is limited to macrophages lodged among the cordons of leucopoietic cells, a finding that is in keeping with Bogoraze and Cazal's (1944) claim (disputed by Bolognari, 1951, see Section 3.3.9) that the white body includes two distinct sets of cells, leucoblasts and 'nephrocytes'. A little carbon deposit is found in the anterior salivary glands, but none elsewhere (Stuart, 1968).

Bacteria are cleared from the intact circulation of *O. dofleini* at about the same rate as carbon particles. Bacteria injected into an isolated section of an afferent branchial artery were not. Bayne (1973), apparently unaware of Stuart's (1968) earlier work on *Eledone*, claimed that this showed that the amoebocytes found in the blood could not be responsible for bacterial clearance, but his experiment is unconvincing. No counts were given of the number of amoebocytes present, and the bacterial count itself fluctuated widely, falling to about half in the first hour after injection and then recovering. No antibacterial agent was found in *Octopus* serum, even of animals previously injected with bacteria, and there was no improvement in the rate of clearance of successive doses injected into the intact bloodstream (Bayne, 1973).

Although octopus blood plasma appears to lack any mechanism for agglutinating foreign particles such as bacteria, yeast or vertebrate red blood cells, it does include a factor that renders these particles attractive to the phagocytic blood cells. Red blood cells first soaked in *Eledone* serum, washed, and then presented to *Eledone* macrophages in culture medium were devoured, while untreated cells were not. Antibodies from rabbit serum, produced by injections of *Eledone* serum, agglutinated red cells dipped in *Eledone* serum, so the effect is due to the attachment of some component from the latter, just possibly haemocyanin since this is the commonest protein present (Stuart, 1968).

There is general agreement that the wandering amoebocytes originate in the 'white body', a conspicuous structure enclosed in one of the blood sinuses that wrap around the optic lobe and optic nerves, cushioning the eye within its orbit (Boycott and Young, 1956a). The white body consists of strings of cells suspended from the thin collagenous capsule that forms the wall of the sinus. Most of the spongy mass of tissue is made up of leucocytes in various stages of development, and many mitotic figures are seen, particularly in young individuals. In addition there is a network of supporting fibres and a system of blood vessels that is particularly well developed in octopods compared with other cephalopods (Bolognari, 1951). Necco and Martin (1963) studied the multiplication of white body cells *in vitro*. They concluded that about one cell in a thousand enters into mitosis per hour; there is an unusually long telophase.

There is no fibrous clotting of cephalopod blood. Wounded animals do not bleed to death because most of their blood vessels are muscular, and because the amoebocytes clump to plug small holes. Electrophysiologists in particular have found a disconcerting tendency for octopuses to shut down the blood supply to any locally exposed nervous tissue; the cells stop transmitting impulses and a first assumption must be that the tissue is dead. After an hour or two (arms) the vessels relax, flow begins again and activity returns (Boycott, Lettvin, Maturana and Wall, 1965; Rowell 1966). Small leaks are stopped by amoebocytes, by a mechanism that seems to be common to all molluscs. The loose cells adhere to damaged tissue, and in turn form an attractive substrate to which further amoebocytes adhere by their pseudopodia; a network is formed entangling still further blood cells. The protoplasmic bridges shorten and plug the wound. In a secondary process, fibrocytes enter and lay down connective tissue to form a more permanent repair (Drew, 1910, Drew and DeMorgan, 1910, Cowden and Curtis, 1973).

3.3 Excretion

3.3.1 *Osmotic and ionic regulation*

Cephalopods all live in the sea and their bloods are nearly isosmotic with seawater. Tissue fluids, including the transparent, almost protein-free contents of the large eyes and the endolymph of the statocysts (Amoore, Rodgers and Young, 1959; Robertson, 1953) are all within 1% of the concentration of the blood plasma. The same is true of the fluids in the kidney, discussed below. Cephalopods, in general, do not appear to osmoregulate (Potts and Todd, 1965; Robertson, 1964).

The only known exception to this generality is found in the buoyancy mechanism of *Nautilus, Sepia,* and *Spirula,* which depends upon extraction of ions from a fluid, at first isosmotic with seawater, contained in their chambered shells. Ions are removed by active transport, water follows and gas comes out of solution under the reduced pressure, partly filling the impermeable shell and lightening the animal (Denton, 1974).

Ionic regulation is by contrast widespread. The concentrations of the common ions so far examined all differ markedly from seawater. Table 3.2 shows their concentrations in the blood plasma of *Eledone*, quoted in terms of the differences from a sample that has been dialysed against seawater across a collodion membrane. 100% would represent the passive equilibrium value for any given ion, differing from that of seawater because of the plasma proteins. Calcium tends to form an indiffusable calcium-protein complex and its high value is at least in part attributable to this. Potassium plasma values may be exaggerated because of amoebocytes that have disintegrated during collection (Potts and Todd, 1965). There is no reason to believe that the estimates

Table 3.2 Concentration of some common ions in the blood plasma of *Eledone cirrhosa* as per cent concentrations in plasma dialysed against seawater (after Robertson, 1964).

Na	*K*	*Ca*	*Mg*	*Cl*	SO_4
97	152	107	103	102	77

Table 3.3 Major inorganic constituents of blood, pericardial fluid (PCF), urine and rectal fluid (RF), of *Octopus dofleini* (from Potts and Todd, 1965).

		mM/l (mean)	*Range (and number of determinations)*	*mM/kg water (mean)*
Na	SW	407		410
	Blood	371	(361–387) (4)	402
	PCF	399	(393–404) (4)	405
	Urine	385	(353–410) (6)	395
	RF	400	(394–406) (2)	412
K	SW	9.1		9.2
	Blood	10.3	(9.9–10.9) (3)	11.2
	PCF	9.1	(7.8– 9.5) (4)	9.2
	Urine	14.6	(11.7–15.1) (8)	15.0
	RF	10.3	(9.7–11.1) (3)	10.6
Ca	SW	8.9		9.0
	Blood	8.2	(8.4–7.8) (3)	8.9
	PCF	8.3	(8.4–8.0) (3)	8.4
	Urine	4.6	(2.9–6.8) (6)	4.7
	RF	8.3	(7.8–8.7) (2)	8.6
Cl	SW	475		479
	Blood	447	(410–466) (6)	484
	PCF	476	(470–482) (3)	483
	Urine	442	(400–461) (3)	453
	RF	472	(472) (1)	486
SO_4	SW	24.5		24.7
	Blood	18.2	(26.9–19.9) (4)	19.7
	PCF	19.0	(18.7–19.3) (4)	19.3
	Urine	35.4	(31.2–42.0) (5)	36.3
	RF	31.9	(31.2–32.6) (2)	32.9

of the other ions are incorrect and since they are free to diffuse, the differences found can only mean that the animal is regulating some or all of them (Robertson, 1964).

Urine concentration indicates that potassium and sulphate are actively secreted into the renal sacs, while calcium is withdrawn (Table 3.3). Sodium is very variable; with ammonia as the principal nitrogenous excretory product, it tends to be displaced in animals such as the *O dofleini* used by Potts and Todd (1965) which were fed regularly. Robertson (1949) reported much higher values for *Eledone* but his animals had been starved for several days and sodium in the renal sac fluid was close to seawater level. Chlorine, again, seems to be determined passively, it is relatively low in urine and high in blood, balancing the active secretion of the divalent sulphate and phosphate.

3.3.2 *Kidney structure and function*

Cephalopods have kidneys built on essentially 'standard' molluscan lines. There is ultrafiltration into a pericardial cavity followed by resorption of metabolites and secretion of wastes into a duct that empties into the mantle cavity.

Fig. 3.15 shows the gross anatomy of the duct system in *Octopus.* Quite a lot is also known about the fine structure of this system. This will not be surveyed here; Turchini (1923) is a useful review of early work at light microscope level, while more recent cytochemical and

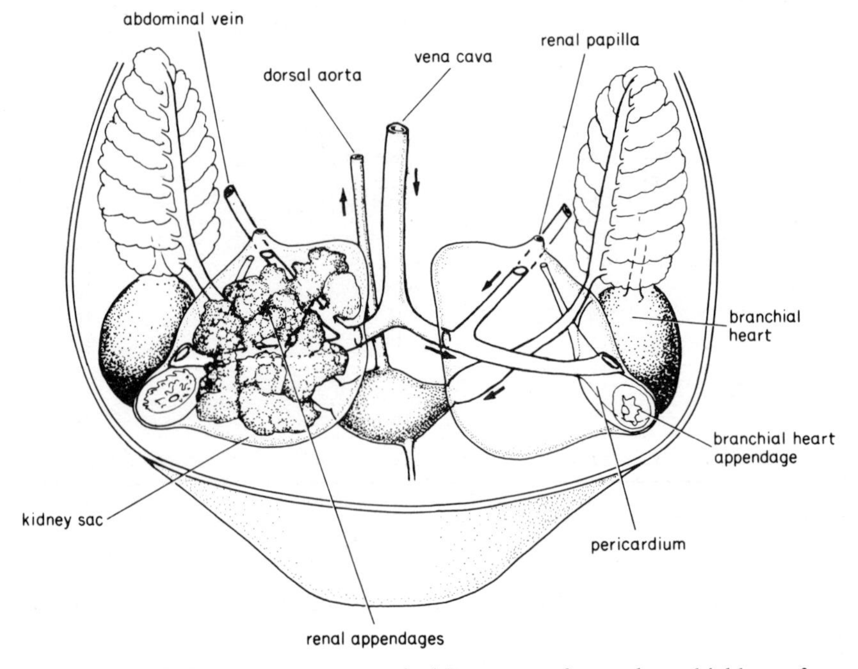

Fig. 3.15 The kidney system. On each side pressure from a branchial heart forces an ultrafiltrate of the blood through the branchial heart appendage into a pericardium. The filtrate passes through a ciliated duct into a further coelomic cavity, the kidney sac. Here it comes in contact with the renal appendages, diverticulae of the lateral vena cava and abdominal vein, which further change its constitution before discharge into the mantle cavity through an aperture in the renal papilla.

EM studies have been reported upon by Schipp and Boletzky (1975), Schipp, Boletzky and Doell (1975) and by Witmer and Martin (1973). The pericardium covers only a part of the gill heart on each side, surrounding the branchial heart appendage, a structure generally considered to correspond with the pericardial gland of other molluscs. The pericardium has a long ciliated duct which opens into a further coelomic space, the kidney sac. This in turn opens into the mantle through a renal papilla. The opening of the pericardial duct lies close to the opening of the renal papilla, a rather strange arrangement in view of the extent to which the urine is now known to be modified within the kidney sac.

In many cephalopods, though not, apparently, in the adult male *O. dofleini* that have been used for most recent experiments on cephalopod kidney function (Potts 1967), there is a duct connecting the gonadial and pericardial coeloms. Young (1967) describes it, from a juvenile *O. vulgaris,* as heavily vascularized and served by an abundant nerve supply. The function of the genito-pericardial duct is unknown, but it might, for example, be important in pregnant females, where there is a very abundant blood flow to the ovary. This comes from the systemic heart and must be at relatively high pressure, so that the fluid found in the ovisac at this stage may well include metabolities that could profitably be resorbed through the walls of the pericardial duct, if not through the walls of the genito-pericardial canal itself.

3.3.3 *Branchial heart appendages*

The filter on which the whole system is based is attached to the side of the branchial heart. The lumen of the heart is continuous with the lumen of the appendage and this interdigitates with a much-branched infolding from the surface of the appendages. There is thus a large area of close contact between the blood space and the coelomic space of the pericardium. Between the two spaces lie two sorts of cell, the podocytes and the lacuna-forming cells. The podocytes are characterized by very numerous end-feet apparently supporting the basal membrane lining the blood space, which often lacks a lining of cells on the blood side. The intercellular spaces between the feet connect with much larger spaces between the lacuna-forming cells (Witmer and Martin, 1973).

The podocytes contain numerous dictyosomes but there is otherwise little sign of secretory activity in *Octopus* (Witmer and Martin,

1973), unlike decapods, where the branchial heart appendage includes further secretory cell types (Schipp and Boletzky, 1975; see also Section 3.3.8 below).

3.3.4 *Pericardial duct*

In *Octopus dofleini*, the pericardial duct is wide enough to insert a cannula. The fluid that can be collected is an ultrafiltrate of the blood (Table 3.3), and has the same freezing point depression and concentration of injected inulin as deproteinized blood (Harrison and Martin, 1965). There is plenty of pressure to drive such a system. Colloid osmotic pressure is only 4.3–5.0 cm of water and the systolic blood pressure, measured in the branchial heart of *O. dofleini* at rest, ranges from 25–50 cm (diastolic about 15 cm). No measurements have been made from within the spongy branchial heart appendage, but since this is in direct connexion with the lumen of the heart the pressure must be very similar there (Johansen and Martin, 1962).

The volume of ultrafiltrate thus produced is considerable, about 10% of the body weight per day in catheterized animals. Blood has been estimated to constitute a little less than 6% of the body weight (see Section 3.2.1 above) so that the whole volume of the blood is being filtered every 12 h or so (Harrison and Martin, 1965).

The structure of the branchial heart appendage includes muscle fibres (Witmer and Martin, 1973) and a very rich nerve supply (Young, 1967) which together suggest that the animal has some control over the filtration rate, as indeed it must if filtration is not to increase greatly when blood pressure and heart-beat frequency rise in exercise (see Section 3.2.4).

A more direct indication of control is Potts' (1965) observation that urine production by the two sides of the kidney can vary considerably, so that for several hours one will produce urine at up to 10 ml h^{-1}, with a barely measurable flow from the other. Production may then be switched to the previously inactive side.

The pericardial canal is not simply a passive transporter of fluid. *In vivo* studies have shown that it actively resorbs glucose (Harrison and Martin, 1965) and amino acids (Martin, 1965). Split and unrolled, it can be used *in vitro* to form a membrane between chambers and will then transport glucose against a concentration gradient (Martin, 1965). The system is poisoned by phlorizin (Fig. 3.16).

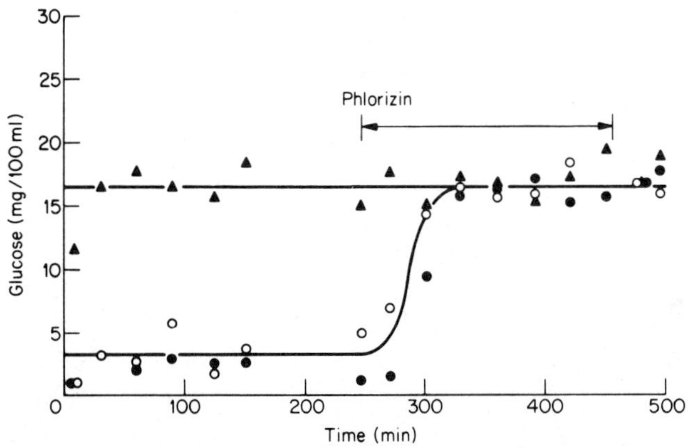

Fig. 3.16 The relationships of the normal concentration of glucose in the blood and pericardial fluid before and after the administration of phlorizon. ▲, blood; ●, right pericadial fluid; ○, left pericardial fluid. Phlorizin given from 245 to 450 min at a dosage of 10.3 mg kg^{-1} (after Harrison and Martin, 1965).

3.3.5 *Renal sacs*

The pericardial ducts empty into the renal sacs. These much larger coelomic spaces are normally filled with a pale yellowish fluid, often cloudy with suspended matter kept stirred by peristaltic movements of the renal appendages. The latter are proliferations of the walls of the vena cava and of the abdominal veins that run to it from the digestive system. In *O. vulgaris* they are bright orange in colour, apart from the purple branchial hearts the most conspicuously coloured objects in the mantle cavity. They are hollow, soft and contractile, with a columnar epithelium of kidney tissue separating blood and urine. This layer has many of the characteristics of the renal epithelia of vertebrates; there is an apical brush border, a much infolded basal labyrinth, many mitochondria and lysosome-like dense bodies; it is clearly an organ likely to be concerned in active transport outwards from the blood-spaces (Schipp, Boletzky and Doell, 1975).

As well as regulating the concentrations of some ion species (Table 3.3), the renal appendages pass nitrogenous and other wastes to the renal fluid, without at the same time adding to its volume, which is entirely dependent upon inflow through the pericardial duct. At least some of the processes involved are active. Thus for example *p*-aminohippuric acid injected into the bloodstream accumulates in the urine at concentrations of up to 170 times its level in the blood. Urea and phenol red are extruded similarly. Secretion is inhibited by 2, 4-dinitrophenol or benemid (Fig. 3.17; Harrison and Martin, 1965).

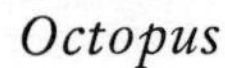

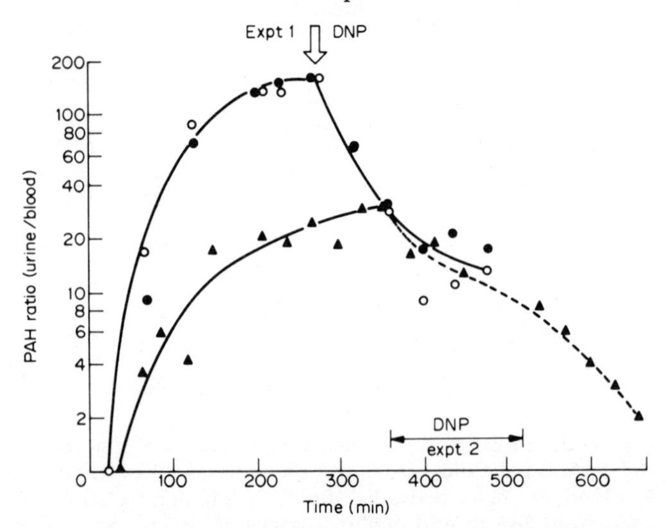

Fig. 3.17 The urine/blood ratios of *p*-aminophippuric acid (PAH) concentrations for each sampling period during perfusion experiments before and after administration of 2, 4-dinitrophenol (DNP). Expt. 1: ●, right renal sac urine; ○, left renal sac urine; DNP given at 300 min in a single dose at 5.2 mg kg^{-1}. Expt. 2: ▲, right renal sac urine; DNP given from 360–510 min, total 7.3 mg kg^{-1} (after Harrison and Martin, 1965).

3.3.6 *Ammonia*

Nitrogenous wastes are voided mainly as ammonia. Potts (1965) found that, in contrast to the pericardial ultrafiltrate, the contents of the renal sacs are normally acid (pH 5.2–5.5). The effect of this would be to trap ammonia diffusing across the renal appendages as NH_3^+, and he was able to show a quantatitive relation between the passage of ammonia and experimentally manipulated changes of the pH of the fluid in the renal sacs. The ammonia content of the renal appendages is, in any case, unusually high, associated with a steady loss of glutamate and amino acids as the blood passes through this part of the kidney. The breakdown to ammonia is so vigorous that there is a net increase in ammonia content as the blood traverses the appendage-bearing parts of the vena cava. Some of the product diffuses down a concentration gradient and is trapped in the acid urine, a larger fraction of the whole runs through the branchial hearts and is lost, presumably again by diffusion, through the gills (Potts, 1965).

Emmanuel and Martin (1956) have estimated that 73% of the non-protein nitrogen in filtered urine is present as ammonia. Together with the presumed much greater loss through the gills, this leaves no doubt that octopus, unlike most molluscs, is overwhelmingly an ammonotelic organism (Potts, 1967). It is interesting in this context that *Nautilus*, the only surviving representative of the ancestral cephalopods, is also now known to be ammonotelic. In this tetrabranchiate there are no gill hearts and the blood is seemingly pumped through the gills by the combined action of the renal appendages and the pericardial glands (≡ branchial glands plus branchial hearts). The pericardial glands let through an ultrafiltrate but the sacs into which they discharge are nowhere confluent with the renal sacs and there is no obvious resorptive structure equivalent to the pericardial duct. The renal sacs themselves are generally crammed with granules of calcium phosphate which may act as a reserve to supply the periodic rapid growth of new partitions in the shell (Martin, 1975).

3.3.7 *Other nitrogenous wastes*

A variety of other nitrogenous substances have been identified from octopuses. Delaunay (1931) has reviewed the earlier work, mainly on *O. vulgaris* and cites (as proportions of total non-protein nitrogen in the urine) urea (2–5%), uric acid (1.4%), purines (25%) and free amino acids (12.5%). Emmanuel (1957) reports some 45 distinct organic compounds, separable chromatographically, from *O. dofleini*. He, with Martin (1956) failed to find uric acid in filtered urine, which suggests that the material is excreted as solid material, extruded directly from the kidney cells, as it is in some gastropods (Bouillon, 1960). In general, there are considerable differences of opinion as to what is and is not present in octopod urine. Potts (1967) in a review of excretion in molluscs generally has stressed the extent to which the identification of substances and in particular any estimate of their relative quantities has depended upon the techniques available at the time. The literature and particularly some of the older work must therefore be viewed with caution; apparent species differences are as likely to be due to techniques as real.

The fluid from octopus kidneys always includes large numbers of Mesozoans, and it is difficult to assess the possible contribution of these organisms to the list of waste products or to the concentrations of free ions in the fluid. It appears, for example, that they concentrate copper which is otherwise, surprisingly, absent from the urine (Emmanuel and Martin, 1956).

3.3.8 *Other sites of excretion: the branchial hearts*

Purines accumulate as yellow granules and as larger violet-coloured concretions within the cells of the branchial hearts; the concretions appear to increase in size as the animals get older. Ammoniacal carmine injected into the bloodstream at first colours the yellow granules and is later (animals sacrificed after one month) found only in the larger concretions. Turchini (1923) considered that these were in all probability permanent.

Decapods' hearts contain only the small yellow granules, and Turchini correlated this with a supposed difference in function of the branchial heart appendages in the two groups. The appendage is more complex in decapods than octopods, and Turchini was particularly impressed by the absence of signs of secretory activity in the ephithelial lining of the organ in octopods. He showed, moreover, that injected dyes accumulated in the gland of *Sepia* but not in *Octopus,* and concluded that in octopods the branchial heart appendage was degenerate and played no part in excretion.

3.3.9 *White body*

The 'white body' in the orbit has also been claimed as a kidney of accumulation. Cazal and Bogoraze (1943) showed cytochemically that purines were present in the white body and asserted that a particular class of cells was responsible for the accumulation of these wastes. Bolognari (1951) denied the existence of morphologically distinct 'nephrocytes' and suggested that the large cells identified as such were in fact leucoblasts. The supposed nephrocytes show many mitotic figures and Cazal and Bogoraze's distinction between these and leucoblasts, based on the presence or absence of granules of (mainly) xanthine breaks down as the animal gets older. In adult octopods, nearly all of the cells in the white body, including those plainly destined to become free amoebocytes, contain yellow granules of this material.

Again, there are differences between octopods and decapods; in both cases the white bodies make blood cells, but in decapods the cells do not contain the yellow granules of nitrogenous wastes. Bolognari (1951) has pointed out the similarity of the granules in the white body of *Octopus* and the granules in the branchial hearts of *Sepia,* and suggests that both are 'kidneys of transformation', handling the breakdown of protein wastes *en route* to the renal appendages.

Now that more is known about excretion in the octopods, and in particular that very large volumes of ultrafiltrate pass through the branchial heart appendages, it would perhaps be wise to re-examine the whole 'kidney of accumulation' concept, for why should an animal bother to accumulate soluble rubbish that is so readily discharged?

3.3.10 *Hepatopancreas*

The liver part of the hepatopancreatic complex produces large quantities of 'brown' and 'grey' bodies, assembled in the liver cells and voided with the faeces. The process of digestion is discussed in Chapter 4. 'Brown bodies' are typical of animals that have fed and will, for instance, include carmine particles that have been taken in with the food; they are produced in very large quantities and form the coloured matter to be seen in the mucous strings that the animals produce. 'Grey bodies' are most numerous during hunger and periods of secretion; they appear to be the cast off ends of secretory cells (Bidder, 1957). The nature of the wastes eliminated in brown and grey bodies has not been studied. Nor has the material derived from the white 'pancreas' which is located at the base of the liver, close to the hepatocaecal duct (Section 4.3.5). Either could include nitrogenous wastes.

3.3.11 *Mucus*

Octopuses, because of their method of feeding (4.3 below), take in comparatively little solid matter from their food, but it includes fish scales and small pieces of crustacean cuticle. The indigestable rubbish is compacted and, like the wastes from the liver, voided in long mucus strings, which are blown away through the funnel. This, and the continual production of mucus at the body surface generally must involve very large losses of mucopolysaccharides and may be important as a sink for nitrogenous materials; so far no attempts have been made to quantify the loss (Delaunay, 1931).

3.3.12 *Phagocytes*

In a number of other molluscs, phagocytic cells collect wastes in the blood and then wander out of the body. In cephalopods solid matter, such as carbon particles injected into the bloodstream, is concentrated by phagocytic cells lodged in the posterior salivary glands, the gills and elsewhere. The common opinion has been that the carbon remains there permanently (Section 3.2.16 above). Froesch (personal communication) has, however, recently shown that phenol red injected into the blood is accumulated by amoebocytes that then migrate through the anterior chamber organ (Fig. 6.17) into the space between the lens and cornea. This space is open to the sea, being enclosed only by two flaps of skin; it could represent an important route for the export of particulate and other rubbish.

CHAPTER FOUR

Feeding and digestion

4.1 Habits and hunting; prey recognition and capture

Octopus vulgaris will feed and grow rapidly on a diet of dead fish in aquaria, and I have seen it eating fish (killed by an explosion) in the sea. So it is potentially a scavenger. It does not, however, seem at all likely that scavenging is important since the animal plainly, and preferentially, attacks live prey. In captivity, food that fails to move is often ignored until touched.

In the laboratory, the octopus will eat crabs and lobsters, bivalves and gastropods, and fish if it can corner them. The animals will also eat each other if not well-matched in size or preoccupied with sex. In the sea, the discarded remains of crab and mollusc shells scattered around inhabited holes testify to a similar range of prey; Altman (1967), for example, lists eight species of molluscan shells and three sorts of crab carapace found in middens at one site in Malta. She also points out that the animals seem to pay no attention to live fish in the sea. Kayes (1974) quotes and confirms a fisherman's report, also from Malta, that the perch *Serranus scriba* often follows *Octopus* on its foraging expeditions and thus gives away the position of the cephalopod, since the fish is comparatively conspicious.

4.1.1 *Prey capture in captivity: crabs and other small moving objects*

In an aquarium, where space is restricted, the most normal pattern of food gathering is a visually directed jet-propelled pounce onto any small object dropped into the tank. The octopus sits in its home, in a flowerpot or amongst a pile of bricks, and watches. It seems always to be awake and alert. If anything unfamiliar moves, the head is raised and the animal faces the disturbance, sideways on, looking out of one eye; quite often the head is raised and lowered several times, as if the octopus were ranging by parallax.

Crabs and other small moving objects are typically trapped under the interbrachial web, following the jet-propelled forward leap; the crab is covered by the outspread arms and gathered in towards the mouth.

The leap itself is commonly preceded by a smooth approach, with the animal sliding out of its home towards the prey. If the prey stops moving, so does the octopus, which may then make jerky movements as though undecided whether or not to attack; it is possible that these have some adaptive value in tending to flush crabs and other crustaceans themselves perhaps undecided whether to move or not. There are complex colour changes, apparently related to the expectations of the octopus, which typically darkens to a deep reddish brown on sighting potential prey; colour changes and their control are discussed in Chapter 10. Having caught the crab, fish or other food dropped into its tank, the octopus normally returns home to eat it. If any further food is available it will collect that *en route* and, indeed, dash out from home to collect more crabs whatever the state of its first victim, to the point where it is unable to retain the struggling elements of its dinner beneath the web.

As soon as it is no longer occupied with actually gathering in the prey, this is killed by a poison produced from the posterior salivary glands, a matter to be dealt with in the section that follows. An octopus will kill several crabs before settling down to eat any one of them.

4.1.2 *Feeding in the sea*

It is difficult to tell how these habits relate to food gathering in nature. The problem is that octopuses can almost certainly see quite as well underwater as a SCUBA diver, who is in any case rendered conspicuous by his bubbles. Most observations, moreover, have been made in broad daylight and relate to animals sitting in their holes or foraging close by. A typical, and nicely illustrated, account is given in Cousteau and Diolé (1973); Lane (1957) is a useful compendium of the older references. It seems that *Octopus vulgaris* is an opportunist, prepared to feed at any time, but most likely to be found foraging away from home between dusk and dawn; that, at least, is when Mediterranean fishermen search for them with lamps, and when they are most likely to be 'out' if spot-checks are made on their homes (Fig. 4.1; Altman, 1967; Kayes, 1974). During these expeditions, the animals will collect a variety of

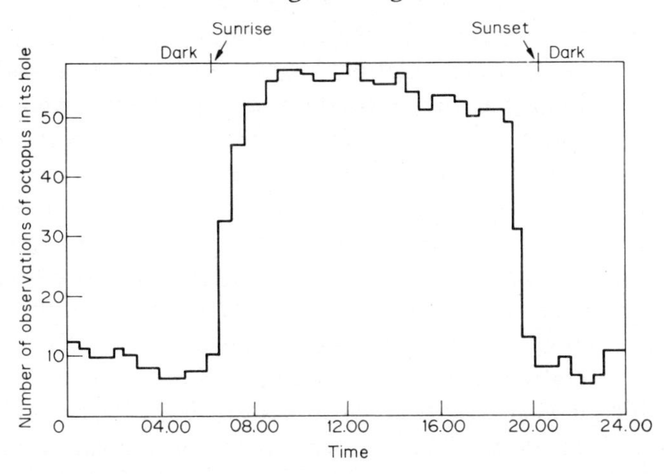

Fig. 4.1 A summary of 7000 observations of octopus holes made at L'Ahrax Bay, Malta, during the period 4th–15th September 1973. Observations of holes known to be inhabited were made every hour; series lasting less than 24 hours have been excluded (from Kayes, 1974).

prey, and it hardly seems probable that the accumulation of several hours hunting is all brought home to be devoured there.

Much of the food of the animals must be gathered by touch. Bivalves move little, and crabs are often hidden in crevices. An octopus, abroad by day, moves over the rocks feeling into cracks and under stones; from time to time it moves swiftly across to gather something that it has detected with one of the arms. *O. cyanea,* observed moving freely over a reef from an observation tower (one of the few series of records where the octopus almost certainly did *not* observe the observer) systematically moved from one dead coral head to the next, enveloping each in turn with the web, while it groped below with the armtips. Recapture on the way home showed it to be carrying several small crabs (Yarnall, 1969).

4.1.3 *Learning to recognize food*

We know a great deal about learning to recognize shapes by sight or touch under laboratory conditions (Chapters 8 and 9), but almost nothing about the extent to which *Octopus* has to learn to recognize its prey in the sea.

A major problem is that *O. vulgaris* has a planktonic phase. Itami, Izawa, Maeda and Nakai (1963) succeeded in rearing the newly hatched animals through this stage, feeding the little octopuses first on shrimp larvae (*Palaemon serrifer*), then on fragments of crab meat and finally on whole small crabs as they settled out of the plankton and onto the bottom. They did not report details of the behaviour of the animals, and other attempts to rear *O. vulgaris* have failed (see Vevers, 1961).

Octopus cyanea, like *O. vulgaris,* has a planktonic stage and we have a little information on the behaviour of the young of this species during the period immediately following settlement. The little (half-gram) octopuses at once establish individual homes, which they defend against their fellows. From their homes they sally forth to pursue and capture small crabs which they take home to eat, their behaviour in every visible manner identical with that of fully-grown individuals (Wells and Wells, 1970). These observations imply innate recognition since the octopuses had, presumably, never seen crabs before. But the same animals had already spent a considerable period growing in the plankton where, it must also be assumed, they were feeding on other crustaceans. We have no means of assessing the contribution made by learning during this period.

Some octopods lay large eggs and omit the planktonic phase. Of these, *O. briareus* (Hanlon, 1975), *O. joubini* (Boletzky and Boletzky, 1969) and *Hapalochlaena maculosa* (Tranter and Augustine, 1973) have now been bred and reared successfully in captivity. But again no attempts have been made to identify the cues that they use to identify food, or the effect of experience on individual behaviour.

Experiments of this sort have, however, been made on prey recognition by the cuttlefish, *Sepia. Sepia* lays large eggs that hatch into young that rapidly settle in sand on the bottom, and then behave very much like adults. They will readily attack the small swimming crustacean *Mysis,* which swarms in places where the adults lay their eggs. The attack, which involves orientation towards the prey and a run in to the correct distance for a sudden strike with the long arms, is made accurately first time, and does not appear to change with practice. Prey is recognized by sight and the little animals are at first highly selective. They will turn to observe other moving objects, but attack only mysids.

Rather surprisingly, the effect of experience is to widen the range of things that will evoke an attack even when that experience is limited to catching mysids. Models of mysids are at first ineffective, but become progressively more attractive over the first ten or twenty days after hatching; the animals finally reach a condition where they may stab at almost any small moving object. It seems that *Sepia* hatches with the

CNS programmed to recognize its likely first meal, and that the programming soon fades, leaving the animal free to learn to recognize a wider variety of prey (Wells, 1958; 1962). Slightly older cuttlefish appear to have to learn to attack crabs circumspectly and preferably from the rear (Boulet, personal communication) and there is plenty of evidence that older cuttlefish, like octopuses, can learn to discriminate (Sanders and Young, 1940; Messenger, 1973a).

In the absence of information to the contrary, it seems quite probable that *Octopus* will be found to behave like *Sepia.* In both genera, the vertical lobe (of proven importance in visual learning, see Chapter 12) grows very considerably relative to the other parts of the brain during the first few weeks after hatching (Wirz, 1954). In the cuttlefish at least, this correlates with the development of a capacity to learn when not to attack (Wells, 1962; Messenger 1973b).

4.2 Killing and eating the prey

4.2.1 *Cephalotoxin and crabs*

Crabs are killed by means of a secretion produced in the posterior salivary glands. The material is extruded into the pocket formed when the crab is gathered into the interbrachial web and appears to penetrate without any wound being made in the carapace of the victim (Ghiretti, 1959; 1960, but see Ballering (1972) who claims that the toxin from *O. apollyon* will not pentrate in this way). Crabs grabbed by octopuses and removed as little as a minute later show muscular fibrillation, and then go limp. A wide variety of pharmacologically active substances is produced by the salivary glands, including dopamine, tyramine, octopamine, 5-HT, histamine, acetylcholine and taurine (references, see Barlow, Juorio and Martin, 1974). Several of these substances have been identified as the poison that kills crabs and for a long time most authors held that tyramine was the agent responsible (Ghiretti, 1960). But tyramine in the quantities present in octopus saliva does not kill crabs. An acetone extract of the saliva, which contains all the amines, produces hyperexcitability when injected, but the crab never relaxes, and it later recovers. The real poison, 'cephalotoxin', appears to be a glycoprotein, retained by dialysis and destroyed by boiling. Neuromuscular preparations of crustacean limbs are not affected by it, so the action is on the central nervous system (Ghiretti, 1960; Songdahl and Shapiro, 1974).

4.2.2 *Bites and man*

People are occasionally bitten by *Octopus vulgaris.* I have been bitten ten or twenty times myself and nearly all of the research workers that I know, working regularly with this species, have been bitten at one time or another, normally while carrying the animals between aquaria. I only know personally of one instance where the bite produced any reaction beyond surprise; a student, who had never handled the animals before (so there was no possibility of a previous sensitizing bite) was bitten on the forearm, which swelled as it might to a bee sting, and remained painful overnight. The swelling disappeared by next day. It is only fair to point out that the animals rarely break the skin. The beaks are comparatively blunt and in the majority of cases it is unlikely that saliva is injected. With the range of substances that this contains (see above, and in the next section) it would be quite remarkable if no local reaction were produced.

Lane (1957) and Russell (1965) have surveyed the matter of octopus bites and give a number of references. The consensus is clearly that octopuses, in general, are not dangerous, although there is at least one species, the small blue-ringed octopus (*Haplochlaena maculosa*) of Australia which bites freely and can be fatal (Dulhunty and Gage, 1970, 1971; Flecker and Cotton, 1955).

4.2.3 *External digestion*

Crabs, paralysed by cephalotoxin, are broken apart and the flesh removed. The carapace, the joints of the limbs and the endophragmal skeleton are discarded clean and unbroken. Very few cuticular fragments are found in the gut.

It seems very unlikely that such efficient cleaning could be achieved simply by nibbling with the beaks and radula, and it is widely believed that some degree of external digestion must be involved. An octopus takes half an hour or so to dispose of a crab that it has caught. During this period the crab is held close to the mouth, in a bag formed by the interbrachial web and it would clearly be possible for it to spit into the cavity and soften up the food before attempting to swallow the bits.

Opinion differs about the extent of the supposed external digestion. Bidder (1966), in reviewing digestion in cephalopods generally, concludes that the body contents of crabs are reduced almost to soup before being swallowed and contrasts the situation in *Octopus* with that in *Sepia* and the squids where quite large chunks of flesh and skeleton are chopped up and swallowed so quickly that there can be no question of predigestion. Altman and Nixon (1970) believe that external digestion is probably limited to loosening the muscle attachments and point out that recognizable fragments of gills, eggs, hepatic caecae and millimetre cubes of muscle can all be found in the crop, apparently unaltered in animals that have recently fed.

Table 4.1 Enzymes from the salivary glands of *Octopus vulgaris* (from Russell, 1965).

Activity	*Salivary gland*		*Reference*
	Posterior	*Anterior*	
Tyramine oxidase	X		Blaschko *et al.* (1952–53)
Tryptamine oxidase	X		Blaschko *et al.* (1952–53)
5-Hydroxytryptamine oxidase	X		Blaschko *et al.* (1952–53)
Proteolytic	X		Ghiretti (1953)
Hyaluronidase	X	X	Romanini (1954)
Mucinolytic	X	X	Romanini (1954)
Dopa decarboxylase	X		Hartman *et al.* (1960)
Histamine oxidase	X	weak	Arvy (1960)
Succinic dehydrogenase	X	X	Arvy (1960)
Phosphatase	X	weak	Arvy (1960)
Adenosine-triphosphatase	X	weak	Arvy (1960)
Butyrylthiocholinesterase	X	O	Arvy (1960)
Acetylthiocholinesterase	X	O	Arvy (1960)
Acetylnaphtholesterase	X	X	Arvy (1960)
Alpha naphtholase		weak	Arvy (1960)

Proteases are included among the enzymes that have been identified from the salivary glands (Table 4.1) but they hardly seem to be present in the range and quantities that would be expected if external digestion were to proceed on any considerable scale. There are, for example, tryptic but no peptic enzymes (Ghiretti, 1950; 1953). There is in fact no direct evidence for or against external digestion. Rather surprisingly, nobody ever seems to have collected a series of crabs at different periods after seizure to examine the state of the flesh inside them.

4.2.4 *Beaks and radula*

Octopus has a typical molluscan radula and parrot-like beaks of tanned protein (Fig. 4.2). The brown beaks are hard and resilient, but not particularly sharp. The radula has long cusps pointing backwards and is clearly more suitable as a rake than as a grinding tool. In older animals, the earlier teeth show signs of wear (Nixon, 1969a, b; 1973). Beaks and radula are embedded in muscles which form an almost spherical structure, free to rotate in the socket formed by the muscular bases of the eight arms.

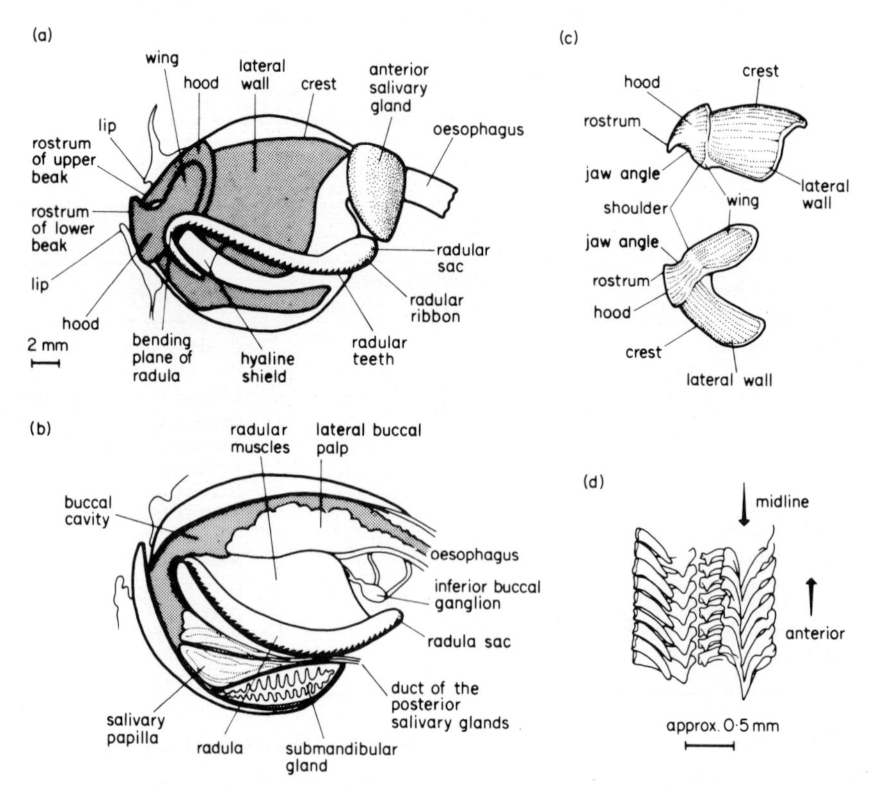

Fig. 4.2 The buccal mass, beaks and radula. (a) shows the relation between the beaks and the other structures, (b) is a longitudinal section, (c) and (d) give further details of the beaks and radula (after Altman and Nixon, 1970; Nixon, 1969b).

Altman and Nixon (1970) have tried to assess the role of beaks and radula in extracting the meat from crabs, by surgical damage and subsequent examination of crab remains discarded by the octopus. Removal of the radula had rather little effect: the more delicate cleaning operations were sometimes neglected and there was a small reduction in the amount of food taken. Removal of the beaks, especially of the lower beak, upset the capacity to clean out the joints; sometimes the animals were even unable to kill their crabs.

It is difficult to interpret these observations. There must inevitably have been a great deal of interference with the operation of structures adjoining or attached to the parts actually removed. The authors themselves suggest that removal of the lower beak probably interferes with the capacity to secrete through the salivary papilla (Fig. 4.2) so that a failure to loosen muscle attachments by external digestion is a perfectly reasonable alternative explanation of their observations; Young (1965b) obtained very similar results by cutting the nerves to the posterior salivary glands.

4.2.5 *Boring into shellfish*

The use of the radula is plainly important in dealing with gastropods and bivalves. *Octopus vulgaris* drills holes in shells in the Bahamas, and it is plain from the shapes of the holes and the scrape marks (and from the noises made while drilling) that the holes are cut out by the radula. There is no evidence of chemical softening as in some shell-boring gastropods (Arnold and Arnold, 1969; Wodinsky, 1969). A wide range of molluscs is attacked: Wodinsky (1969) lists 28 species, ranging from chitons to bivalves.

In the Bahamas, the commonest prey are species of *Strombus.* The conch is grasped with the arms, oriented so that the spire is towards the mouth, and drilled, usually within the first 180° of the spiral away from the lip. Sinking the hole proceeds at about 1.25 mm per hour. Having broken through into the cavity between shell and mantle, the octopus injects a large quantity (over 4 ml in some cases, Wodinsky, 1969) of a viscous secretion which so weakens the gastropod that it can be dragged out of its shell. From the crab experience, one might expect the secretion to include material that could loosen the attachment of the columella muscle.

There are a number of interesting matters arising from the considerable body of data amassed by Arnold and Arnold (1969) and Wodinsky (1969). For one thing, it appears that this particular behaviour may be learned. Arnold and Arnold showed that individual octopuses preferred to drill in particular places on the spire (Fig. 4.3) and Wodinsky carried out an interesting experiment in which octopuses from Bimini, accustomed to drill gastropods, were presented with oysters (*Crassostrea virginica*) from Florida. In three weeks they made no attempt to drill these, while accepting gastropods (*Tectarius musicatus*) imported from the same area. *O. vulgaris* bores oysters off Japan (Fujita, 1916; quoted in Wodinsky, 1969), but not apparently in the

Bahamas. It is interesting that in Naples, where octopuses have been kept in the aquarium for more than a century, there have, until recently, been no records of octopuses drilling shells; in the Bahamas

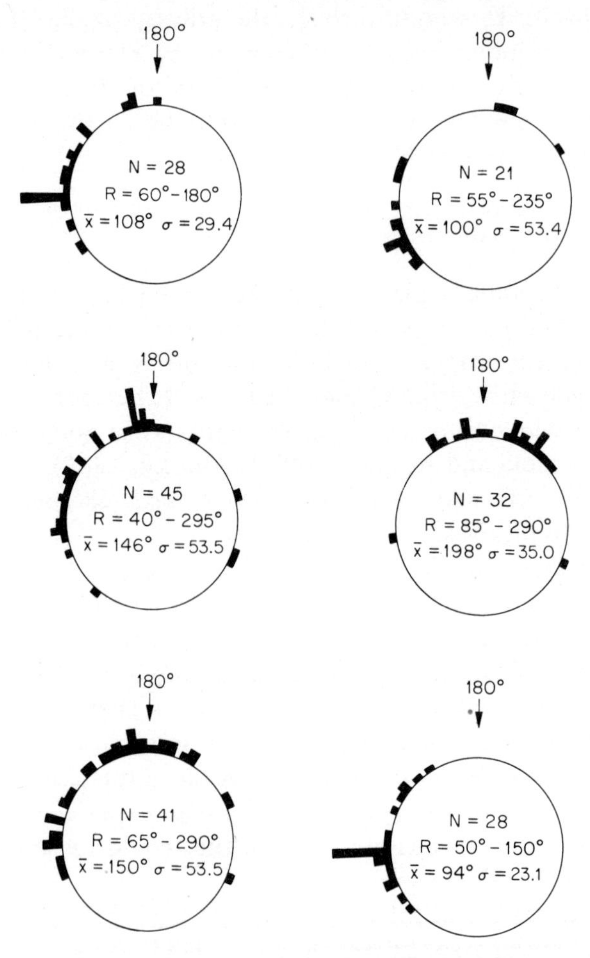

Fig. 4.3 Histograms of position of the borehole on the spires of shells bored by six representative *Octopus*. Each individual has a marked tendency to choose a given area in relation to the lip (from Arnold and Arnold, 1969).

they will even drill holes in shells occupied by hermit crabs (Wodinsky, *loc. cit.*). Wells and Wells (1956, 1957c), for example, found that, while soft objects made of cork or plasticine were often bitten, there were no recognizable attempts to drill these, or the bivalves that they used in tactile training experiments. If an octopus was unable to open a bivalve because the shells had been cleaned out and stuck together, it would wrench at it for an hour or more and then discard it intact. More recently, however, it has been proved that the Naples octopuses both can and do drill holes; Nixon (personal communication) found that *Mytilus* is regularly drilled when offered, typical penetrations being by a single hole made close to the umbo. *Murex* is also attacked, and often several holes are made. Earlier failures to observe this behaviour could mean that not all Mediterranean octopuses perform in this way, or that they do it less readily than their relatives in the Atlantic. The matter, and in particular the effect of individual experience, plainly deserves further study.

4.3 Digestive system

4.3.1 *The buccal mass and salivary glands*

Within the sphere of the buccal mass, or closely applied to it, are two sorts of gland. The submandibular (= sublingual or subradular) gland is a large organ in most octopods (absent in cirromorpha), but its function is entirely unknown. Lying at the back of the buccal mass and wrapped around it are the paired anterior salivary glands, with openings at the front of the lateral palps (Fig. 4.2). The anterior salivary glands produce an abundant mucus (Gennaro, Lorincz and Brewster, 1965) and they have also been reported to secrete a dipeptidase (Sawano, 1935) and hyaluronidase (Romanini, 1952, see Table 4.1). Their function is problematic; the hyaluronidase perhaps helps to liquify the liberal but otherwise rather viscous secretion from the posterior salivary glands.

The posterior salivary glands lie behind the head, in the neck between this and the abdomen. Nerves and ducts connect them to the buccal mass and they clearly function as a part of this apparatus so far as digestion is concerned. Their structure has been described by Ducros (1971) and Matus (1971).

The role of the posterior salivary glands in killing crustacean prey, and in loosening the muscle attachments of crabs and molluscs has already been considered above. In addition, the glands appear to play a part in excretion, possibly acting as kidneys of accumulation and/or as an avenue for the elimination of fluid wastes (Section 3.2.16). At least some of the wide range of pharmacologically active substances that are released into the saliva also pass into the blood, where they seem to affect heartbeat and the peripheral circulation, as well as central control of the state of expansion of the chromatophores (Section 6.2.4). So the glands apparently have a third, endocrine, function as well as playing a part in both digestion and excretion.

Food is forced into the oesophagus by the combined action of the lateral buccal palps and the radula. The latter is repeatedly pushed forward to lick at the food. As it moves forward, the lateral teeth (Fig. 4.2) fan outwards; the process is reversed as the radula retracts, raking fragments backwards into the oesophagus. The lateral buccal palps come forward to meet in front of the radula as this is retracted. Waves of contraction carry down the lateral palps and continue along the oesophagus, thrusting the food back by peristalsis (Young, 1971). Since ingestion is still possible after destruction of the radula (Section 4.2.4) peristalsis alone will apparently suffice to suck in food that is perhaps already softened by external digestion by this stage.

4.3.2 *Nervous control of eating*

The posterior salivary glands have a dual innervation. One set of nerves runs from the superior buccal ganglion at the front of the supraoesophageal brain; these nerves descend towards the subradular ganglion and then double back along the salivary ducts (Fig. 2.7, p. 22; Young, 1965b). The other originates at the subradular ganglion; these nerves run to the circular muscle surrounding the secretory tubules and must be the immediate cause of salivary discharge. The group from the superior buccal ganglion runs to the secretory cells themselves (Ducros, 1971; 1972a, b, c; Martin and Barlow, 1971).

Young (1965b) succeeded in cutting the nerves to the posterior salivary glands on both sides of a single animal, while leaving one of the interbuccal connectives intact. This animal attacked crabs, tore them apart and then succeeded in removing most of the flesh, an observation that could clearly be relevant to the whole question of external digestion. It will be noted, however, that the operation as Young carried it out (by a cut close to the brain) does not eliminate the salivary gland innervation completely; subradular control of the smooth muscle of the glands remains and the animal was presumably still capable of discharging secretion, though perhaps in smaller quanties than usual since control of the secretory cells themselves had been eliminated.

In a further five animals, the interbuccal connectives were cut as well as the salivary nerves from the brain. These octopuses no longer poisoned their crabs and although they sometimes tore them apart they seemed unable or disinclined to remove the flesh. Evidently, the inferior buccal and subradular ganglia are not, by themselves, able to organize eating. Some connexion with the superior buccal ganglion must remain. The rest of the supraoesophageal brain is unnecessary (Young, 1965b).

4.3.3 *Crop and stomach*

Most of the gut hangs from the roof of the mantle cavity surrounded by a series of coelomic membranes (see Chapter 2). Behind the buccal mass, it consists of a crop, stomach and intestine, serially arranged, with a diverticulum, the spiral caecum. A large digestive gland is connected to the caecum through a pair of ducts which open into the second whorl of the spiral. Fig. 2.4 in Chapter 2 shows the general layout. The gut is muscular and very distendable, as might be expected; it contracts upon dissection of the fresh animal, so that its shape in life and the distribution of its contents at any stage in digestion are difficult to establish with certainty. The description that follows is based on Bidder (1957, 1966) who also reviews previous work in the field.

Food passes by peristalsis down the rather narrow oesophagus, through the ring of central nervous tissue between the supra- and suboesophageal parts of the brain. As the oesophagus reaches the level of the mantle it widens to form a crop which may be expanded to form a pocket in animals that have recently fed. There are no glands feeding into the crop and it is, presumably, mainly a storage organ. Fankboner (personal communication) has shown that labelled leucine is taken up by the walls of the crop, so that even this part of the gut is potentially absorptive if the saliva can produce a significant breakdown of proteins.

From the crop, food passes into the stomach, a second glandless chitin-lined sac. The stomach musculature is arranged to form two opposable pads lined with somewhat thicker cuticle and the organ is plainly built to mix incoming food with other gut contents. In specimens that have been opened for examination, the stomach is typically empty because it contracts on dissection or fixation. Occasionally, however, it is found to be filled with a dark brownish-purple fluid presumably – on account of its colour – derived from the hepatopancreas. The fluid contains proteases at a pH of about 5.6 (Sawano, 1935) and must be responsible for the preliminary digestion that takes place in the stomach before its fluid contents are passed on into the caecum. By the time bones and fragments of cuticle leave the stomach they are clean and flesh-free.

4.3.4 *Caecum*

The partially digested mush in the stomach is filtered as it leaves. Fine particles and fluids can pass into the spiral caecum, bones, fish scales and bits of cuticle are prevented from doing so by a fringe of processes, the last part of the cuticular lining of the foregut; they pass directly into the intestine.

The caecum is a thin-walled spiral sac of complex internal structure. The paired hepatopancreatic ducts join and run down the columella of the spiral, opening into the second whorl. A groove runs back from this opening, along the columella but inside the caecum, to the junction with stomach and intestine. It can be covered by a fold; hepatopancreatic juice can be delivered to stomach and caecum independently. The rest of the caecum is largely filled with a series of ciliated leaflets, converging on a second groove that spirals round to terminate in the intestine. No detailed description of these structures or their function is available for *Octopus*, but Bidder (1957) states that 'the action of the caecum is that described for *Loligo*' and refers to her own detailed account of the structure in *Loligo* and *Alloteuthis* (Bidder, 1950). The sludge from the stomach is passed into the caecum in instalments. It includes partially digested proteinaceous materials, fat droplets and, presumably, carbohydrates. The ciliated leaflets sort this material into fluid and particulate matter which is caught up in the mucus that they produce, wound into a rope and carried along the central groove out of the caecum and into the intestine. Quite apart from its function as a sorting organ, the caecum is important in absorption. Boucaud-Camou, Boucher-Rodoni and Mangold (1976) found that ^{3}H-labelled glycine was readily taken up by the walls of the caecum, which showed the highest specific activity of any part of the gut. Enriques (1902) had already shown that the caecum takes up fats.

4.3.5 *Digestive gland*

In octopods, the fluid is then pumped up into the digestive gland by contraction of the caecum. This is quite unlike the situation in *Loligo* where the flow from the hepatopancreas is all one-way with digestion and absorption entirely through the walls of the caecum and intestine.

Digestion is evidently by instalments, since material derived from a single meal, a crab injected with a suspension of carmine, can be found throughout the digestive system if the octopus is examined 4–6 h later. (Table 4.2). Since material still in the crop at this stage is practically

Table 4.2 Distribution of carmine in the gut (after Bidder, 1957).

Animal	*Interval after meal hours and minutes*	*Crop*	*Caecum*	*Mucus strings in intestine*	*Hepatopancreas lumen*	*Hepatopancreas cells*
a	00.25	Fluid	–	–	–	–
W	01.40	?	+	–	–	–
β	03.30	?	+	+	–	–
γ	04.35	Full	+	+	+	+
Z	04.55	?	+	++	++	–
T	05.40	Full	+	+	–	+
δ	06.00	Full	+	+	+	+
E	10.40	?	?	?	+	BB
G	10.44	?	?	?	+	+BB
N	14.00	?	+	+	BB	BB
Q	14.40	?	–	++	–	+
P	14.55	?	–	+	BB	+BB
V	18.00	Empty	–	–	–	–
J	24.00	Empty	–	+	BB	BB
M	3 days	?	–	–	BB	BB

BB = Brown bodies; indicating excretion of wastes; ? = no observation made; + = carmine present; – = carmine absent.

undigested, while clean rubbish is already present in the intestine, the digestive gland must either be importing and exporting material simultaneously, or cycling. Histological evidence implies the latter, at least for the brown 'liver' part of the hepatopancreas. The appearance of the liver varies very considerably from one sample to the next, even among individuals killed at the same time after a meal, a fact that accounts for a number of conflicting reports on its function (see Bidder, 1957). Each liver cell runs through a cycle of changes. It can secrete, and at this time is found charged with droplets. It then passes into an absorptive phase, becomes ciliated and takes up fluids (including carmine from dyed meals) into vacuoles. Finally each cell becomes excretory, and accumulates a single 'brown body' which is discharged into the lumen and ultimately turns up in the faeces (Table 4.2). At any one time nearly all of the liver cells will be found in the same phase.

In the samples examined so far the state of the liver cells has shown no obvious correlation with the presence or absence of caecal fluid in the lumen. Thus Bidder (1957) found individuals with liver cells in the secretory and excretory phases apparently ready to discharge their contents into freshly-arrived, carmine-stained fluid from the caecum. This presumably means that the liver proceeds independently of the caecum, a strangely inefficient way of organizing a digestive system. An alternative explanation of these observations, that the presence of food at apparently 'inappropriate' times is a further consequence of the muscularity of the gut so that fluid is forced up when the animal is killed, is improbable since no particulate matter is found in the mix (Bidder, personal communication).

The function of the 'pancreatic' part of the hepatopancreas is not at all clear. It is quite different in outward appearance, milky white in contrast to the brown or green (in fish-fed animals) colour of the liver in which it is embedded. It does not produce 'brown bodies' or other solid wastes (Arvy, 1960). It does, however, absorb ^{3}H-glycine from food and shows a high specific activity that might be related to a rapid passage of amino acids from the gut lumen to the bloodstream (Boucaud-Camou, Boucher-Rodoni and Mangold, 1976).

Sawano (1935) and Arvy (1960) have reviewed histochemical and biochemical work on the liver and pancreas. Products from the two differ in minor ways, but no clear functional differences were identified.

Rather mysteriously, extracts from the liver of *Octopus* include a cellulase (D'Aniello and Scardi, 1971). Cellulase is evidently widespread in cephalopod livers, which is odd because so far as we know, all cephalopods are and always have been carnivorous. Furia, Gianfreda and Scardi (1972) suggest, without much apparent conviction, that the enzyme may help to digest unicellular algae ingested with their normal crustacean or molluscan prey.

4.3.6 *Intestine and rectum*

Coarse particulate matter arriving directly from the stomach, finer rubbish sorted out in the caecum, and the mass of 'brown bodies' derived from the liver all accumulate in the intestine. The mucus in which they are carried from the caecum is augmented by mucus secreted from the walls of the intestine itself, creating a tangled mass of strings that weave together to form a rope enclosing fish scales, brown bodies and chips of cuticle, orange, red brown, or silvery in appearance depending upon the food that the animal has taken in the last day or so. At intervals long lengths of this faecal rope are blown out through the funnel. It is sufficiently coherent to remain intact despite the sometimes quite violent mantle contractions that accompany its release. The ropes can be found floating in aquaria for hours after discharge.

4.3.7 *Time scales*

The experiments summarized in Table 4.2 were carried out at Naples in June–July when the temperature in aquaria would have been about 24 °C. Carmine, fed with a crab meal, was found in the caecum after one and a half hours, and in mucus strings in the intestine two hours later. The crop was still quite full six hours after the meal and appreciable quantities of carmine were still traceable in the liver and intestine next day. This is considerably slower than the rate in squids, where the enzyme flow from the hepatopancreas can be continuous and digestion completed within 4–6 hours (Bidder, 1966). Interestingly, the size of a meal taken by an octopus has little effect on the time taken to clear it, despite the need for digestion by instalments. The time taken seems to depend only upon the temperature. *Octopus*, as we have seen, took about eighteen hours at 24 °C to digest a crab, *O. cyanea* required 12 at 30 °C and *Eledone cirrosa*, which normally lives in deeper, cooler water 15 at 20 °C and 30 hours at 10 °C.

The comparatively long time taken to void the last remains of a meal give a somewhat false impression of the efficiency of the cephalopod digestive process; what matters is how long it is before a substantial part of the meal has been digested. The instalments system and the tendency to deal with a given proportion of a meal in a given time, more or less irrespective of its absolute size means that the effective rate of digestion can be far higher than the crude clearance data would suggests. If the rate of digestion is calculated in the form:

$$\frac{\text{Ingested food as \% of the weight of the octopus}}{\text{Duration of digestion in hours}} = x$$

Octopus cyanea at 30 °C yields a value of $x = 0.5$ while *Eledone* at 10 °C and 20 °C shows 0.14 and 0.29. These values are high compared with values derived similarly from littoral fish, which range around 0.1. Pelagic fish (tunny) and squid (*Illex*) digest more rapidly with values of 0.7 and 0.6. In general it appears that cephalopods are at least as efficient in this respect as their fish competitors, while the benthic species dispose of their meals more rapidly (Boucher-Rodini, 1973).

In view of the very high conversion rates achieved by *Octopus* (Section 5.3.1) it is clear that the digestive system is an efficient means of converting environment into *Octopus.*

4.3.8 *Control of the digestive process*

The gut has an abundant nerve supply, forming a plexus that includes nerve cells (Alexandrowicz, 1928a). This plexus is fed by nerve trunks from two sources, the visceral nerves from the palliovisceral lobes at the hind part of the suboesophageal brain, and the sympathetic nerves from the inferior buccal ganglion. Both sets are paired, the latter unequally, with the right nerve larger than the left (Young, 1971). Visceral and sympathetic nerves converge on the gastric ganglion, situated close to the point at which crop, stomach, caecum and intestine meet (Fig. 4.4). Stimulation of the gastric ganglion or of the visceral nerves causes violent contractions of the gut from the crop downwards, with an accelerated flow from the liver which Falloise (1906) assumed to be liberation of stored secretion. Stimulation of the sympathetic nerves at their departure from the inferior buccal ganglion initiates peristaltic movements mainly of the oesophagus. Cutting the nerves does not stop feeding (Young, 1971).

Beyond this, nothing is known about the nervous control of digestion. It would not be impossible to cut the visceral nerve supply to the gastric ganglion (which might control some aspects of digestion locally without reference to the CNS) and even to sever certain of the nerves radiating from this. But it would be a major problem to interpret the consequences in the event of any result more complex than an outright cessation of function by some part of the gut.

The animal is evidently sensitive to the presence of food in the crop. After feeding there is an increase in the tendency to attack objects seen, which rises to a maximum after about half an hour, and then declines progressively. The rise and fall correlates nicely with the weight of the crop over the same period (Young, 1960c). Other factors affecting food intake are considered in Section 5.3.1.

A single report has indicated that there may be hormonal as well as nervous control of gut function in *Octopus.* Ledrut and Ungar (1937) isolated a material from the caecum that had the same physiological effects as secretin when injected into the dog. Injected into the bloodstream of octopuses the material induced hypersecretion by the hepatopancreas, as did also extracts of mammalian secretin.

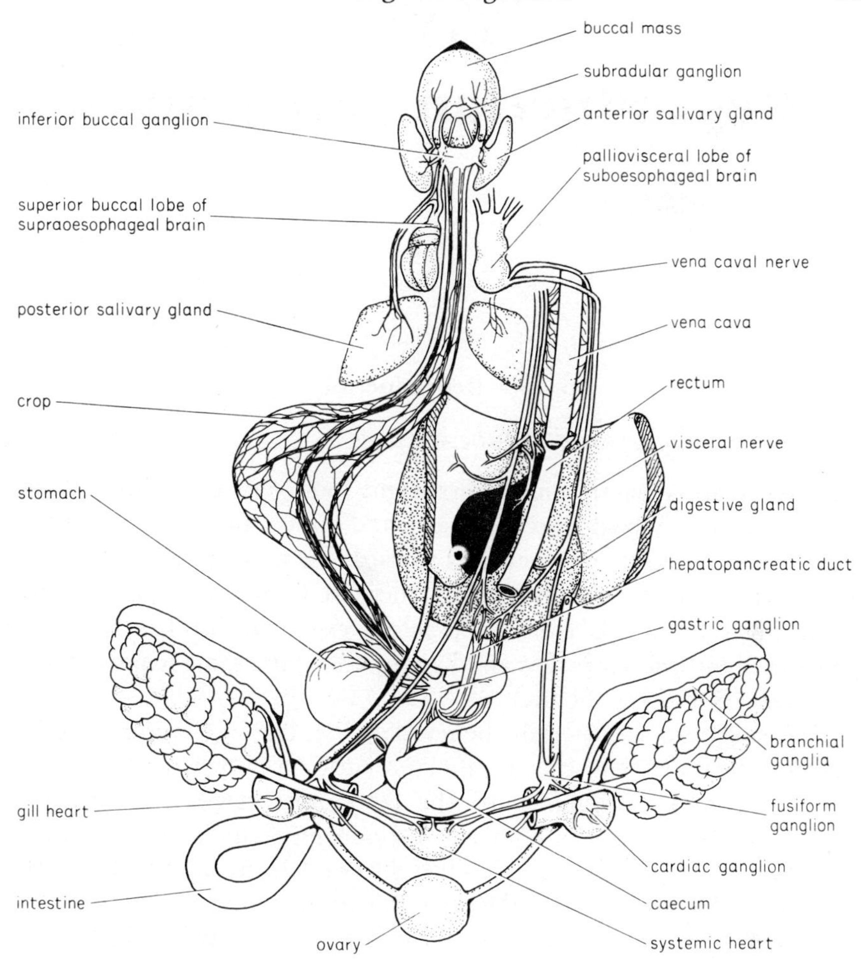

Fig. 4.4 Innervation of the gut. Drawn from the ventral surface, cf. Fig. 2.4 which shows the gut in a dorsal view at an earlier stage in dissection. Semi-diagrammatic, showing half of the supraoesophageal brain on the one side and half of the suboesophageal on the other (after Young, 1971).

CHAPTER FIVE

Reproduction and growth

5.1 Reproduction

5.1.1 *Sexual dimorphism and the sex ratio*

The sexes are separate and at first sight look very much alike. Males have a hectocotylized third right arm and mature individuals generally have a few notably larger suckers towards the bases of the second and third pairs of arms (Section 2.1.2). Otherwise there is no reliable means of sexing the animals from an external examination. The two sexes grow at about the same rate and live for about the same length of time (Section 5.3.1 *et seq*, below).

The sex ratio is generally believed to be 1:1 in *Octopus vulgaris*, despite the fact that this is not the ratio obtained in commercial catches. Samples brought into the laboratory at Naples for use in behavioural experiments regularly include more males than females; in summer 1974, for example, there were 102 males to 78 females (131:100, Buckley, personal communication) while a sample of 465 collected over the three year period 1954–7, yielded 251 males (117:100, Wells and Wells, 1959). In a larger sample of 900 animals caught at Banyuls, Mangold-Wirz (1963) found 143 males to every 100 females. The preponderance of males could be due to differences in behaviour. The males mature earlier and may constitute the more mobile fraction of the population, and we know that mature females tend to seek out holes in the rocks, where they remain to guard their eggs. There is an annual migration inshore in the spring, at which males appear to precede females (Mangold-Wirz, 1963). Any or all of these factors could increase the ratio of males to females caught in predominantly shallow-water traps and trawls.

To date, nobody has tried to analyse the sex ratio of a sample of octopuses at hatching. But this has been done for *Sepia*; Montalenti and Vitagliano (1946) found 64 males and 69 females in 133 embryonic and newly hatched cuttlefish. Together with the available information for other octopods and decapods (see Mangold-Wirz, 1963), none of which show unaccountable departures from a 1:1 ratio, it seems fair to guess that male and female *Octopus vulgaris* are equally common at birth and have about the same chances of surviving to maturity.

5.1.2 *Male reproductive system*

The male genital system consists of an unpaired testis and duct, opening into the left side of the mantle cavity. The testis is a compact mass of cells surrounding a series of channels that converge on a ventral outlet from the coelomic sac surrounding the gonad. There is an abundant blood supply. The walls of the radiating ducts within the testis are formed from germinal epithelium throughout their length; the germ cells proliferate and the spermatocytes differentiate as they move inwards. Free sperm migrate into the testis sac and from there into the male duct, where they are packed into spermatophores.

When the mantle is folded back, the coiled male duct and the testis can be seen enclosed in a drop-shaded capsule, fat and cream-coloured at the testis end. In animals over about 150 g white spermatophores can usually be seen in the anterior more transparent parts of the duct. The system is shown *in situ* in Fig. 3.11, p. 41. Dissected out it is found to include three blind-ending diverticulae and a storage compartment, Needham's sac, in which the finished spermatophores accumulate prior to mating (Fig. 5.1).

The relative sizes and shapes of the components of the male duct vary from one species to the next. Descriptions are available for *O. vulgaris* (Brock, 1878; Belonoschkin, 1929a; Marchand, 1907) *O. bimaculoides* (Peterson, 1959) and *O. dofleini* (Mann, Martin and Thiersch, 1970). The terms used to described the elements of the system differ from one author to the next; in the account that follows, I have adopted the terminology used by Mann *et al.* (1970), with the further division of the spermatophoric gland proposed by Belonoschkin (1929a)

The sperm leave the testis sac through a ciliated pore into the first section of the duct system, a narrow convoluted tube, the proximal vas deferens. They emerge from this compacted into a 'sperm rope', with the sperm heads aligned and rather little fluid between them. As the rope leaves the vas deferens it is coiled into a spiral. In the following, wider, part of the duct, the first section of the spermatophoric gland, the coiled rope is embedded in a mucilaginous secretion. In

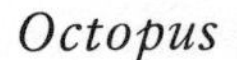

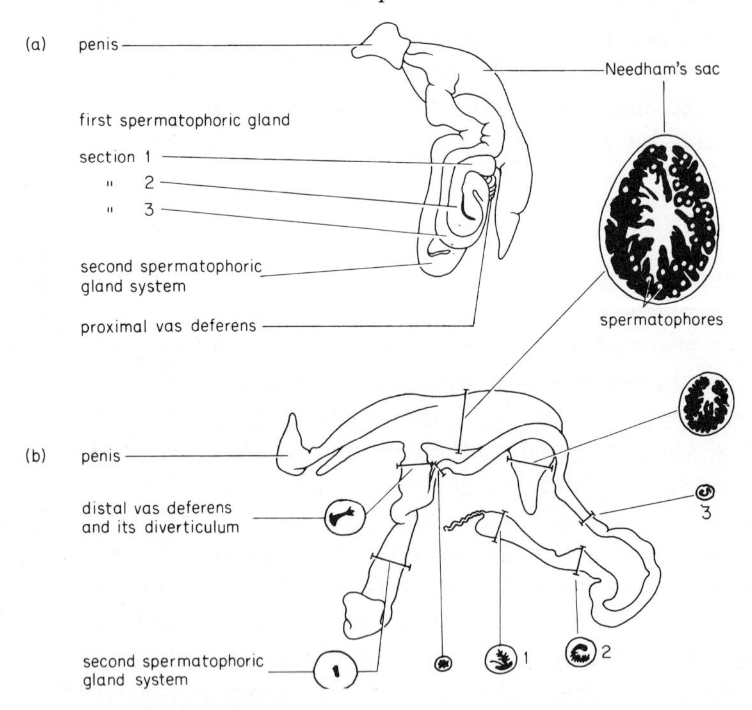

Fig. 5.1 The male duct of *Octopus.* (a) after removal of the connective tissue capsule and testis, which would feed into the proximal vas deferens; (b) displayed following further dissection, with cross-sections of the ducts. The upper cross-section of Needham's sac is drawn to a larger scale to show the stacked spermatophores.

O. bimaculoides, degenerating cells are shed from the tubular epithelium and these too seem to become incorporated into the sperm rope (Peterson, 1959).

The sperm rope with its mucin coating is carried through the 'intercalary piece' that ends the first section of the spermatophoric gland. As it passes, a membrane (the sperm membrane or middle tunic) is added, sealing off the rope. Movement of individual sperm now ceases until the sperm are liberated from the spermatophores after copulation (Belonoschkin, 1929a).

In the next two sections of the first spermatophoric gland further sheaths are added, together with a gel column, the main core of the ejaculatory apparatus, which at this stage lies posterior to the rest of the spermatophore (Mann *et al.*, 1970).

The spermatophore continues into the second spermatophoric gland, a blind ending sac. It seems probable that the ejaculatory apparatus is

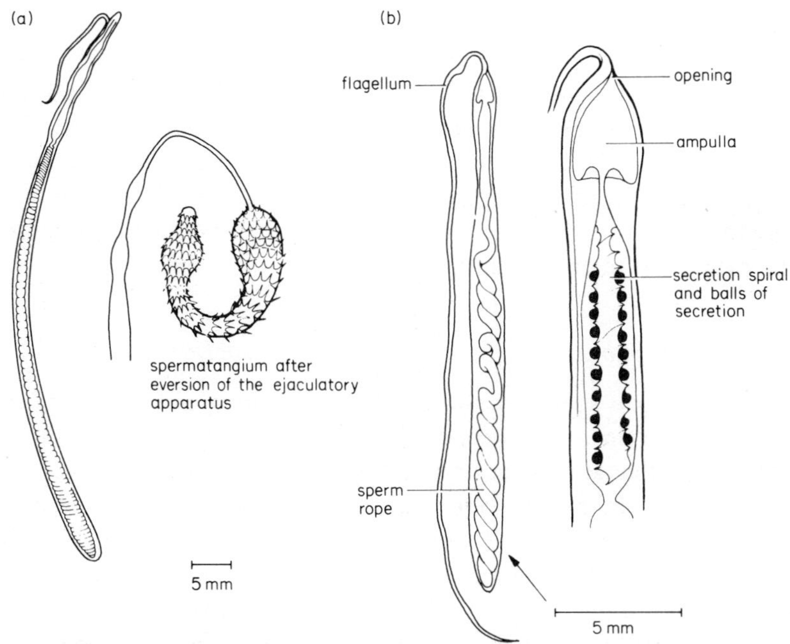

Fig. 5.2 (a) Spermatophore of *Eledone cirrhosa*. The everted spermatangium breaks off and is later found in the ovary. (b) Spermatophore of *Octopus vulgaris*. In this species the spermatophore splits close to the origin of the flagellum and the sperm are liberated into the oviduct. ((a) after Fort, 1937 and (b) after Meyer, 1911).

completed here. After a while the spermatophore emerges, cap end first (Fig. 5.2) and passes through the distal vas deferens to lodge in Needham's sac where it is stored. According to Peterson (1959) a 'general finishing and hardening or dehydration' occurs here. It will be noted that the spermatophore is now once again reversed, with the cap end facing away from the male opening; before it is placed in the groove of the hectocotylus, the packet will be shunted into yet another blind sac, the diverticulum of the penis. Mature males usually have a single spermatophore lodged here, facing cap end outwards, correctly oriented towards the hectocotylus and ultimately towards the oviduct of the female.

An *Octopus vulgaris* of 500–700 g that has not recently mated will normally carry about fifty spermatophores. No estimates of the number of sperm this represents have been made. In *O. dofleini*, which stores 10 or so much larger (1 m long!) spermatophores, each contains about 10 g of sperm (Mann *et al.*, 1970).

There is obviously much to be learned about the male duct system and the control of spermatophore production. The system is ciliated and muscular, it is innervated from the fusiform ganglion (Young, 1967) and it contains a variety of gland cells. Between them, these secrete a morphologically complex structure. Synthesis is known to be under hormonal control, apparently by the same optic gland gonadotropin that controls the onset of sexual maturity and the development of the ovary and testis (Chapter 6).

5.1.3 *Spermatophores and the spermatophoric reaction*

Typical *O. vulgaris* spermatophores are 2–3 cm long and about 0.5 mm across at their blunt, sperm rope end. They are a little smaller in animals of 150–200 g that are just beginning to produce sperm. Their structure, as seen in the light microscope is outlined in Fig. 5.2. The spermatophore is tubular, turgid and springy. Inside two distinct divisions are visible to the naked eye. One, the posterior or 'male-oriented' part is opaque white, largely filled with a tightly coiled rope of compacted sperms. The other, anterior, region is more translucent; it ends in a thin thread which is also carried back, stuck along the body of the tube. This 'female-oriented' region is found on examination to be an invaginated folded tube, the 'ejaculatory apparatus', the anterior end of which forms a cap. These features vary considerably from one species to the next, and because spermatophores are tough and readily preserved, spermatophore structure can be used as a taxonomic character (Marchard, 1913). As with the male duct system, there is also considerable variation in the names used to label the recognizable parts. Here, I have again adopted those used by Mann *et al.* (1970). Fort (1937) is a review of earlier work and includes a table relating the terms used by different authors.

Mann *et al.* (1970) have explored the biochemistry of the spermatophore and its contents in *O. dofleini.* The outer tunic is proteinaceous (not 'chitin-like', as usually stated in the literature) and elastic. It can be pulled out to nearly twice its resting length, and includes a high proportion of proline, as is usual in the elastic structural proteins of other animals. Inside, the sperm rope is surrounded by a transparent viscous fluid, the spermatophoric plasma. Sperm and fluid can be separated by centrifugation. Both are highly concentrated materials containing more than 30% dry weight of organic matter; about a fifth of this is present as carbohydrate reserves, amino sugar in the plasma and glycogen in the sperm. Free glucose is present, but in very variable quantities, as it also is in the blood (Goddard, 1968). In front of the sperm rope lies the gelatinous rod of the ejaculatory apparatus, low in glygogen but high in other organic materials, separated from the sperm rope by a region filled with an amber-coloured syrupy 'cement' (a bad term, since there is no indication that it has anything to do with cementing in octopuses; in *Loligo* it sticks the bag of extruded sperm to the female (Drew, 1919)).

In mating, the spermatophore, held in the muscular penis, is transferred to the guide and carried down the groove in the hectocotylized arm (Section 2.1.2). The arm tip is inserted into the mantle cavity of the female. Quite what happens then has never been observed, but if the pair is separated a few minutes after the passage of the first spermatophore, the spermatophore is found lodged, filament end forward, in one of the oviducts.

It then discharges the sperm into the oviduct. This 'spermatophoric reaction' will take place *in vitro* after about 30 minutes (*O. dofleini*) if the spermatophore is kept in seawater. Matters can be accelerated by gently tugging at the cap thread, as probably happens *in vivo* when the spermatophore is pushed or carried into the oviduct, which shows vigorous peristaltic movements at this time.

The most complete account that we have is that of Mann *et al.* (1970), for the very large spermatophores of *O. dofleini.* Discharge begins with the uptake of seawater at the hind end of the spermatophore. The sperm rope in this region begins to uncoil and move away from the blunt end of the capsule, pushing the ejaculatory apparatus before it. The spermatophore breaks just behind the cap (in other species, including *O. vulgaris* and *Eledone*, the cap evaginates, Fort 1937, 1941) and the entire ejaculatory apparatus is extruded. The spermatophore continues to swell and there is a considerable increase in length and volume. The ejaculatory apparatus evaginates slowly, pushed forward by the advancing sperm rope and sperm plasma. Quite suddenly there is an explosive increase in the rate of evagination and the whole apparatus and the front part of the outer tunic of the spermatophore balloons out to form a bladder, into which the rest of the sperm rope passes. Fig. 5.3 summarizes some of the changes that can be measured, 60 minutes being the time at which the sperm bladder began to form in this instance. The driving force for extrusion of the sperm bladder appears to be dilution of the spermatophoric plasma by seawater coupled with the considerable elasticity of the outer tunic of the spermatophore; perhaps suprisingly there is no measurable decrease in osmolarity during the spermatophoric reaction (Mann *et al.*, 1970). It will be noted also that the ejaculatory apparatus behaves in a remarkable manner, twisting, evaginating and uncoiling as it does so in a way reminiscent of coelenterate nematocysts, so that that this too may play a part, adding active expansion to passive inflation of the bladder.

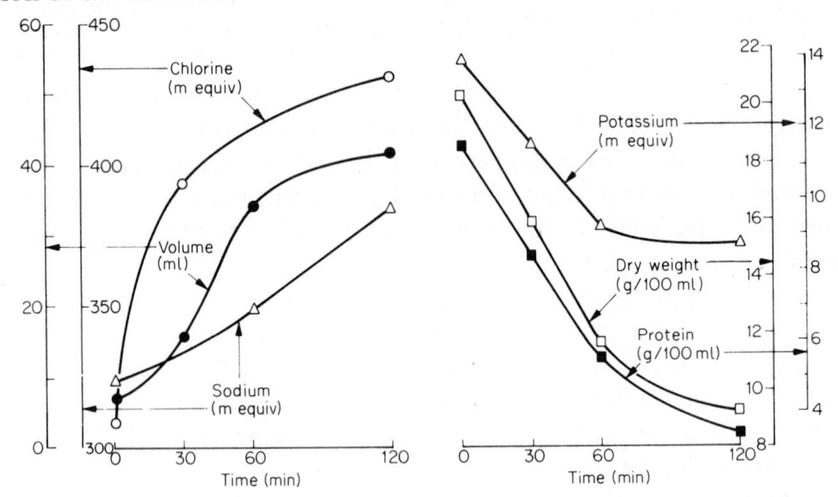

Fig. 5.3 Changes in volume and the contents of the spermatophoric plasma of *O. dofleini* during the spermatophoric reaction. The sperm bladder began to form 60 min after placing the spermatophore in seawater (from Mann, Martin and Thiersch, 1970).

The fate of the sperm bladder apparently differs from one genus to the next. In *Octopus* it bursts, liberating sperm into the oviduct which carries them toward the oviducal glands by peristalsis. In *O. vulgaris,* they accumulate in the spermathecal sections of the oviducal glands where they are presumably stored until the eggs are ready (Belonoschkin, 1929b; Wells, 1960b). In *Eledone cirrosa*, the whole sperm bladder (Fig. 5.2) breaks off and is carried up into the ovary, where several may be found in maturing animals (Mangold-Wirz, 1963; Orelli, 1962).

5.1.4 *Female reproductive system*

The female has a single ovary and two oviducts, one opening on either side of the midline about halfway along the mantle cavity (Fig. 2.2 p. 15). As with the male there is a gonadial coelom, linked to the pericardium of each branchial heart by a narrow duct.

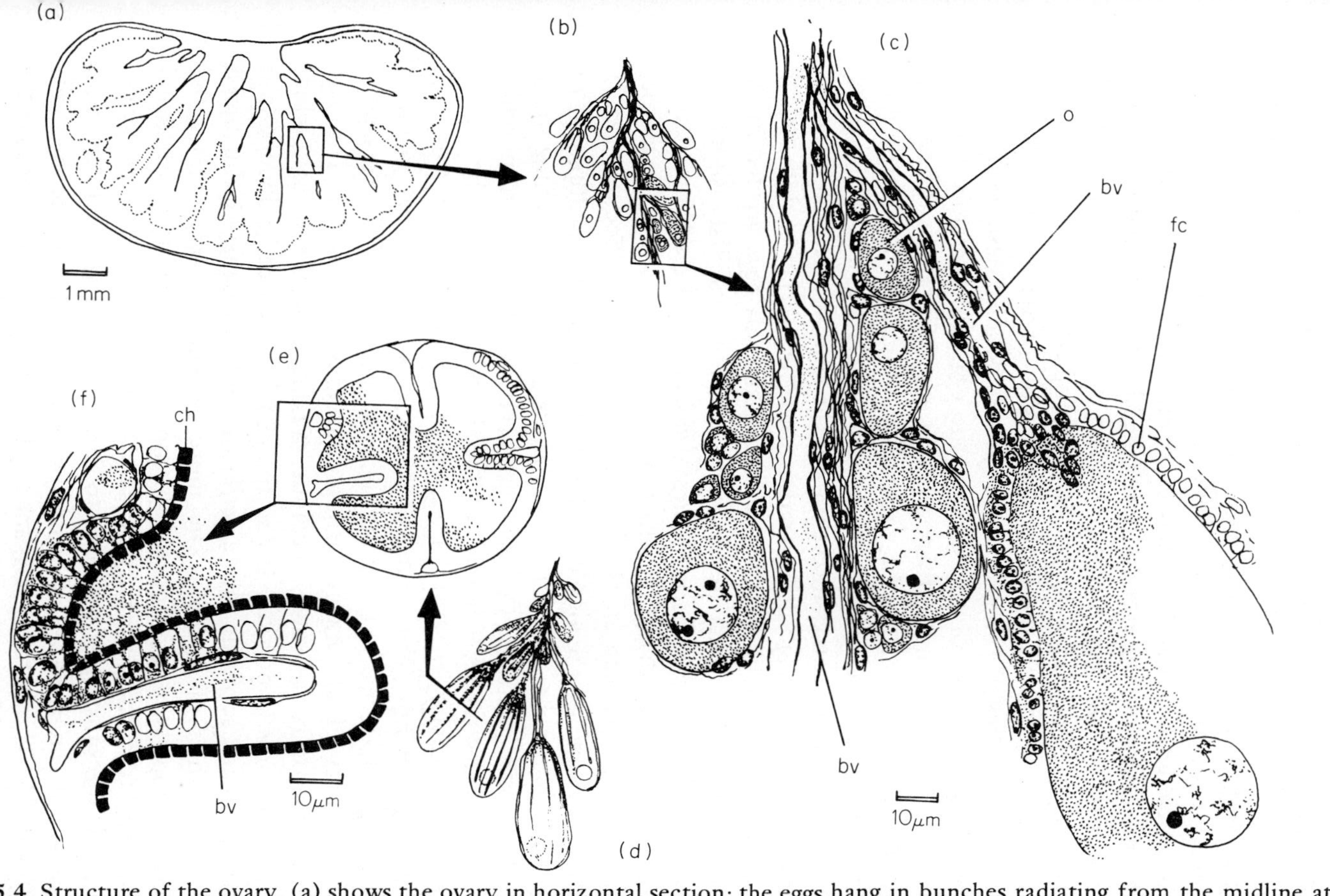

Fig. 5.4 Structure of the ovary. (a) shows the ovary in horizontal section; the eggs hang in bunches radiating from the midline at the anterior end. (b) and (c) show further details. As the ovary matures the follicle cells divide and enlarge until they form a series of folds invading the oocyte cytoplasm (d, e, f). An extensive capillary blood system envelops each egg. The chorion and the main bulk of the proteinaceous yolk are secreted simultaneously. bv — blood vessel. ch — chorion. fc — follicle cell. O — oocyte.

The germinal epithelium hangs in a series of strings, radiating from the front of the gonadial coelom, between the two oviducts. The ovarian artery and vein enter here and there is a complete system of capillaries which eventually envelopes each individual egg. The ovarian tissue includes oogonia, follicle cells, contractile and connective tissue elements that move the eggs and form the blood vessels and the fibrous framework of the egg strings. As the eggs develop, each oocyte becomes enveloped in a double layer of follicle cells, the inner layer cuboid and secretory, the outer a thin skin of flattened epithelium (Fig. 5.4; Buckley, 1977 gives a much more detailed description). The former secrete most of the yolk in the egg, and probably also the chorion which hardens to form a tough egg shell before the eggs are laid. During development, the eggs move in the ovisac as muscle fibres in the egg stalks and around the eggs themselves contract and relax. The ovarian artery comes directly from the systemic heart and there is considerable scope for ultrafiltration into the ovarian coelom, which may be the means by which the very numerous eggs are fed in the early stages before each develops its individual capillary supply. It could also be the reason for the heavy vascularization and innervation of the genitopericardial duct which may form an additional excretory and resorptive system in the maturing female (see Section 3.3.2).

Development of the ovary is always paralleled by development of the oviducts and the oviducal glands that lie about halfway along them. As the animal matures, the oviducts swell and stiffen with muscular and connective tissue. The ends of the ducts fan out onto the roof of the mantle cavity to form a series of ridges radiating from the oviducal openings. The oviducal glands enlarge as sexual maturity approaches. Externally, the glands become ridged, until they resemble the segments of an orange, grouped around the oviduct. The part nearest the ovary is brown in colour, in contrast to the brilliant white of the more distal portion. Inside the gland is seen to be divided into two parts each secreting into a series of ducts coverging on the lumen of the oviduct (Fig. 5.5). The more proximal (ovary side) gland secretes droplets of a sulfonated mucopolysaccharide, the more distal a mucoprotein; they are secreted at pH's of 6.4 and 6.6 respectively. *In vitro*, the substances combine to form a precipitate that can be re-dissolved if NaOH or HCl are added to bring the pH outside the limits 9.0 and 5.5. The latter is relevant to the situation *in vivo* because the mucus in the oviduct of mature animals is acid, with a pH of 5.2, compared with 6.8 in immature octopuses; during their passage down the oviduct the mixed gland materials are prevented from polimerizing by the oviducal mucus. The mucus, it seems, must also add some further constituent, since the final product, which is used to stick the eggs together in strings attached to the wall or roof of the animal's home, is stable only after contact with it (Froesch and Marthy, 1975).

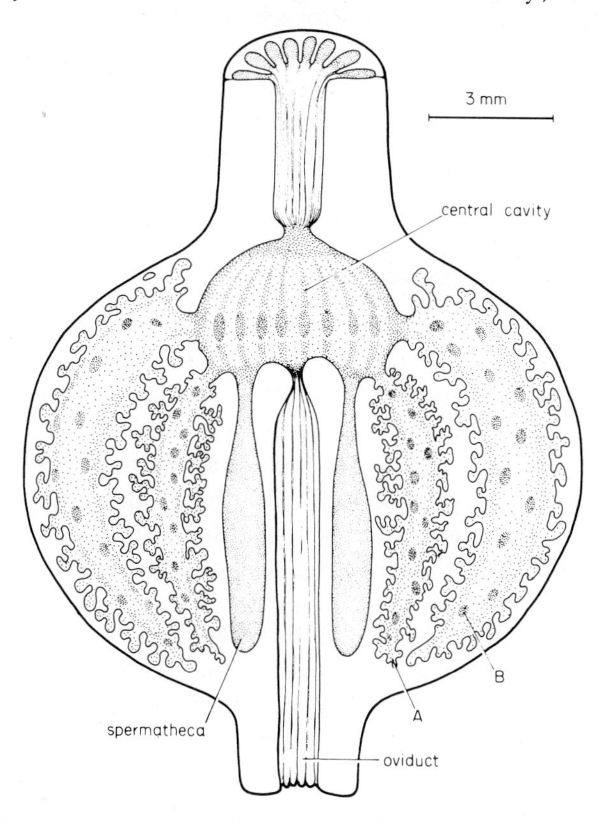

Fig. 5.5 Diagrammatic view of the oviducal gland of *Octopus vulgaris.* The gland has been opened by a longitudinal cut. It shows three sets of compartments which open into the central cavity: the spermathecae next to the proximal part of the oviduct, the proximal (A) and the distal (B) parts of the gland (from Froesch and Marthy, 1975).

In addition to the two glandular portions, the oviducal glands contain a third series of convergent lumina, opening into a ring of thin-walled spermathecae, the destination of sperm released from spermatophores placed in the oviducts. Animals as small as 180 g have been observed with spermathecae packed with spermatozoa. The sperm penetrate deeply into the walls of the sacs with their spiral acrosomes and then become immobile. They remain viable for long periods. Mangold and Boletzky (1973) cite an instance where a female, isolated in January, laid fertile eggs in May, four months later, and the small size of some of the mated animals suggests that sperm may even be stored overwinter in this manner, in anticipation of a spring spawning (Belonoschkin, 1929b; Froesch and Marthy, 1975).

Not all octopods store sperm in spermathecae. In *Eledone*, for example, the bladder of sperm discharged by the spermatophore is carried up directly into the ovary (Section 5.1.3 above).

Endocrine control of the state of the ovary and its ducts is discussed in Chapter 6, below.

5.1.5 *Sexual behaviour; displays and mating*

An adult male octopus that has become accustomed to its aquarium will emerge from his home to approach any other octopus introduced into his tank. If the intruder sits still, the occupant will flush darkly, and often approach with the third right, hectocotylized arm extended towards the newcomer. If the latter moves, the male may spread its interbrachial web, appearing much larger than usual, and it may pale to produce the 'dymantic' coloration, with the eyes accentuated by dark rings against the flat white saucer of the outspread web (see Section 10.3). A 'confident' octopus remains dark. Once tactile contact is made the male will typically rush forward and try to envelope the visitor in his web, probing at the same time with the hectocotylus.

What happens next seems to depend on the sex of the visitor. Both males and females will attempt to fend off the occupant and may break away more than once before they are finally cornered. There does not seem to be any specific display that would distinguish between the sexes of *O. vulgaris* and the approach of an established male of this species appears to be the same irrespective of the sex of the intruder. Packard (1961) has pointed out that males have some exceptionally large suckers close to the base of arm pairs 2 and 3 and suggests that these may be used to advertise their sex to females, perhaps particularly those that are larger than the male himself (Fig. 5.6a). If this is so the display is unusual in being directed at the female. Other cephalopods, both decapods like *Sepia* (Tinbergen, 1939) and *Loligo* (Drew, 1911) and octopods such as *O. cyanea* (Wells and Wells, 1972a) seem to aim their often quite striking displays at discouraging the male opposition rather than seduction of the opposite sex; any failure to return the

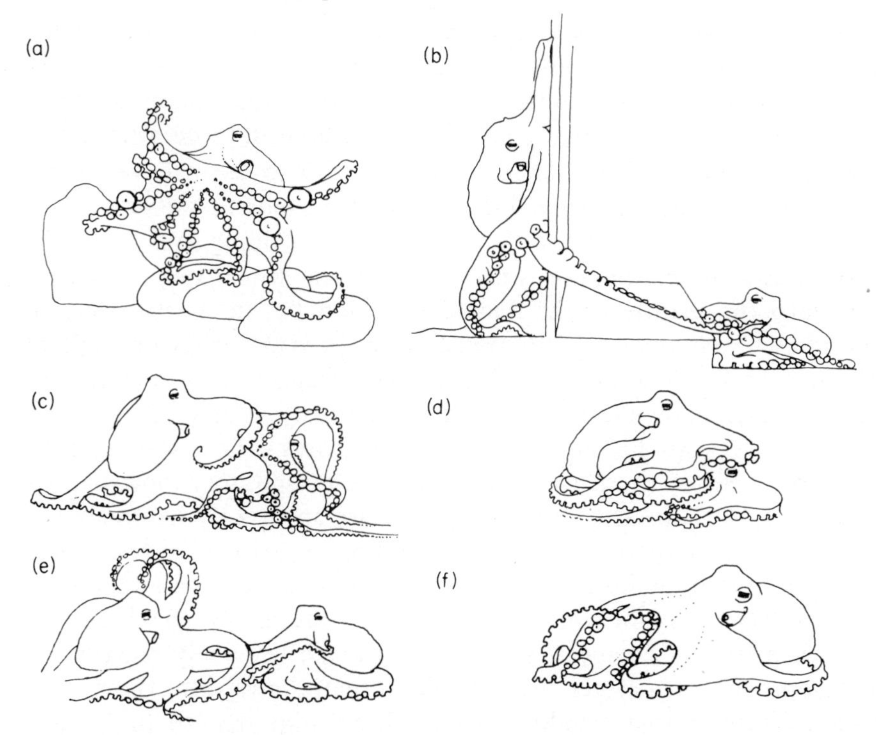

Fig. 5.6 Sexual displays and mating in *Octopus vulgaris*. (a) shows a male displaying the large suckers on his second and third right arms; Packard (1961) suggests that this allows other octopuses to identify the sex of the displayer. (b) Copulation around the edge of the partition dividing two aquaria, male on the left. (c) – (f) Positions of male and female in copulation. There is considerable variability, both in the behaviour of the same individual at different times, and between individuals. (a) and (b) from photographs in Packard (1961) and Wells and Wells (1959). (c) – (f) after Wells and Wells (1972a).

display may be followed by attempted rape. Packard (1963) himself admits that he has never observed the sucker display in animals of the same size and since such animals quite clearly do identify each others' sex and rapidly at that it can only be supposed either that there are subtle visual clues that we have not so far detected (Wells and Wells (1972a) report observations on 161 matings) or that sex recognition is chemotactile in this species.

In any case, the behaviour of the animals alters as soon as a contact has been made. Males fight, and when the smaller cannot escape, he will generally be killed, and eaten. A female typically submits to the demands of the male, who will then sit on or beside her caressing her head and abdomen with the hectocotylus, and having successfully inserted this into her mantle cavity proceed to pass spermatophores (Fig. 5.6). Copulation may continue for an hour or more. During this time the pair sit quietly. At intervals of 20 s or so the male will hump his back and place the funnel down over the guide at the end of the groove in the hectocotylus. A moment later a ripple runs down the groove as a spermatophore is carried along the arm at some 20 cm s^{-1} and into the mantle of the female. Occasionally, the hectocotylus is withdrawn and repositioned, transferring, perhaps, from one side of the mantle to the other. The female may struggle in an unconvinced sort of way, and then settle down again. Eventually, it may be supposed, the male exhausts his supply of spermatophores - the discarded cases are blown out of the mantle by the female's respiratory movements - and the animals part. By this stage they are apparently habituated to each other and there is no further fighting, at least for a few hours.

We do not know how far these aquarium observations are representative of *Octopus* behaviour in the sea. Presumably the female is normally free to escape if she wishes, though her likely reaction on the approach of a larger male would be withdrawal into the nearest hole, a retreat that would probably leave him free to probe her.

We know, from aquarium observations like that recorded in Fig. 5.6b that the female does not always run away even when she is able, and we may suppose that the animals sometimes settle down in pairs as *O. cyanea* apparently does (van Heukelem, 1966; Yarnall, 1969; Wells and Wells, 1972a). But I have never observed or heard of this in *O. vulgaris*; octopuses encountered while SCUBA diving seem always to be alone.

5.1.6 *Oviposition and care of the eggs*

Females lay their eggs in strings, attached to the rooves of their homes in the rocks. If no home is available in an aquarium they will stick the egg bunches to the sidewall of their tank. The operation seems to be preceded by an attempt to clean off the area by continually running over it and plucking with the suckers. Rather surprisingly, no detailed description of egg laying is available for *O. vulgaris,* but Arakawa (1962) has given one for *O. luteus* where the results seem to be similar. Just before egg laying, the female attaches herself to the substructure by means of the large suckers at the base of the arms. The area around the mouth is raised to make a small cone-shaped cavity, and the funnel inserted into this from behind. After two or three vigorous exhalations, which would serve to blow away detritus, a few eggs, together with a glutinous material – presumably the secretion of the oviducal glands – are passed through the funnel to the small suckers immediately surrounding the mouth. These suckers manipulate the eggs, pressing them onto the rock and one another, sticking them together by the stalks, so that they hang freely in bunches from a common stem. The presence of numerous discs of cast-off sucker cuticle in the glue uniting the eggs in the long strings laid by *O. vulgaris* implies that the same detailed manipulation occurs here (Froesch and Marthy, 1975).

Octopus vulgaris will lay many thousands of eggs, in strings of several hundred, an operation that may take a week or more. By comparison with many other species, the eggs are very small (1 × 3 mm) and very numerous (150 000 in a typical clutch, Heldt, 1948) – *O. bimaculoides,* for example lays only a few tens of eggs, but each measures 17.5 × 9.5 mm (Akimushkin, 1965).

Having laid her eggs, the female stays to brood them. During this period, which may last from 4 to 6 weeks at 22–23 °C (Boletzky, 1969; Vevers, 1961) she rarely leaves the eggs and rarely feeds, becoming very emaciated. The armtips are passed repeatedly between the egg bunches and jets of water directed through them from the funnel. Vevers (1961) records that the respiratory rate increases during brooding, and several authors (references see Vevers, *loc. cit.*) have reported that rubbish and intruding animals (including those eaten at other times) are pushed away to arms length. Shortly after the eggs hatch, she dies.

One is, of course, once again dependent upon observations made in aquaria, where conditions may not be ideal. But it does seem to be generally true that octopuses, of whatever species, die after spawning whether in captivity or in the sea. Reports from laboratories (Borer, 1971; van Heukelem, 1973, and a number of older records summarized in Nixon, 1969c) all suggest that death is universal. Studies of large numbers of animals from populations of *Octopus* and *Eledone* and smaller samples from other species (Mangold-Wirz, 1963) yield only a few females that, from their age and condition, could possibly have bred more than once.

Male *O. vulgaris* seem to grow at about the same rate as females (Section 5.3 below) and since the average weight of males in samples collected at random is not noticeably larger, it seems likely that most of these also die before the end of their second year of life. In *Octopus cyanea*, which has a very similar life cycle, studies of individual males show that these, like the females, eventually cease to feed, lose weight, and die (Van Heukelem, 1973).

5.2 Embryology

5.2.1 *Cleavage and gastrulation*

A review which concentrates on physiology and behaviour is no place to survey the quite considerable literature on octopod embryology, the more so because the studies made to date have been predominantly descriptive. Cephalopods are only just beginning to attract the attentions of experimental embryologists, and nearly all the work so far has been done with squids (see Arnold, 1971; Arnold and Williams-Arnold, 1976; Marthy, 1970, 1972, 1975). The two sections that follow are included because a total omission of embryology would leave a noticeable gap in what might otherwise claim to be a more or less all-round introduction to the literature on *Octopus* biology. These sections are however deliberately kept short – a minimal account with a minimal reference list. A more detailed review, with a more complete list of references, is given in Wells and Wells, 1977.

The egg is presumably fertilized as it passes through the oviducal gland. Development begins at once with the formation of a germinal disc of clear cytoplasm at the apical end of the egg, beneath the terminal micropyle. The first cleavage furrow appears at 9–14 hours, depending upon the temperature, and the process continues for the next two days to form a discoblastula of some 1200 cells, capping one end of the yolk (Naef, 1928). Cephalopods, unlike other molluscs, show no signs of spiral cleavage and it is a moot question whether the eggs are to be regarded as mosaic or regulative; the normal pattern of embryogenesis is rigid, but it seems that the fate of specific cells is not (Arnold, 1971).

Gastrulation begins with the formation of a syncytial yolk epithelium as nuclei from the edge of the blastodisc migrate away to form a thin layer of clear cytoplasm that eventually envelopes the yolk and grows under the blastodisc. Experimental investigations of the role of this layer have yielded conflicting results. Arnold has shown that if the cells that come to overlie it are removed others move in and form the appropriate organs; if the yolk epithelium is removed as well, there is no such repair, while the two layers transplanted together will form the organs *in vitro.* The yolk epithelium seems, however, to derive its inductive properties from the egg cortex since the results of damage to this before the nuclei spread out from the blastodisc, or of displacement of the blastodisc itself, all seem to lead to the conclusion that the egg cytoplasm itself is patterned. As Arnold (1971) points out, this may mean that the follicle cells are ultimately responsible for establishing the pattern of the embryo.

In apparent contradiction to these results Marthy (1972, 1975) succeeded in rearing dwarf, but otherwise seemingly normal embryos from blastodiscs separated from the rest of the egg, and from eggs (both of *Loligo* and *Octopus*) divided in two by a ligature, isolating the animal pole from some 60% of the yolk. He has suggested that the apparent anomaly may be explicable in terms of the nutritive function of the yolk syncytium; local interference may disturb the formation of an active yolk-resorbing syncytium and subsequently starve the cells that come to lie above the damaged area (Marthy, 1973, 1975).

5.2.2 *Organ formation and absorption of the yolk*

Endoderm and mesoderm at first develop together as a layer of cells delaminated from the more peripheral blastomeres (Sacarrão, 1961). This spreads inwards under the developing embryo, which is simultaneously generating an outside layer, the extraembryonic ectoderm, that grows down to further enclose the yolk. Free amoeboid cells are formed at the edge of the blastodisc, and come to lie in the perivitelline space between the chorion and the cellular membranes surrounding the yolk; they are perhaps phagocytic (Sacarrão, 1945).

One of the earliest rudiments to develop is the 'shell gland', an ectodermal invagination at the hind end of the mantle, a reminder of the earlier history of the octopods, now destined to produce only the two narrow chitinous rods that lie in the mantle above the bases of the gills (Spiess, 1972).

Soon afterwards a whole range of organ system rudiments begin to become visible from the outside of the embryo and from this point on the animal is recognizably cephalopod (Fig. 5.7). It also begins to move, first simply rotating along its long axis within the egg, and then undergoing two longitudunal inversions, in which the embryo first moves towards the stalked end of the egg, and then reverts to its original position, hind end facing the apex through which it will

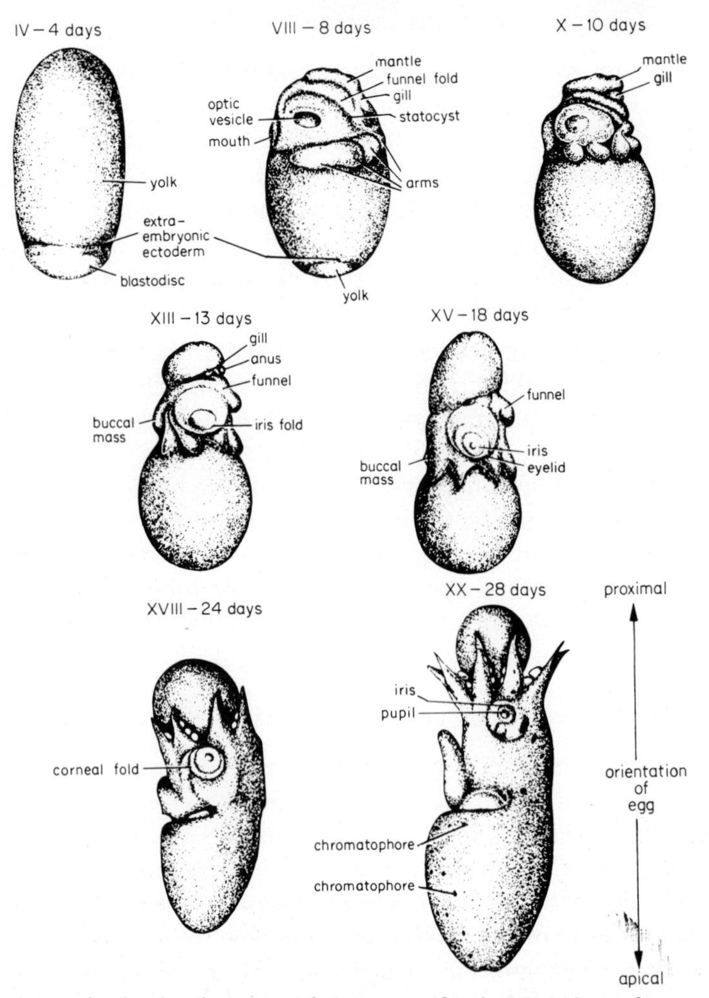

Fig. 5.7 Stages in the development of *Octopus vulgaris*. IV, 4 days after spawning. The extra-embryonic ectoderm begins to spread down over the yolk. VIII, rudiments of the arms, mantle, funnel, eyes and statocysts can be seen. Inside, the blood system has just begun to develop. The embryo inverts itself in the egg. X, optic vesicle is fully formed and pulsation of the outer yolk sac begins. XIII, the funnel begins to overlap the gill, and suckers are developing on the arms; inside, the elements of the coelomic system, including gonads and kidneys are becoming identifiable. XV, further development of the mantle and eyes, internal differentiation of the main elements of the CNS. XVIII, rapid reduction of the remaining yolk sac as the animal prepares to hatch; the animal again turns round in the egg and a 'hatching gland' is formed at the extreme hind end of the mantle. XX, the hatching gland weakens the egg shell and the animal escapes, swimming up into the plankton. Development takes about 28 days at 22°C (after Naef, 1928).

eventually escape. The first inversion occurs at stage VII (in Naef's 1928 classification – see Fig. 5.7), the second at stage XVIII shortly before hatching. The reason for the first of these strange manoeuvres is entirely unknown, the second presumably brings the embryo back into an appropriate position for a un-impeded escape from the egg (Boletzky, 1971).

The formation of the blood system begins at stages VIII–IX and from then on yolk absorption is always through the blood system rather than the gut; there is an extensive extraembryonic blood sinus, between the membranes around the yolk (Boletzky, 1975). Absorption steadily reduces the size of the external yolk sac, aided by muscular contractions of the buccal region which gradually drive the remaining yolk inside the animal; when *Octopus* hatches, it still has a considerable supply of yolk inside it, mainly in the digestive gland. There is a corresponding delay in the development of the gut (Sacarrão, 1945). The animals, presumably, do not feed during the first few days of their life in the plankton.

The development of the coelom system, kidneys and gonad have been described by Marthy (1968).

As the octopus approaches the moment of hatching, various organs in the skin are completed. These include the first few chromatophores (Fioroni, 1970), a hatching gland ('Hoyle's organ') which secretes a material responsible for dissolving the apical section of the chorion (Fioroni, 1962a) and the rather mysterious series of pits and bristles known as Kölliker's organs. These last are structures found only in octopods (and not in all of these, *O. briareus*, for example, does not have them). Each consists of a pit containing a small bundle of chitinous rodlets, lying beneath a layer of ectodermal cells in places where the animal will later have skin papillae (Fioroni, 1962b). *In vivo* observations show that the bundles can be evaginated and spread repeatedly (Boletzky, 1973). Their function is unknown.

5.3 Postembryonic growth and life span

5.3.1 *Longitudinal studies of growth rate*

The young of octopuses that lay large numbers of small eggs typically spend a period in the plankton after hatching. This period has been estimated as up to 3 months in the English channel – cold water, at the northern limit of the range of *O. vulgaris* (Rees, 1950; Rees and Lumby, 1954) – but it is normally assumed to be much shorter. Itami,

Izawa, Maeda and Nakai (1963) recorded settlement after 33–40 days at 25 °C in the laboratory. During the planktonic phase the arms grow in size relative to the rest of the body (Fig. 5.8) and the bristles (Kölliker's bundles) present at hatching disappear; by the time the young settle on the seabed they have grown from about 3 mm in total length to a little over 1 cm and closely resemble adults in their general appearance.

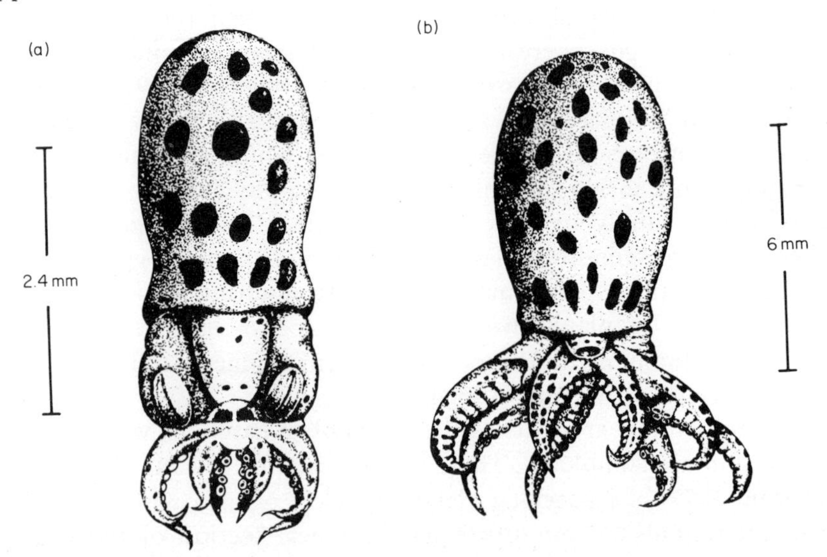

Fig. 5.8 *Octopus vulgaris* 'larvae' from the English Channel. (a) is soon after hatching (cf. stage XX in Fig. 5.7), (b) is a later stage at which the animals are beginning to settle down on the bottom. These are drawn from fixed specimens; the arms would be relatively larger in life, particularly in (b) (after Rees, 1950).

The growth of the newly settled animals is very rapid indeed. Itami *et al.* (1963) fed theirs on shrimp and crab meat and later on small live crabs and obtained a tenfold average increase in size within 40 days at 25 °C (Fig. 5.11). Mangold and Boletzky (1973) reared two similar animals at 20–30°C and found rates only a little slower than this.

Nixon (1969b), Mangold-Wirz (1963) and Mangold and Boletzky (1973) have summarized the information available for larger animals. Two sorts of evidence can be used to estimate the probable growth rate and lifespan of octopuses in the sea. The most direct arises from longitudinal studies on the growth of individual animals in aquaria; complimentary evidence comes from sampling natural populations.

Fig. 5.9 shows the results of several longitudinal studies carried out at Naples, Banyuls, and in the aquarium of the Zoological Society in London. In Naples the circulating water temperature falls from 25 °C in summer (July–August) to 15 °C or less in midwinter and this is

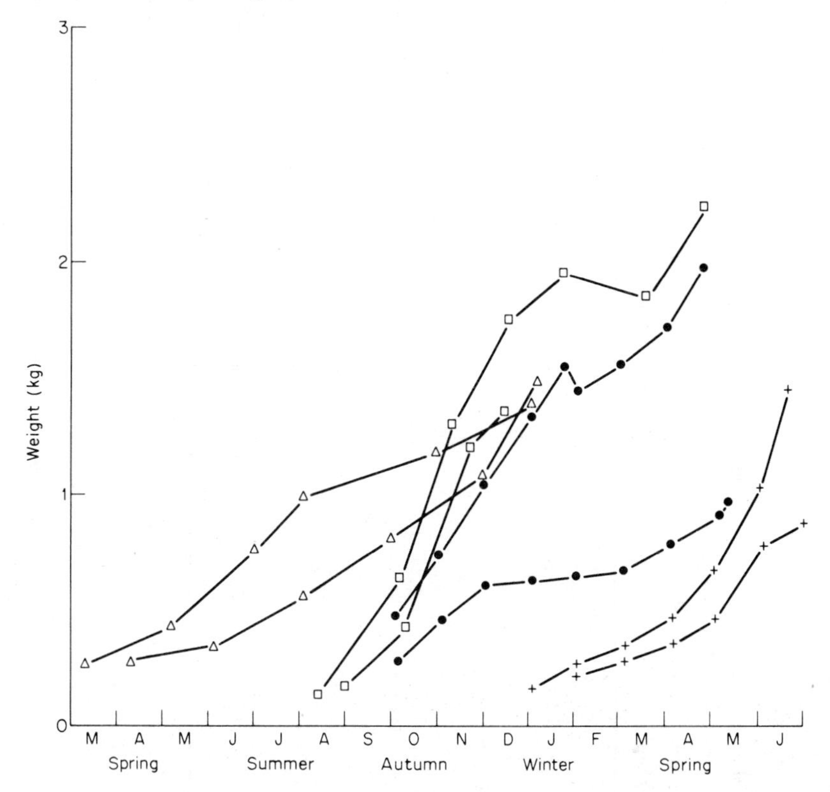

Fig.5.9 Growth rates in longitudinal studies. ● Nixon (1966), extremes from a sample of 12, Naples: + Mangold and Boletzky (1973), from a sample of 13, Banyuls; □ Lo Bianco (1903) 2 males, Naples; △ Vevers, 2 males, London (from Nixon, 1969c).

clearly reflected in the rate of growth of the animals, which ate less and in some cases even lost weight during this period (Nixon, 1966). Mangold and Boletzky (1973) reared octopuses in Banyuls at constant temperatures of 10, 15 and 20 °C and found, as expected, that both food intake and growth fell with temperature (Table 5.1). The matter is complicated by animal size; little octopuses eat more and grow faster. But they have a higher metabolic rate (Fig. 3.4) so that their conversion of crab meat to octopus is relatively poor though still remarkable at an average of around 50% (Table 5.1).

Table 5.1 *Octopus vulgaris.* Daily feeding rate, food conversion, and daily growth rate in 11 individuals raised at 10°C, 15°C and 20 °C. In the last three columns, the underlined values represent the means for whole periods of observation; the weekly minima and maxima are given in brackets (from Mangold and Boletzky, 1973).

Temperature (°C)	*Initial body weight* (g)	*Final body weight* (g)	*Period* (days)	*Total food intake* (g)	*Average weight increase per day* (g)	*Daily feeding rate* (%)	*Food conversion* (%)	*Daily growth rate* (%)
20	112.1	449.3	29	538.5	11.6	6.6 (6.2–8.4)	46.1 (25.6–85.9)	4.14 (4.20–5.08)
	178.0	420.0	29	522.3	8.2	6.0 (5.4–7.0)	46.1 (25.6–85.9)	3.11 (1.79–4.66)
	240.1	493.7	41	634.4	6.2	4.2 (2.3–7.7)	40.0 (24.8–63.1)	1.68 (1.14–2.61)
	587.3	1267.0	29	1165.5	23.6	4.7 (3.8–5.0)	58.3 (27.9–90.6)	2.73 (2.14–4.19)
15	81.5	111.0	19	82.1	1.6	4.5 (2.5–9.5)	35.9 (27.0–57.5)	1.91 (0.69–2.74)
	152.5	250.1	27	219.2	3.6	4.0 (2.0–6.5)	44.5 (27.7–81.4)	1.79 (0.85–2.29)
	233.0	347.0	24	154.6	4.8	2.3 (1.9–2.5)	73.5 (42.6–87.3)	1.69 (1.42–2.12)
	343.2	450.0	18	163.4	5.9	2.3 (2.1–2.6)	65.5 (30.0–82.8)	1.50 (0.87–1.73)
10	79.5	152.5	62	124.6	1.2	1.7 (1.4–2.7)	58.4 (27.1–92.5)	1.01 (0.47–1.42)
	185.4	311.2	65	293.7	1.9	1.8 (1.5–2.5)	42.8 (25.6–48.1)	0.78 (0.35–1.28)
	379.1	461.2	21	124.4	3.9	1.4 (0.8–2.0)	65.9 (46.9–78.8)	0.93 (0.48–1.57)

Daily growth rate was calculated as $\frac{w_2 - w_1}{tW}$. $W = \frac{w_1 + w_2}{2}$ where w_1 = the initial weight; w_2 the final body weight and t = time in days.

Food conversion = $\frac{w_2 - w_1}{\text{Food Intake}}$ and Feeding rate = $\frac{\text{Food Intake}}{tW}$.

If octopuses are starved, they lose weight and become visibly emaciated. The total loss can be considerable and quite rapid at high temperatures. Starvation does not appear to debilitate the animals, which remain alert and capable of swift movement after a loss of more than 40% of their body weight (Table 5.2).

Two other longitudinal studies have been reported. Wood (1963) recorded that a single individual in Florida grew from 55 g to 3.5 kg in 10 months, but gives no intermediate weights. Itami (1964) branded and released 1600 animals of from 200 to 500 g off Japan. 15 of these were recovered; those recaptured after 5 to 7 weeks in the sea had trebled their weight at 19–25 °C. These rates of growth are clearly compatible with those found in European laboratories; Itami's experiment, moreover, suggests that food availability is probably not a limiting factor in the sea.

5.3.2 *Sampling studies of growth rate*

Assuming that growth in the sea is even approximately the same as that found in the laboratory, it is clear that octopuses of breeding size (1 kg and above, in females, see Table 5.3) could well have been spawned in the previous year. It is more difficult to be sure that this

Table 5.2 *Octopus vulgaris.* Changes in body weight due to starvation.

Temperature (° C)	*Period of starvation (days)*	*Initial weight (g)*	*Final weight (g)*	*Loss (%)*	*Initial weight (g)*	*Final weight (g)*	*Loss (%)*	*Initial weight (g)*	*Final weight (g)*	*Loss (%)*
20	6	493.0	440.0	4.7	56.2	48.1	14.4	1500	1445	3.7
15	6	450.7	432.0	4.1	74.1	68.0	8.8	1520	1483	2.4
10	6	420.4	412.1	2.0	72.1	68.5	5.0	1400	1378	1.6
20	30	460.0	419.0	8.9						
15	30	450.0	421.0	6.4						
10	30	470.0	450.0	4.3						
25	38	175.0	110.0	37.0						
25	38	200.0	125.0	37.5						
25	32	350.0	200.0	42.8						
25	36	260.0	130.0	50.0						
25	36	240.0	130.0	46.0						

Data on the animals kept at 10–20° from Mangold and Boletzky (1973). Those at 25° from Wells and Wells (1975) or O'Dor and Wells, (in preparation). The last three animals kept at 25 °C were under stress in the course of an experiment on the mobilization of food reserves. They were injected with ^{14}C and ^{3}H–labelled leucine and had the tips of some of their arms removed during the starvation period. Allowance has been made for the weights of these samples. A trio of controls of comparable weight, treated in the same manner, but fed on fish at a rate of about 10 g per day showed weight gains ranging from 21–33%.

does indeed occur and whether or not it is likely to be the normal situation. Growth is very dependent upon temperature in both embryonic and postembryonic stages. Below 13 °C, in the shallow sea off Banyuls in March or early April, or at greater depths (octopus spawn has been collected down to 85 m) late in the year, the eggs could take up to 3 months to develop, and, as Rees and Lumby (1954) have pointed out, planktonic development can take as long again. The animal might still weigh less than a gram six months after it was spawned.

Mangold-Wirz (1963) analysed the size and sexual condition of a sample of 900 octopuses collected in trawls and traps from the Port-Vendres/Banyuls region in the Western Mediterranean. Her findings are summarized in Table 5.3 which is taken, with minor modifications from Nixon (1969c). Mangold-Wirz's original data is given in terms of dorsal mantle length and Nixon has converted this into weights so that the sample findings can be compared with her own longitudinal studies.

The sampling data suggest that a population of octopuses moves inshore from deeper water in February and March. These animals are

Table 5.3 Summarizes the information given by Mangold-Wirz (1963) for *O. vulgaris* caught in the Bay of Banyuls and Port-Vendres region of the Mediterranean (modified from Nixon, 1969c).

Date	*Sex*	*Dorsal mantle length (mm)*	*Deduced total body weight (g)*	*State of gonads*	*Probable age (months)*
Spring spawned					
November		50	70	Undeveloped	5–6
March	Male	90–100	270–400	Developing	9–10
	Female	90–95	270–325		
July	Male	130–145	600–800		13
	Female	100–135	400–650		
Summer spawned					
December		30	20		3
April	Male	57–69	60–115	Undeveloped	8
	Female	55–80	54–180		
June	Male	80–83	190–250		
	Female	76–100	160–370		
August	Male	106–110	460–540		12
	Female	95–120	330–700		
February	Male	140–160	1200–1800	Ripe	18
	Female	135–150	1000–1500	Almost ripe	
March	Male	150–170	1500–2250	Ripe	19
	Female	145–160	1300–1800	Maturing	
June	Male	180–200	2700–3700	Ripe	
	Female	165–175	2000–2500	Ripe eggs laid	22

quite large, 1 kg or more and both sexes are ripe, or ripening. The males precede the females. The females spawn in May and June and then stay with their eggs until these hatch later in the summer; the large males apparently retire to deeper water in July.

It seems likely that the small animals collected in November–December are the offspring of these spring immigrants; those spawning early being responsible for the c. 70 g animals found in shallow water in November, while the later spawners (octopuses guarding their eggs are still found in October) generate the small c. 20 g animals that can be collected in shallow water at the end of the year (Mangold-Wirz, 1963).

If this view is correct, the large animals that move in from deep water in February are unlikely to be as young as 9 months, since this would imply a rate of growth well above that found in the summer-time longitudinal studies, during a period when the animals have retired to deep, cold water. It is more probable, as Mangold-Wirz suggests that the large animals are at least 18 months and perhaps up to two years old when they spawn.

In addition to the breeding animals, a large population of smaller, immature octopuses is found inshore during the summer months. In March–April these range from 60–400 g, probably representing, respectively, the late and early spawnings of the previous year. The larger animals have gonads that are just beginning to develop towards maturity in the spring and it is possible that the most precocious of these spawn later in the same year, at an age of a little over 12 months. The rest, averaging some 700 g by the end of summer, retire to deeper water in winter and reappear to spawn in the year after. In Table 5.3 these middle-sized animals are represented as two populations, early and late-spawned, although they really represent a continuum derived from eggs hatched at any time from May to November. We know, besides, from the longitudinal studies, that there are enormous individual differences in growth rate, even among animals kept and fed under identical conditions, so that it is strictly impossible to make an accurate estimate of the age of any individual *O. vulgaris* in a temperate climate from its size (Mangold-Wirz, 1963; Mangold and Boletzky, 1973).

5.3.3 *Lifespan*

Evidence has already been presented (Section 5.1.6) that female octopuses rarely, if ever, survive spawning so that the overwhelming

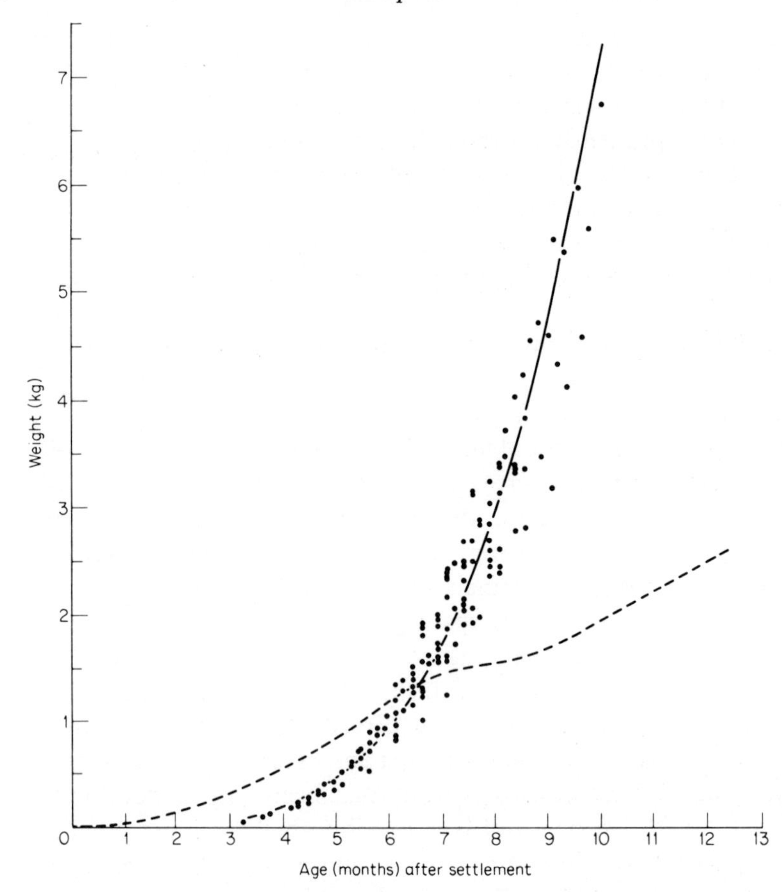

Fig. 5.10 The growth curve of *Octopus cyanea* at 25.6°C obtained by plotting weights of 29 individuals (14 ♂♂. 15 ♀♀) at 15 day intervals is shown. The solid line represents the fitted curve described by body weight (g) = $(1.28 \times 10^{-6}) \times$ age in days$^{3.92}$. The dotted line represents the growth of *O. vulgaris* (redrawn from Nixon, 1969b) where there is a seasonal depression of feeding (and hence growth rate) by water temperature (from Van Heukelem, 1973).

majority die off by the end of their second year. The same seems to be true of males; if it were not so one would expect to find a substantial population of animals weighing (on projection from the growth curves derived from longitudinal and sampling studies in years 1 and 2) 5 kg and more by their second breeding season. No such population is found, although occasional individuals of up to 25 kg have been reported (references, see Nixon, 1969c).

More direct evidence on this matter has been obtained by van

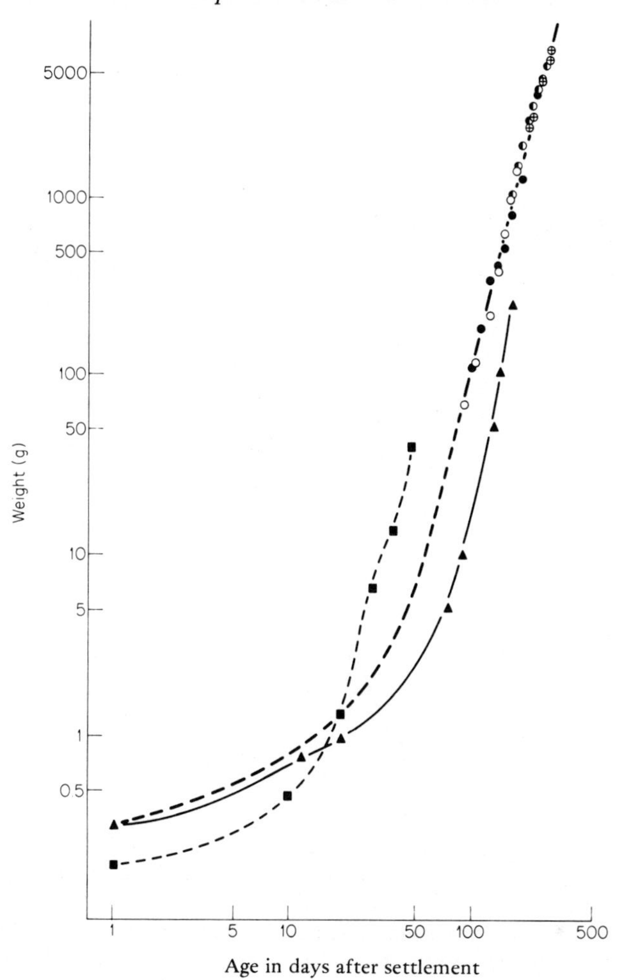

Fig. 5.11 The growth of four *O. cyanea* (2 males ○ and ● and 2 females ◐ and ⊕ on an *ad libitum* diet of crabs for 90 day periods) is shown on a double logarithmic grid. Triangles indicate the most rapid growth of newly settled *O. cyanea* (redrawn from Wells and Wells 1970). Squares indicate the average growth of laboratory reared *O. vulgaris* (Itami *et al.* 1963). The heavy dashed line depicts the probable growth of *O. cyanea* if fed *ad libitum* from settlement to 67 g (from Van Heukelem, 1973).

Heukelem (1973) working with *O. cyanea.* His population came from Hawaii, where the water temperature varies comparatively little, between 22 and 28°C. He carried out longitudinal studies on 29 individuals kept in aquaria, and on a further 7 octopuses that were

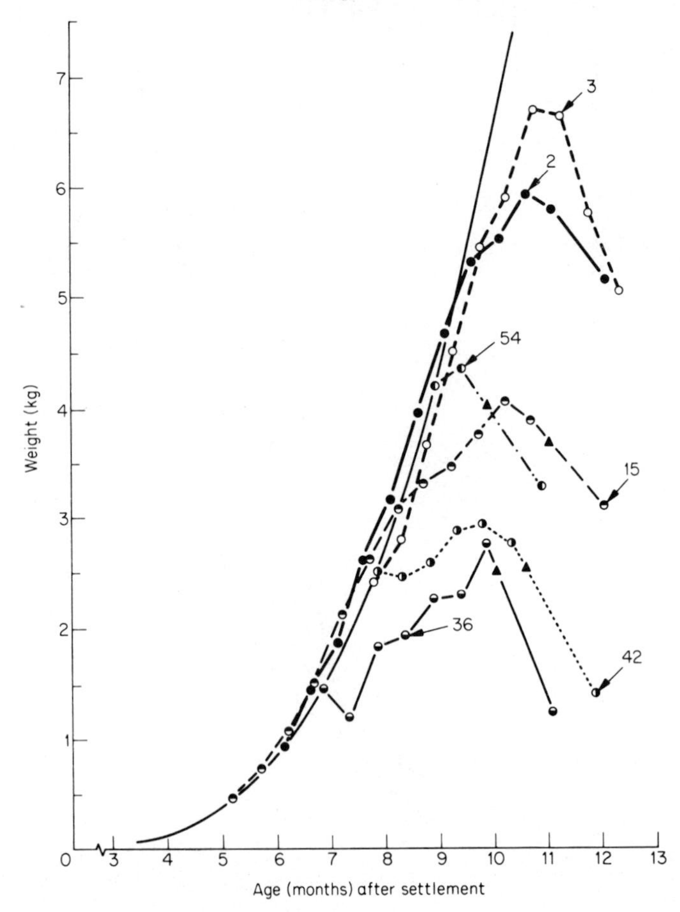

Fig. 5.12 The growth and life-span of female *O. cyanea* aged by their weight at capture. All except nos. 36 and 42 were fed *ad libitum*. Triangles indicate the estimated age when spawning occurred. Nos 2 and 3 died in the act of spawning; their gonads represented 21 and 15% of their total body weights respectively. Food intake decreased and eventually stopped as the gonad matured, hence there was a slowdown in growth and often a loss in weight before spawning. The final point for each animal indicates the age and weight at death (from Van Heukelem, 1973).

branded and released into the sea. The two groups gave very similar results, suggesting that food is not a growth-limiting factor in the wild; the results with the larger, captive, sample are summarized in Fig. 5.10.

Taken together with Wells and Wells' (1970) records of the growth of newly hatched *O. cyanea* and the results of Itami *et al.* (1963) with *O. vulgaris* (Fig. 5.11), these studies suggest that *O. cyanea* grows at much the same rate as *O. vulgaris* in the Mediterranean summertime.

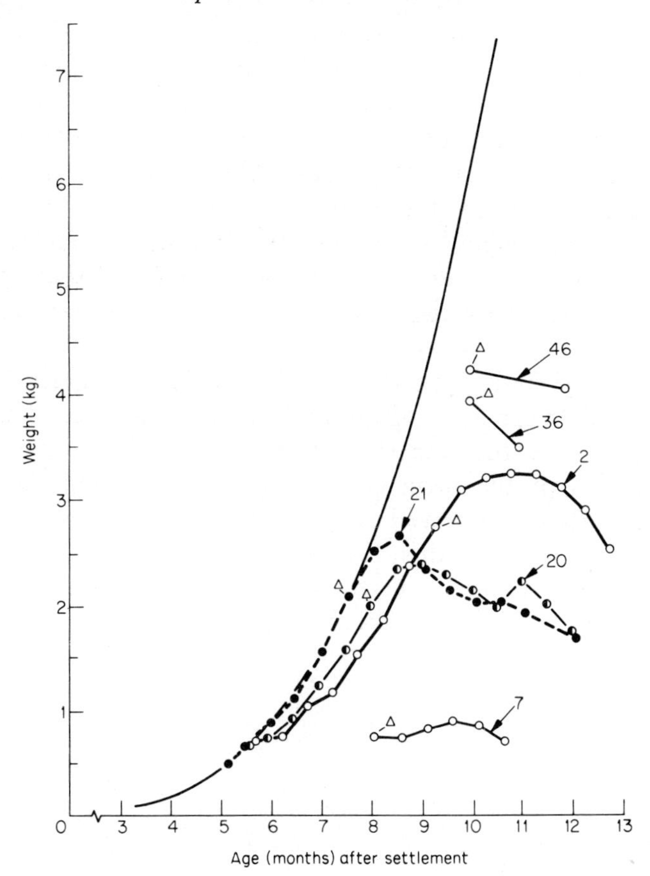

Fig. 5.13 The growth and life-span of male *O. cyanea* aged by their weight at capture. The last point for each animal is the weight at death. Nos. 7, 36 and 46 are examples of males that had enlarged suckers when captured and thus could not be aged. Triangles indicate the weights at which sucker enlargement was first noticed (from Van Heukelem, 1973).

But because there is no slowing due to a winter fall in water temperature, *O. cyanea* achieves its breeding weight of 2 kg and more within a year of spawning. The steady temperature also means that in this species it is reasonable to predict the age of an animal from its weight, since food supply does not seem to be limiting.

Both sexes die after they have bred. Females, as with *O. vulgaris,* stop feeding as their ovaries mature and they are already losing weight by the time they spawn (Fig. 5.12). Sakaguchi (1968) has noted that the activities of the proteolytic enzymes formed in the hepatopancreas and posterior salivary glands decline to about one tenth and one thirtieth of their normal values, respectively, at the time of spawning Males too cease feeding at about the same age and body weight as their mates. Van Heukelem has noted that this seems to be associated with the appearance of extra large suckers (as in *O. vulgaris*, Section 5.1.5) in the mature males (Fig. 5.13). At all events the animals take a decreasing ration of food, even when this is present *ad lib*, and they begin to appear 'flabby' with considerable loss of their capacity to change colour. 'Their general appearance resembled that of females that had hatched their eggs and were about to die' (van Heukelem, 1973).

No very large individuals (> 6.5 kg) of *O. cyanea* of either sex have been reported. In commenting on this and the occurrence of occasional very large *O. vulgaris*, van Heukelem (1973) points out that, given growth curves of the form seen in either species and the range of individual variation observed in *O. vulgaris*, animals even of 10 and 20 kg need be no older than their siblings of one tenth their size. An animal that can double its weight in a month could well achieve these very large sizes within two years of hatching.

CHAPTER

SIX

Endocrinology

An animal as complicated as *Octopus* might be expected to enjoy a hormonal regulation system as complex as that found in mammals, or at least as elaborate as that found in arthropods. In fact, rather little is known about hormonal control of anything in cephalopods. We know something about the hormonal control of gonadial development and this story necessarily fills the greater part of the present chapter. Apart from this, a scattering of putative endocrine organs has been identified, but there is no solid evidence of the function of any of these.

6.1 Optic glands and the hormonal control of sexual maturity

6.1.1 *The structure and innervation of the optic glands*

The optic glands are small rounded bodies found on the upper posterior edges of the optic tracts (Fig. 6.1). In young *O. vulgaris* they are pale yellow in colour, becoming swollen and orange as the animal matures. They have been known for many years. Owen (1832) figures them in a drawing of the brain of *Sepia,* and they are illustrated and labelled as 'glandula ottica' in *Octopus Aldrovandi* by Delle Chiaje (1828). Subsequent descriptions in the literature nearly all reported them as nervous tissue, apparently confounding what was visible to the naked eye on dissection, with the olfactory and peduncle lobes visible in sections under the light miscroscope. Boycott and Young (1956a) have reviewed and clarified the ensuing confusion. There is now no doubt that these bodies are endocrine glands. Their fine structure has been described by Björkman (1963), Nishioka, Bern and Golding (1970) and more recently by Froesch (1974).

The glands are heavily vascularized, blood being supplied through the posterior optic lobe artery. Inside the gland the arterial vessels

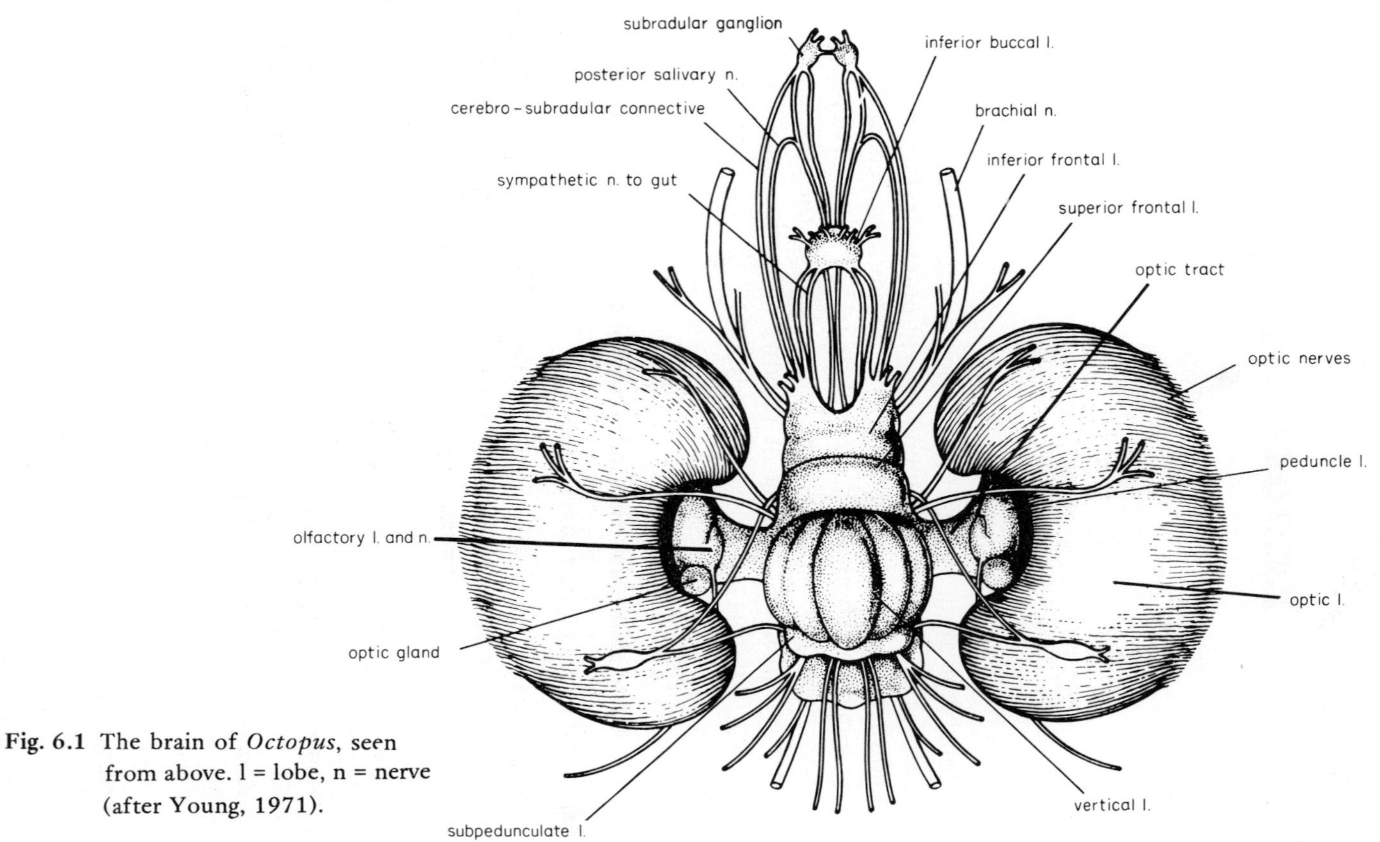

Fig. 6.1 The brain of *Octopus*, seen from above. l = lobe, n = nerve (after Young, 1971).

enlarge to form wide blood spaces joined by a capillary mesh to venous sinuses around the gland and ultimately to the system of great veins and sinuses draining the orbit (Young, 1971). Apart from those associated with blood vessels, there are only two types of cell in the gland. Large stellate cells, with massive (10–15 μm diameter) rounded nuclei and prominent nucleoli, are responsible for the production of the optic gland hormone – they change in size and appearance, though apparently not in numbers, as the gland matures. Among these, and evidently not concerned in secretion, since they remain in the same condition whatever the secretory state of the gland, are the smaller oval nuclei of the supporting cells, probably fibrocytes responsible for the production of a connective tissue framework to the gland (Boycott and Young, 1956a, Wells and Wells, 1959). The cytoplasm of the large stellate cells contains many mitochondria, extensive Golgi apparatus and large numbers of ribosomes. The endoplasmic reticulum (in the resting gland) consists of a sparse system of fine tubules. The fine structure of the supporting cells is similar with fewer mitochondria and ribosomes, 'the cytoarchitecture of the gland suggests that it is of nervous origin. The supporting cells resemble glia cells' (Björkman, 1963).

No trace of the optic glands is visible in the embryonic stages and they cannot be found in the planktonic larvae. They appear in young animals at about the stage that these settle onto the seabed and seem to develop from nerve cells close to the olfactory lobe (Bonichon, 1967).

The optic glands are innervated from the subpedunculate lobe at the back of the supraoesophageal brain. Removal of this lobe is followed by degeneration of at least some of the nerves, which enter the gland from a bundle running along the optic stalk (Wells and Wells, 1959; Froesch, 1974). Both axo-axonal and axo-glandular synapses have been observed, the former apparently disappearing as the gland matures. Clear synaptic vesicles of about 40 nm, with occasional larger (80 nm) dense-cored vesicles are found in the nerve axons; there is no evidence of neurosecretion (Froesch, 1974).

If immature animals are blinded by optic nerves section, the optic glands enlarge and the octopuses mature precociously (see below). This observation led Defretin and Richard (1967) to test the effect of reducing the daylength on the state of the glands in *Sepia.* They kept cuttlefish in tanks in the dark for 22 out of every 24 hours. A control group remained in continuous light. The large stellate cells in the glands of the short daylength animals enlarged, with a massive increase in endoplasmic reticulum and Golgi. Budding from the Golgi were large numbers of dense vesicles 100 nm in diameter, absent in the resting glands and presumably containing the glandular secretion, which must be shed into the bloodstream.

6.1.2 *The function of the optic glands*

In 1956a, Boycott and Young noted that the optic glands were enlarged in a proportion of the animals that they were using for experiments on brain function in learning. Animals in which the optic tracts had been sectioned had large glands and grossly enlarged gonads. Similar effects were observed after destruction of the subvertical lobes.

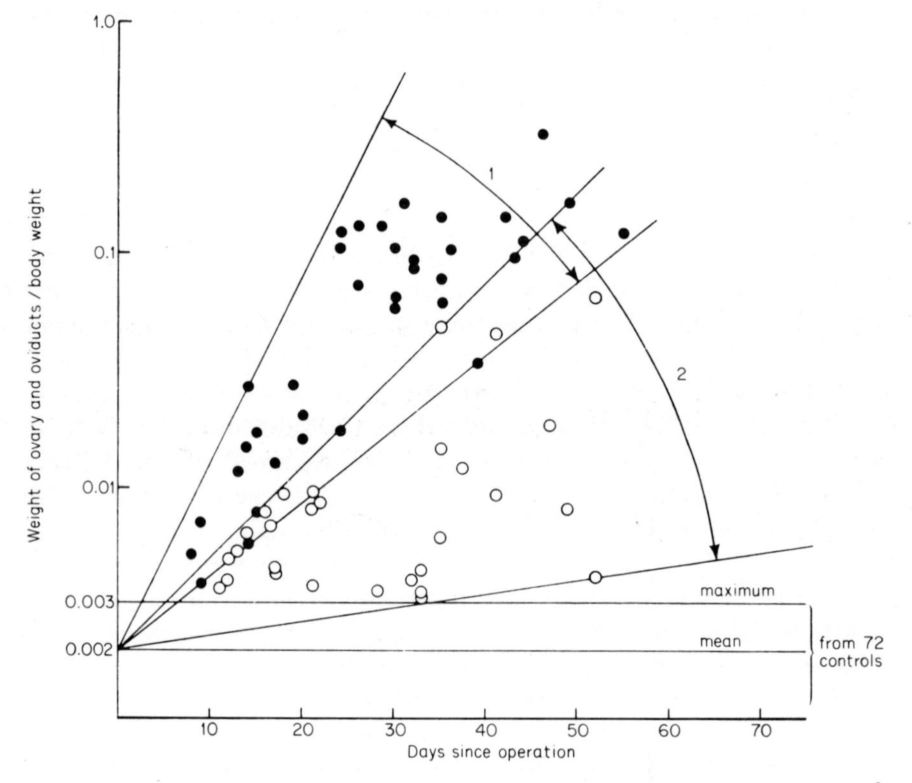

Fig. 6.2 The effect of activating the optic glands. This shows the consequences of (1) removing the subpedunculate lobes (plotted ●) and (2) cutting the optic nerves (○), on the weight of the ovary and oviducts in animals that would otherwise have been immature at body weights of 200–800 g (from Wells and Wells, 1959).

Following up these observations Wells and Wells (1959) were able to demonstrate that any operation cutting the nerve supply to the glands from the subpedunculate lobe (which lies behind the subvertical) will produce enlargement of the glands and subsequent enlargement of the ovary. A similar, though somewhat slower, response is

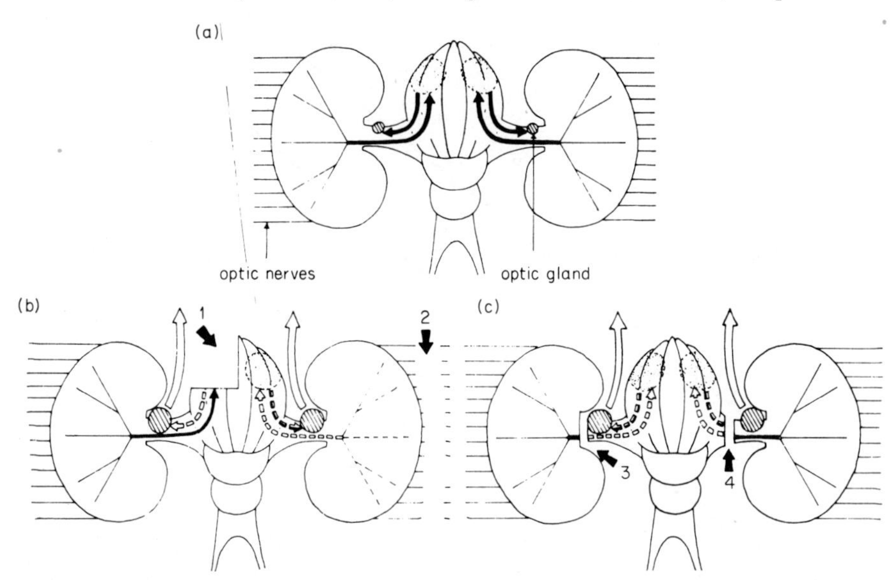

Fig. 6.3 The mechanism of hormal control of gonad maturation in *Octopus.* (a) Situation in an immature, unoperated *Octopus,* where secretion by the optic glands is held in check by an inhibitory nerve supply, (b) Two operations that cause the optic glands to secrete a product causing the gonad to enlarge, being (1) removal of the source of the inhibitory nerve supply, and (2) optic nerves section. (c) Further operations having the same effect upon the gonads, thus eliminating the possibility that there is also an excitatory innervation, being (3) optic lobe removal and (4) optic tract section (from Wells and Wells, 1959).

produced by cutting the optic nerves (Fig. 6.2). The gonad remains small, whatever the operation, if the optic glands are removed. Since the effects of blinding and more central brain lesions were not additive, it was postulated that control of the condition of the glands always ran through the subpedunculate lobe, so that the full chain of control is:

Light (? daylength) ⟶ subpedunculate lobe of the brain ⟶ optic glands ⟶ gonad.

The first two links in this chain are inhibitory, and nervous, the last excitatory and hormonal. The effect of various lesions is explained on these assumptions in Fig. 6.3.

Subsequent work has in general confirmed these hypotheses. Removal of the subpedunculate lobe is always followed by enlargement of the optic glands. If the operation is unilateral, only the gland on that side enlarges. Excitation of one or of both of the optic glands

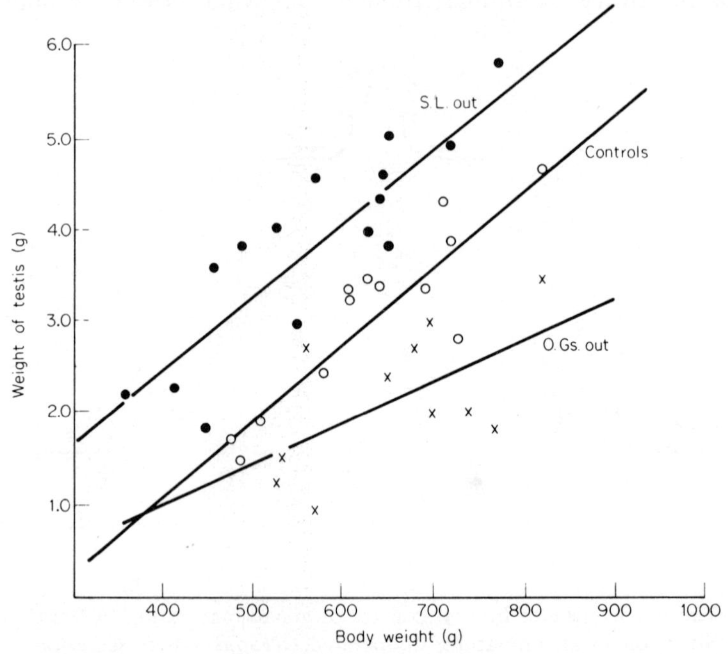

Fig. 6.4 The effect of activating or removing the optic glands from males. The results of activating the glands by removal of the subpedunculate lobes are plotted ●. Control weights are shown ○, and the effect of removing the optic glands from animals X. Regression lines summarising these data indicate that the effect of the brain operation is greatest in the smallest and least mature octopuses, while removal of the optic glands has most effect upon large individuals in a relatively advanced state of maturity (data from Wells and Wells, 1972b).

always produces enlargement of the gonad. The effect is most spectacular in females, where the ovary may increase in size from about 1/500th to as much as 1/5th of the body weight within a month of the operation (Fig. 6.2). But it is also found in males, where the testis, particularly of small animals, can double in weight in a similar period (Fig. 6.4).

Males, as we have seen in Chapter 5, tend to mature earlier than their mates and this allows for a further type of experiment that is not readily made with females, which mature more abruptly at a greater age and size. Removal of the optic glands from males that are already producing sperm is followed by a decrease in the weight of the testis and the eventual cessation of sperm and spermatophore production; the effect is most marked in the largest males (Figs. 6.4 and 6.5). The optic glands in males weighing less than 1000 g are generally small, no larger than those in their immature female counterparts, yellow rather than orange and not obviously secreting. It seems that even glands that are not visibly enlarged must be producing a trickle of secretion. This is sufficient to stimulate sperm production in males but not, it would appear, enough to stimulate yolk production in the ovary, a matter that is discussed below (Wells and Wells, 1972b).

The link between daylength and secretion of the gonadotropin is less certain than the relation between subpedunculate lobe, optic gland and gonad. If *Sepia* is kept under short daylength conditions, it matures precociously (Defretin and Richard, 1967; Richard, 1967), as one might expect from the effects of blinding *Octopus.* Blue light, in particular, seems to have a marked inhibitory effect on gonadial development and this, again, is what one would expect. *Sepia,* like *Octopus,* migrates to deeper water in winter and if the system is to be light-regulated at all, the animals must be sensitive to those wavelengths that can penetrate into the depths (Richard, 1970b, 1971).

Rather sadly (since it is otherwise a tidy story) attempts to replicate the cuttlefish experiments using *Octopus* have yielded only equivocal results: only a few of the animals appear to respond (Buckley, 1977; Mangold, personal communication). The effect of blinding the animals, moreover, could just possibly be unconnected with daylength, since other operations that damage structures in the orbit, or the implantation of foreign bodies into the orbit can also result in optic gland enlargement, albeit much more slowly than optic nerves section (Froesch, personal communication). Starvation has been claimed as a cause of precocious maturation in *Eledone* (Mangold and Boucher-Rodoni, 1973), though again the results have not been confirmed with *Octopus.* Octopuses kept in the laboratory for long periods generally have gonads that are slightly larger than those from animals of the same size fresh from the sea (Wells and Wells, 1975). It would appear that a variety of stresses can trigger optic gland and gonad enlargement in the laboratory; the conservative position at the present time is that we just do not know what controls the onset of sexual maturity in *Octopus* in the wild.

Because optic glands deprived of their nerve supply begin to secrete

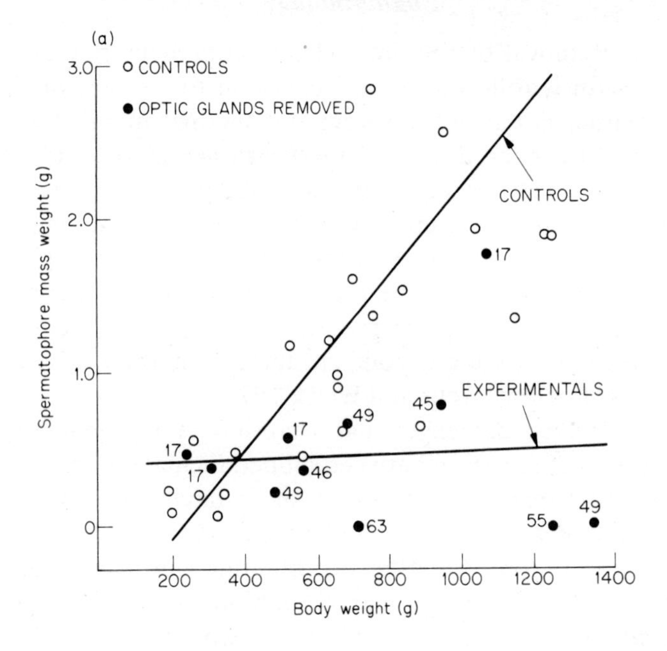

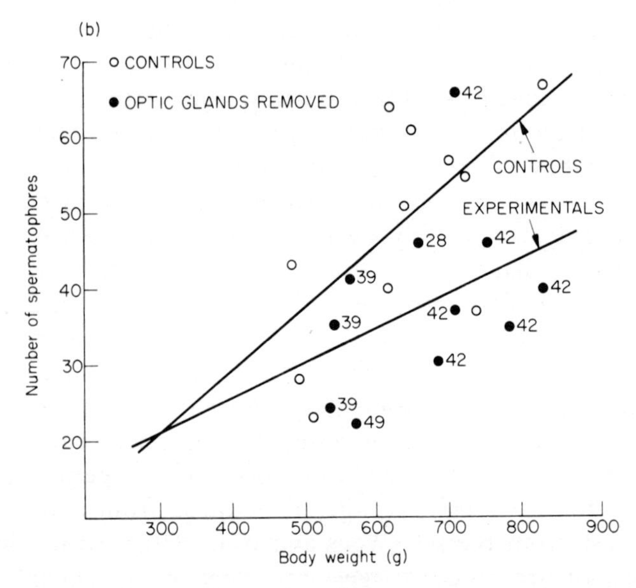

Fig. 6.5 Optic gland removal and its effect on spermatophore production. (a) *O. cyanea*, (b) *O. vulgaris*. In each case the figure beside an experimental result (●) shows the number of days since operation. In *O. cyanea* the mass of spermatophores was weighed, not counted; there are *circa* 200 spermatophores to the gram in this species (from Wells and Wells, 1972b).

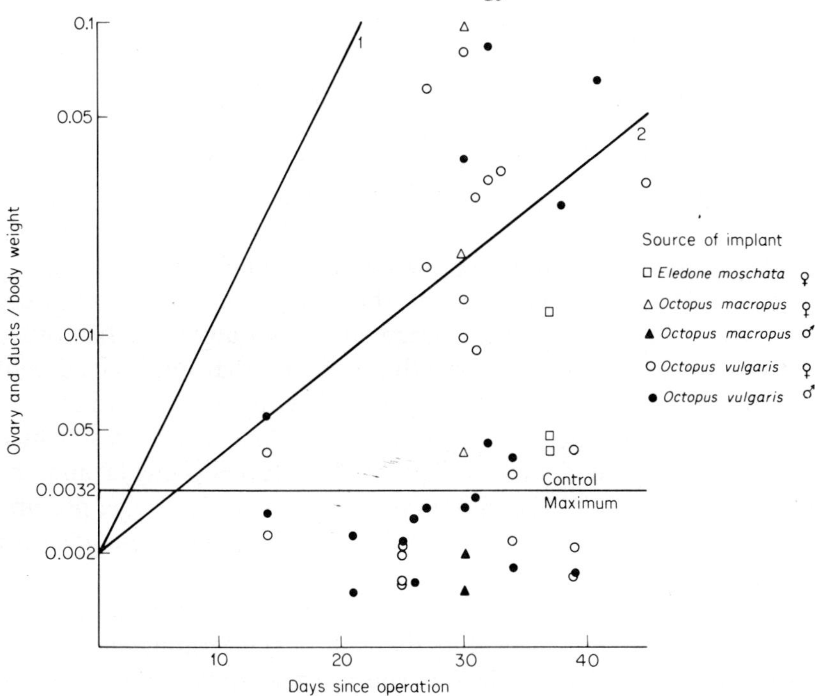

Fig. 6.6 The effect of implanting optic glands into immature females. The ordinate, as in Fig. 6.2, is a log scale, while the lines 1 and 2 indicate, respectively, the maximum rates of enlargement found after removal of the subpedunculate lobes and section of the optic nerves (see Fig. 6.2). The maximum rate of ovarian enlargement induced by implants is comparable with that induced by subpedunculate removal (from Wells, 1976a).

it is possible to implant the glands knowing that they will become active if they survive in the recipient. Glands implanted into the system of blood sinuses behind the eyes adhere to the walls of the sinuses where they attract an arterial blood supply and begin to secrete. A month later (at 25 °C) the gonads of recipients that would otherwise be immature are noticeably enlarged. The sex of the donor, and the condition of the glands (active or inactive) when implanted seems to make no difference to the result. Implants derived from other species and genera (*O. macropus* or *Eledone moschata*) will also survive and induce accelerated gonadial development in the recipients (Fig. 6.6). So far no success has been achieved with implants derived from *Sepia* or *Loligo,* although there is reason to believe that the hormone produced by the optic glands of decapods is very similar to that found in *Octopus* (Section 6.1.3 below, Wells and Wells, 1975).

6.1.3 *In vitro experiments*

Fragments of the ovary or testis of *Sepia* can be kept for several weeks *in vitro,* and this has yielded information, not readily obtainable from *in vivo* experiments with *Octopus,* about some of the stages at which the optic gland hormone appears to be necessary for development. The normal development of octopus eggs has been outlined in Section 5.1.4; *Sepia* too has a germinal epithelium and follicle cells surrounding each ovum. The follicle cells secrete the yolk in the later stages of egg growth (see below). In culture, fragments of young cuttlefish ovary will grow, apparently quite normally, provided that optic gland cells are included in the sample; the germinal epithelium continues to divide to produce oogonia, these in turn produce oocytes, which acquire a coating of follicle cells, the follicle cells divide and the oocytes begin to swell. The system breaks down at this stage, presumably because the nutritive demand of the swelling eggs is too great for the nutrient medium.

Similar fragments of ovary, cultured without optic glands, will survive for as long, or longer, but the cells of the germinal epithelium and the follicle cells fail to divide. The oogonia already present when the culture is set up all transform into primary oocytes during the first three weeks, and these develop normally to the point where their further enlargement depends upon the follicle cells (Durchon and Richard, 1967; Richard, 1970a).

A similar block is found in cultures of testes. With optic glands present, spermatogonia are produced from the germinal epithelium and divide to give spermatocytes and eventually sperm. Without optic glands, the spermatocytes develop normally, but the system eventually slides to a halt because there are no further divisions of the spermatogonia (Richard, 1970a).

In vitro experiments have also been carried out using the eggs from octopuses at a much more advanced stage in development, during the period of intensive yolk production that immediately precedes egg laying. In a series of preliminary experiments, O'Dor and Wells (1973) showed that the ovary itself synthesizes most and perhaps all of the proteinaceous yolk made in the later phases of maturation of the eggs. Labelled amino acid, injected into the vascular system of maturing animals disappears from the blood within minutes (in a typical experiment 97% of a dose of ^{14}C-leucine had gone within 10 minutes) and begins to accumulate in the ovary (within 5 hours, 38% of the total

Table 6.1 Details of ovary and body weights and of labelled protein distribution for animals sacrificed after 5 h (from O'Dor and Wells, 1973).

Group	*Animal wt (g)*	*Ovary wt (g)*	*Protein DPM/Total DPM injected* × *100*		
			Ovary	*Blood**	*Liver*
Control	352	0.6	0.1	3.2	1.9
OG–†	296	18	0.2	2.5	0.6
OG+†	254	45	38	0.5	0.1

* Assuming blood volume = 5.8% of body weight as reported by Martin *et al.* 1958.
† OG+, animal with optic glands activated by removal of the subpedunculate lobe 22 days before. OG–, similar animal, but with the optic glands removed 5 days before this experiment.

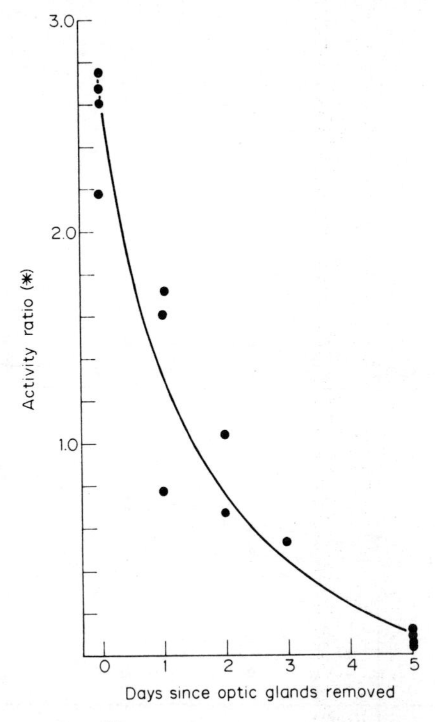

Fig. 6.7 Incorporation of [^{14}C] leucine into protein in eggs *in vivo* after removal of the optic glands. The activity ratio is the dpm in protein per g eggs divided by the dpm injected as amino acid per g body weight (after Wells, O'Dor and Buckley, 1975).

label injected was present as protein in the ovary, Table 6.1). Uptake and synthesis stop if the optic glands are removed; within two days of this operation protein synthesis had dropped to about half of its initial value. After five days it was no longer measurable (Fig. 6.7, Table 6.1).

By analogy with yolk-forming systems in arthropods (Brookes, 1969) and vertebrates (Wallace and Jared, 1969) one would expect yolk protein to be synthesized elsewhere in the body and carried to the ovary via the bloodstream. There is no evidence of this in octopuses. Amino acid injected into the bloodstream never reappears in substantial quantities as protein in the blood (Fig. 6.8). The small quantity of protein that is found there is almost certainly haemocyanin, synthesized in the branchial glands (Section 3.2.14) and is certainly not destined for the ovary, as blood transfusion experiments show (Fig. 6.9, O'Dor and Wells, 1973).

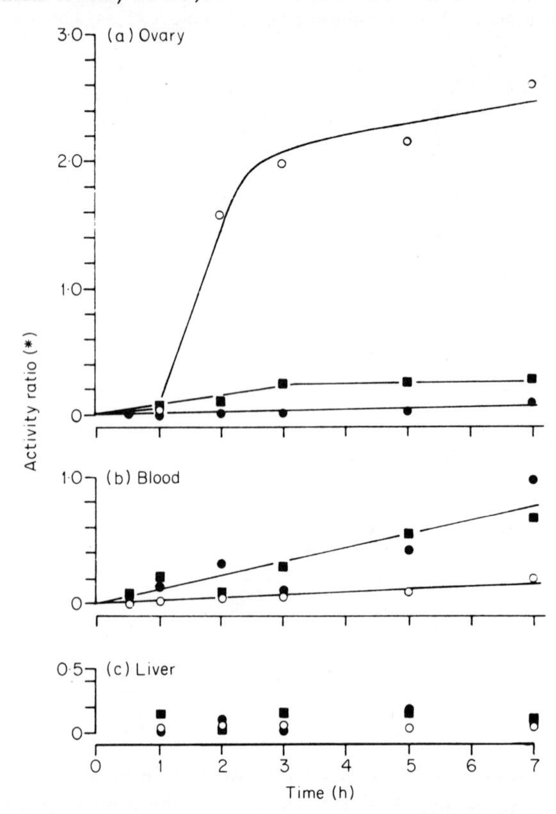

Fig 6.8 The levels of radioactivity in the protein fractions of the ovaries (a), blood (b) and livers (c) of the female octopuses in three different physiological states sacrificed at intervals following injection of [^{14}C] leucine. ■ controls, ○, precociously mature, ●, precociously mature, but with the optic glands removed 5 days prior to injection. * The activity ratio is the dpm per ml of blood or per g of tissue divided by the dpm injected per g body weight (from O'Dor and Wells, 1973).

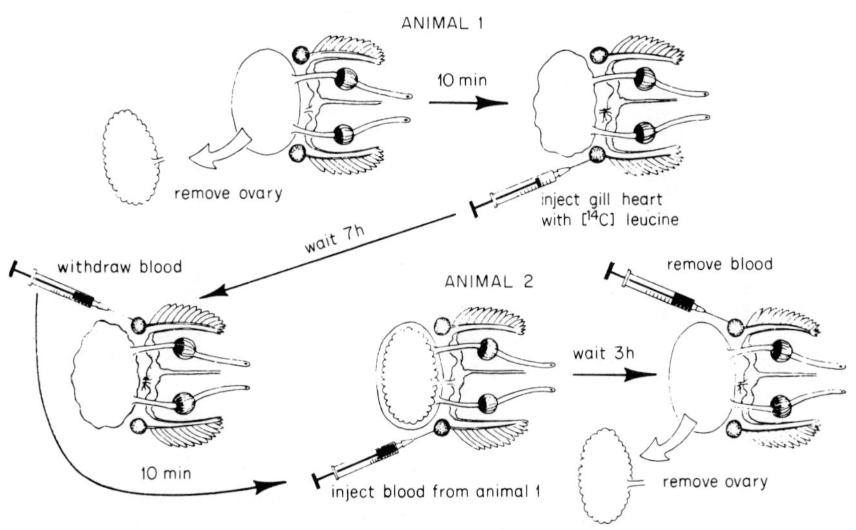

Fig. 6.9 Pattern of transfusion experiments to determine whether proteins found in the blood are incorporated into the ovary. Labelled blood proteins were produced by a precociously maturing animal (1) ovariectomized just before the beginning of the experiment. If this protein was destined for the ovary one would expect it to settle there in the recipient (animal 2). It remained in the blood (from O'Dor and Wells, 1973).

The fact that the ovary synthesizes its own yolk protein has opened up the possibility of developing *in vitro* assay systems for the optic gland gonadotropin. Eggs taken from animals with active optic glands and enlarged ovaries will continue to create yolk protein if incubated in a suitable nutritive medium, and the synthesis can be controlled by adding extracts of optic glands.

Figs. 6.10 and 6.11 summarize the results of some typical experiments. Uptake of label and incorporation into protein are both greatly enhanced by the addition of optic gland extracts. These can be made from fresh or frozen octopus glands or from the glands of squids. Media in which glands have been incubated and then removed are as effective as extracts. As with the implantation experiments, the sex of the gland donor appears to be irrelevant; the hormones produced by males and females and by octopuses and squids are identical, or very much alike.

From a study of the kinetics of labelled amino acid uptake and exchange and the synthesis of labelled protein it would appear that the hormone acts independently on uptake and synthesis (Fig. 6.12; O'Dor and Wells, 1975; Wells, O'Dor and Buckley, 1975).

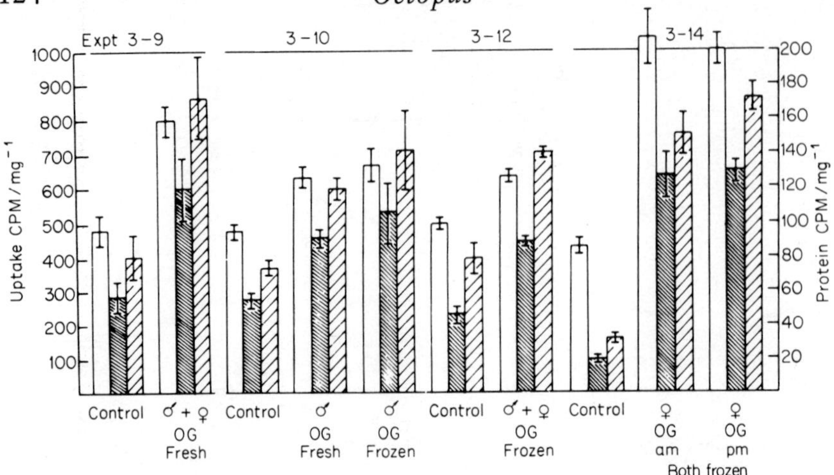

Fig. 6.10 The results of 4 *in vitro* experiments with oocyte/follicular cell complexes from *Octopus*. In each case the effect of adding an optic gland homogenate to bunches of eggs was measured in terms of uptake of [^{14}C] leucine from a nutritive medium (□ left hand scale of counts) and synthesis of labelled protein (water-soluable protein ▧ and total protein ▨; right hand scale). Most of the protein made was water-soluble and this is characteristic of cephalopod yolk (Fugii, 1960). Male and female, frozen and fresh optic glands were all effective in increasing uptake and synthesis. Each experiment was made with eggs from a different ovary and homogenates from a different set of glands. Each column shows an average result from six or eight replicate samples while the vertical lines indicate the standard errors of these means (from Wells, 1976a).

6.1.4 *Protein synthesis in the ovary*

The experiments outlined above show only that the ovary makes yolk protein. They do not show which part of the ovary is responsible.

Early accounts of the relation between ova and follicle cells in *Sepia* (Lankester, 1875; Yung Ko Ching, 1930) indicated the follicle cells as the likely site of yolk secretion. These light microscope studies have been followed by investigations at electron microscope level in *Octopus* (Buckley, 1977; Wells, O'Dor and Buckley, 1975) which leave little doubt that the view of the early workers was correct; in the later stages of yolk production it is the follicle cells of cephalopods and not the oocytes that do the work. The follicle cells are packed with rough endoplasmic reticulum and active Golgi apparatus. Electron-dense material, originating from the Golgi, is apparently exported to the oocyte down finger-like processes traversing the

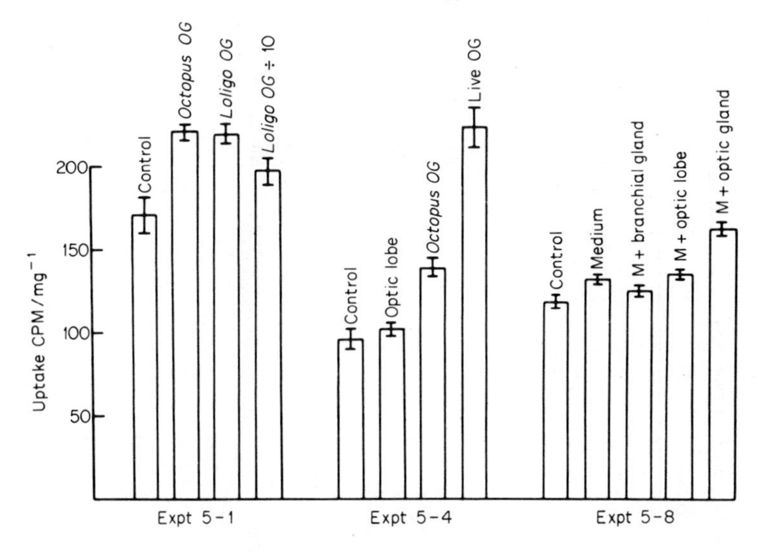

Fig. 6.11 Further *in vitro* experiments with *Octopus* eggs. In experiment 5-1, homogenates of *Octopus* and *Loligo* optic glands were added, at rates of 1 and 2 glands per 7 ml respectively; the second *Loligo* column shows the effect of diluting this homogenate 10 times. Experiment 5-4 shows that an homogenate of optic lobe is without effect, while the presence of a live optic gland in the medium greatly increases uptake; (experiment 5-4 was run for 5 hours after 1 hour pre-incubation instead of the usual 3 + 3). Experiment 5-8 used media in which fragments of optic lobe, branchial gland and optic glands had been incubated overnight. Columns shows means, and standard errors (from Wells, 1976a).

chorion (Plate 1). Corresponding follicle cells from an animal deprived of its optic glands two or three days before have ceased to produce dense granules, contain only much shrunken Golgi and an endoplasmic reticulum that has begun to break up into vesicles. Within five days many of the organelles have disappeared altogether (Wells, O'Dor and Buckley, 1975). In contrast to this the oocyte itself has few organelles and shows no effect of hormone deprivation. Its main phase of synthetic activity comes earlier in development, in the period of carbohydrate and lipid accumulation which precedes the full development of the follicle cell coat. But even at this stage in the proceedings removal of the optic glands has no detectable effect (Buckley, 1977). It appears that the hormone produced by the glands is necessary only at two rather distinct periods in the development of the eggs. Without it, the oogonia fail to divide; this takes place very early in development, soon after the little octopuses have settled on the bottom and

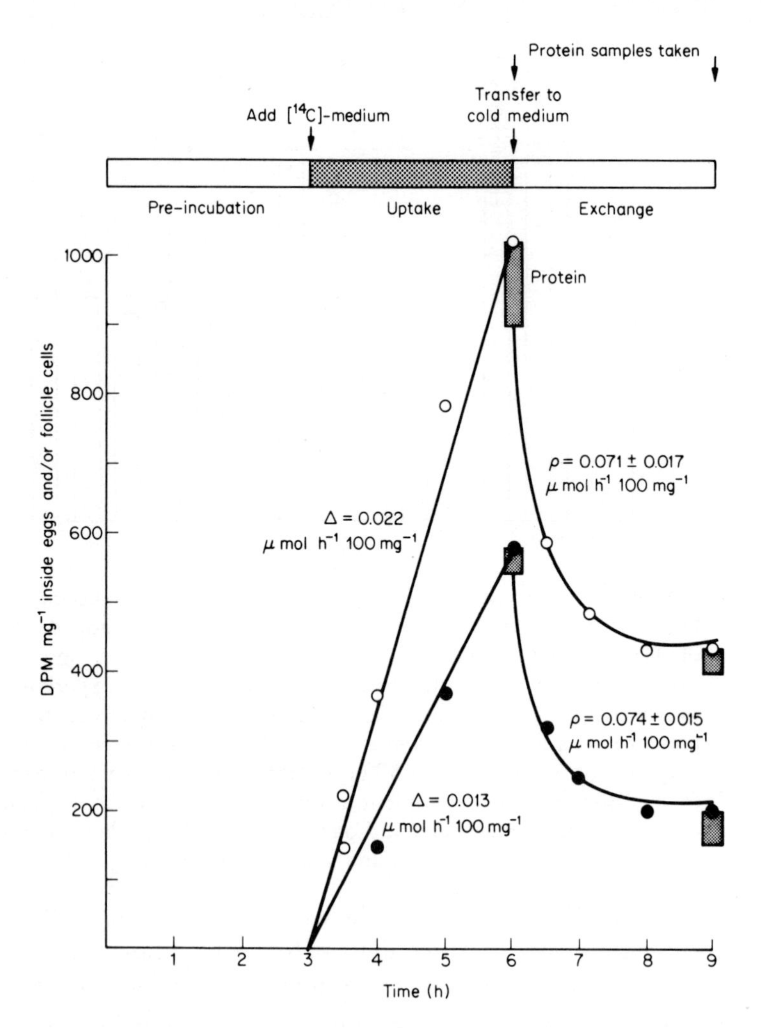

Fig. 6.12 Uptake of labelled amino acid and synthesis of labelled protein by the eggs of octopuses *in vitro*. The donor had her active optic glands removed 2 days before the eggs were taken. The eggs were pre-incubated in seawater with (○) and without (●) optic gland extract. After 3 h labelled medium was added and the uptake of label into the eggs followed by measuring its disappearance from the medium. Three hours later the eggs were transferred into cold medium, and the escape of label again followed by sampling the medium; protein samples were taken at the beginning and end of this period—the fraction of the total count attributable to protein is shown in the stippled columns. Plots ○ and ● show total count values obtained experimentally. Solid lines and the values of Δ (net inward movement) and ρ (exchange) were calculated using the equation developed by Sheppard and Beyl (1951) to describe

the uptake of [^{24}Na] by irradiated erythrocytes; they found rapid exchange superimposed on a steady net uptake. In the first part of this experiment [^{14}C] uptake was linear, indicating that very little cold leucine was available for exchange after pre-incubation in seawater. On transfer to cold medium, uptake continues but is at first masked by exchange—the curves are just beginning to turn upwards 3 h later. Exchange depended on the concentration in the eggs at the beginning of the period in cold medium. It was not affected by the presence of optic gland hormone, so the higher uptake associated with the hormone in the first part of the experiment cannot be attributed to an increase in cell permeability (from Wells, O'Dor and Buckley, 1975).

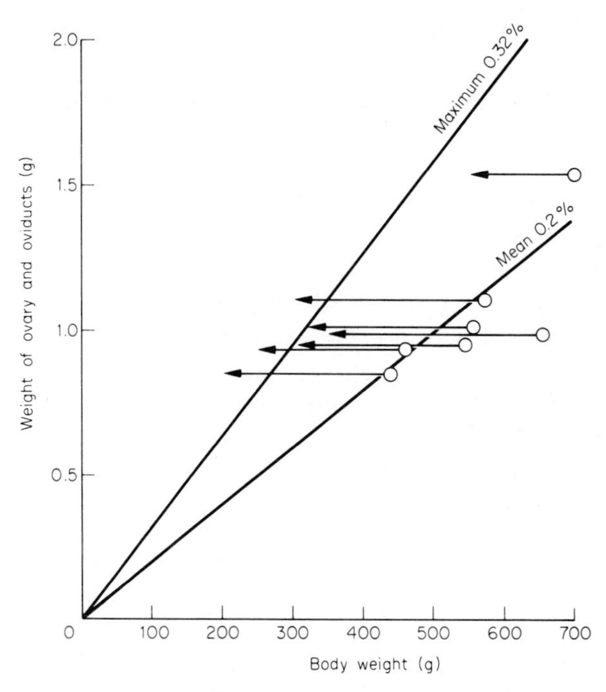

Fig. 6.13 Growth of ovary in the absence of the optic glands. These animals had their optic glands removed one month before they were killed. Points ○ show the ovary weights at death. Arrows show the same weights projected to the animals' body weights at the time the glands were removed. Lines at 0.2 and 0.32% show the mean and maximum weights of the ovaries in controls (see Fig. 6.2). Since it is unlikely that nearly half of the sample had ovaries above the control maximum at the time of operation, the weights at death must mean that the ovaries grow at the same rate as the rest of the body under these circumstances (from Buckley, 1977).

the optic glands have differentiated. There follows a period during which the ovary will grow, apparently at a normal rate, in the absence of the optic glands (Fig. 6.13). The glands become essential again only in the last phases of ovarian maturation, when the animal begins to accumulate the massive quantities of proteinaceous yolk needed to feed the developing embryos.

6.1.5 *Optic glands and the sex ducts*

Besides controlling the condition of the gonad itself, the hormone produced by the optic glands evidently exerts a direct effect upon the oviducts and oviducal glands, and on the complex apparatus responsible for packaging the sperm and assembling the spermatophores.

When the optic glands enlarge, so do the male and female ducts. The testicular ducts increase in weight (Fig. 6.14a) and the number of spermatophores to be found in Needham's sac (Section 5.1.2) more than doubles within three weeks of operations that activate the optic glands (Fig. 6.14b). The effect is greatest in the smallest, most immature males, a state of affairs that can be contrasted with the effect of removing the optic glands (Fig. 6.5) which is most marked in the larger animals. There are corresponding changes to the oviducts and oviducal glands. Both swell, increasing enormously in size and the latter, which are both sperm reservoirs and the source of the glue that binds the egg stalks together at oviposition, show considerable histological changes (Table 6.2; Wells, 1960b; Froesch and Marthy, 1975).

Table 6.2 Growth of the oviducts in the absence of the ovary.

Animal	*Wt. at operation (g)*	*Wt. of ovary removed (g)*	*Wt. at death (g)*	*Wt. of ducts at death (g)*	*Time since operation (days)*
454	350	0.53	420	2.64	20
457	220	0.30	300	2.30	33
461	300	0.36	430	3.02	34
509	220	0.40	260	1.95	32

These animals had the subpedunculate lobe of the brain and the ovary removed. Their optic glands were enlarged at death. The duct weights found then were typical of fully mature animals; in immature octopuses the ducts rarely weigh more than about one quarter as much as the ovary. (Wells, unpublished results).

Castration does not appear to alter the condition of the ducts in males or females. Callan (1940) kept five animals for 3–5 months

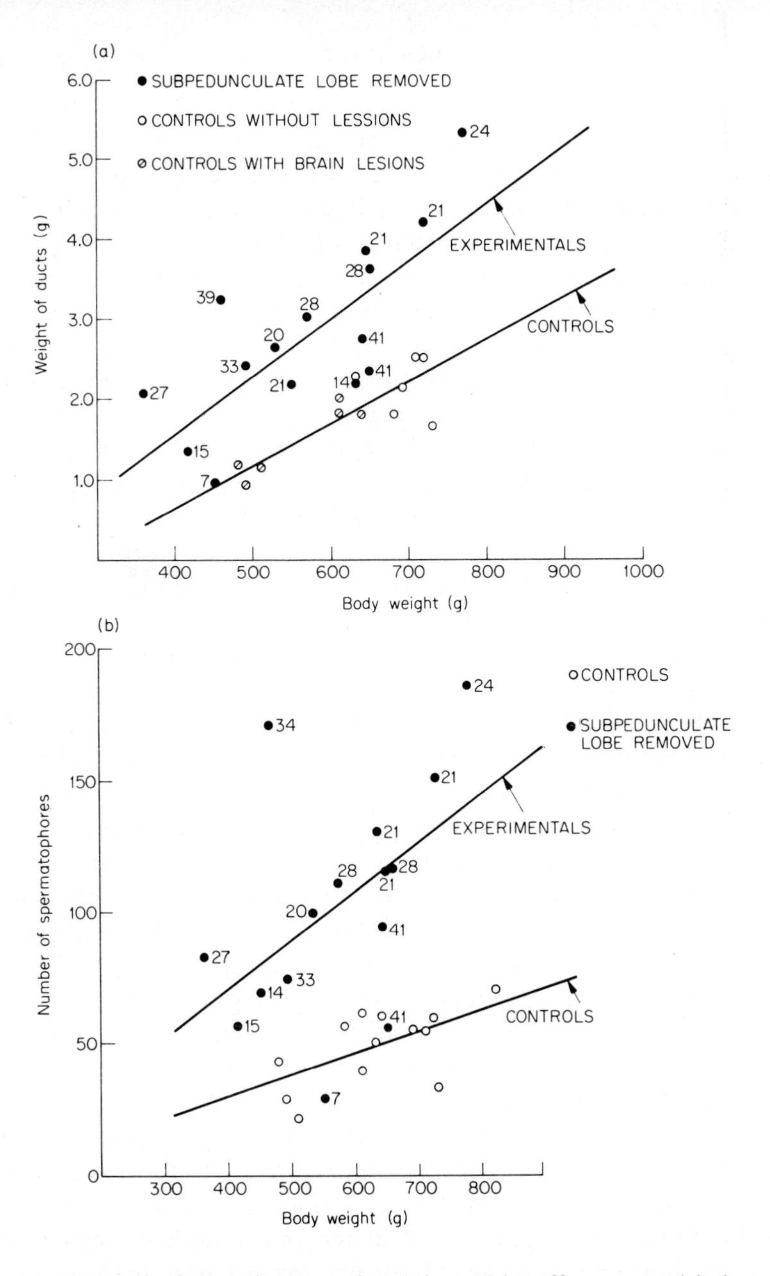

Fig. 6.14 Removal of the subpedunculate lobe and its effects upon (a) the male ducts, and (b) their production of spermatophores. Some of the controls in (a) and all of those in (b) had central nervous operations which did *not* include subpedunculate removal. Figures beside experimental plots show the number of days since operation (from Wells and Wells, 1972b).

after ovariectomy and reported no change in the state of the oviducts or oviducal glands. His animals were presumably immature. He also castrated seven males by removing the testis, and a further four by removing the testis and all the male ducts except the penis. Neither operation affected the condition of the penis. Callan did not report on the condition of the rest of the male ducts in the seven animals from which the testis alone was removed. He was mainly interested in regeneration of the hectocotylus, and found that castration had no effect on this; both sexes regenerated appropriate tips to their third right arms after removal of the gonads.

Subsequent work on the effects of castration has confirmed and extended these findings. Taki (1944) found that removal of the testis was followed by hypertrophy and eventual degeneration of the male ducts and cites this as evidence for the existence of a male sex hormone. This was contrary to Callan's (1940) finding; Callan would presumably have noticed and reported if the male ducts of his animals had enlarged as greatly as those of Taki's, and in an attempt to settle the matter Wells and Wells (1972b) castrated a further fifteen octopuses. These animals were all in the range 250–700 g, with ripe spermatophores in their ducts at operation. The results are shown in Fig. 6.15. Castration produced oedema in eight out of the fifteen animals, the enlarged testicular ducts being gorged with fluid, at least some of which (from its bluish colour) was blood. Our own view is that the seven cases where the ducts remained normal in size are more significant than the swollen eight, and that the oedema was a result of interference with blood and coelomic drainage systems. Castration involves splitting the gonadial coelom, ligating the genital artery and vein, and considerable distortion of the contents of the mantle cavity. It was concluded that the testis does not produce a hormone affecting the condition of the male ducts.

In 1969, Wells and Wells reported a single instance in which *O. cyanea*, castrated by removal of testis and ducts 36 days previously, proved on dissection to have grossly enlarged optic glands. This was a possible indication of gonadial feedback and could have been evidence of a sex hormone. But in 23 further experiments with *O. vulgaris* (fifteen with the testis alone removed and eight with testis and ducts removed (Wells and Wells, 1972b)), no such result was obtained, and it seems probable that the earlier observation lit, by chance, upon a precociously maturing animal. Van Heukelem (1973) has shown that male *O. cyanea* can become fully mature and begin to loose weight

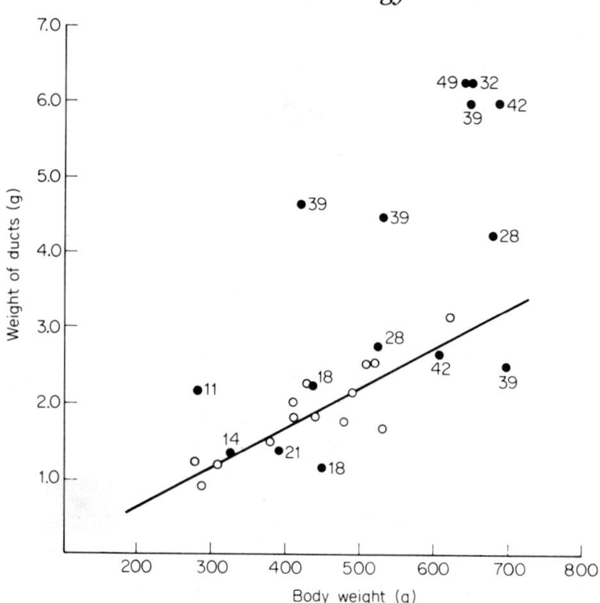

Fig. 6.15 The effect of removing the testis on the male ducts. Filled circles show the experimental animals, with the number of days since castration alongside each point. Open circles are controls (from Wells and Wells, 1972b).

(Section 5.3.3) at a little over 500 g; our specimen weighed 1025 g when caught, and 535 g when killed 36 days later.

Evidence about the relation between the ovary and its ducts is more sparse, but less equivocal than that available for males. Callan (1940, see above) found that ovariectomy had no effect on the ducts in immature animals, and Taki (1944) had the same result. In a series of experiments with very small female animals, Wells (1960b) removed the subpedunculate lobe, activated one or both of the optic glands and induced precocious development of the ovisac, oviducts and oviducal glands in five instances without at the same time causing the ovary itself to enlarge. The ovaries of these small animals, 15–50 g when killed but considerably smaller when operated upon, remained minute, hanging inside large fluid-filled oviscas. It would appear to be possible to select females so small that the ovary is not yet competent to respond to the optic gland hormone. But this does not stop the effect on the ducts.

In a more direct proof of the direct action of the optic gland hormone on the oviducts, four larger (but still immature) animals were ovariectomized and at the same time had the subpedunculate lobe of the brain removed. The animals were killed about a month later. In every case, the oviducts and oviducal glands were grossly enlarged at death (Table 6.2). This can only mean that in females, as in males, no product of the gonad is required as an intermediary between the optic gland hormone and the sex ducts.

6.1.6 *Optic glands, gonadial condition and sexual behaviour*

Activation of the glands is followed by maturation, and removal by regression of the sex organs. One might expect this to have some effect upon sexual behaviour. But this does not seem to be the case, at least for males. Table 6.3 summarizes the performance of 111 males in 161 matings; 49 of these animals were castrated or had operations affecting the condition of their optic glands. There is no evidence that either form of interference has the least effect upon the reaction of males to females placed in their tanks. Even octopuses with their testes and male ducts removed will approach females, insert the hectocotylus and pass imaginary spermatophores down the groove in the hectocotylus into the mantle of the female; the entire sequence of humping the body, placing the funnel over the guide in the web, and the peristaltic ripple that would normally carry the spermatophore proceeds as if the male were intact (Wells and Wells, 1972a). The only operation that will prevent this is removal of the tip from the hectocotylus; evidently some signal from the tip indicates that it has located an oviduct in the mantle and initiates spermatophore transfer. Males will copulate with females having the distal ends of the oviducts removed, but do not pass spermatophores.

Similar observations were made with *O. cyanea,* which has a male colour display; again castration or altering the condition of the optic glands had no measurable effect on sexual performance (Wells and Wells, 1972a).

It should be emphasized that these results were all obtained with mature male animals, all of which had sexual experience before operation. No experiments have been made with juveniles castrated before the onset of sexual maturity.

Van Heukelem (1973) has reported a change in male behaviour that occurs in the last stages of the lifetime of *O. cyanea*, coincident with the development of large suckers in the males (Section 5.3.3). The animals practically cease to feed and become 'noticeably more aggressive' at this stage, two changes that are almost certainly associated with the gross enlargement of the optic glands that seems to occur only in the largest males. These observations have not been quantified, so that the possibility of a direct effect of optic gland hormone on behaviour has not yet been entirely eliminated.

The situation with regard to females has not been studied. Under aquarium conditions, females do not have a great deal of choice in the matter of mating, and preliminary observations (*O. vulgaris*) suggest that their attraction, and male recognition of their sex is not dependent upon the presence of ovary or its ducts. Immature animals, in which both are relatively undeveloped, will mate readily and can be found with sperm in the oviducts and oviducal glands (see Section 5.1.4). Care of the eggs could be under hormonal control, but no experiments have been reported.

6.2 Other endocrine organs

A number of structures have been identified as probable endocrine organs, mainly on structural grounds. Most of these and their relation to the CNS are shown in Fig. 6.16.

There are:

(1) A mass of two to three million neurones that extends from the palliovisceral lobe in the brain to the walls of the anterior vena cava, the 'neurosecretory system of the vena cava' (NSV system) of Alexandrowicz (1964, 1965). The fine structure of this system has been described by Martin (1968).

(2) A similar series of cells, extending from the inferior buccal ganglion to end in the walls of the buccal sinus, the 'juxtaganglionic tissue' (Bogoraze and Cazal, 1944). This too has been seen to contain typical neurosecretory granules under the electron microscope (Barber, 1967).

(3) Neurosecretory cells in the superior buccal ganglion (Bonichon, 1968), in the subpedunculate lobe (Froesch, 1974) and in the olfactory lobe, close to the optic gland (Bonichon, 1967). Neurosecretory cells seem to be rather uncommon in the CNS of cephalopods compared with other molluscs (Golding, 1974).

(4) The posterior salivary glands, which are held to act as endocrine as well as exocrine structures (Sereni, 1930, and other references given in Section 6.2.4, below).

(5) The branchial glands, underlying the gills. Taki (1964) has described

Table 6.3 Timings in the sequence of events in the copulation of normal and operated animals (from Wells and Wells, 1972a).

Year *n = number of animals* *m = number of matings*	*Time to first contact*		*Time to probe from first contact*		*Median and range time*		
	<1min	*>1 min*	*<1 min*	*>1 min*	*From probe to first A and P*	*From first to second A and P*	*From second to third A and P*
Unoperated animals and controls							
1969							
n = 28 m = 37	76%	24%	81%	19%	1 min (<30 s to 13 min)	<15 s (<15 s to 2 min)	15 s –
1970							
n = 34 m = 39	87%	13%	85%	15%	<1 min (<30 s to 17 min)	<1 min (<15 s to 3 min)	<1 min (<15 s to 7 min)
Animals with the optic glands removed							
1969							
n = 5 m = 13	77%	23%	69%	31%	1 min (30 s to 19½ min)	<15 s (<15 s to 30 s)	– –
1970							
n = 5 m = 6	67%	33%	67%	33%	<1 min (30 s to 1¼ min)	<30 s (15 s to 16¼ min)	1 min (<15s to 6¾ min)
Animals with subpedunculate lobe removed on one side							
1969							
n = 7 m = 13	77%	23%	100%	–	1 min (30 s to 17½ min)	<15 s (<15s to 2 min)	<15s –
1970							
n = 9 m = 10	70%	30%	100%	–	1 min (<30 s to 4½ min)	<15s (<15 s to 30 s)	2 min (<15 s to 3½ min)

Castrated animals 1: testis removed							
1969							
n = 7	80%	20%	80%	20%	1½ min	<15 s	<15 s
m = 15					(<30 s to 11 min)	(<15 s to 30 s)	(<15 s to 30 s)
1970							
n = 8	100%	—	90%	10%	<1 min	<15 s	15 s
m = 10					(<30 s to 1½ min)	(<15 s to 30 s)	(<15 s to 9¾ min)
Castrated animals 2: testis and ducts removed							
1969							
n = 2	100%	—	100%	—	1 min	15 s	30 s
m = 6					(<30 s to 2 min)	(<15 s to 30 s)	(<15 s to 1 min)
1971							
n = 6	100%	—	100%	—	1½ min	15 s	15 s
m = 12					(1 min to 4 min)	(<15 s to 30 s)	(<15 s to 1 min)

'Time to probe' = time to insert the hectocotylus after the first tactile contact with the female.
A and P = Arching and pumping. When a spermatophore is passed, the male contracts his mantle in a characteristic manner ('arching'), humping his back and bringing the penis forward into the funnel, which is placed over the guide into the groove down the hectocotylus (see Fig. 6.2). There follows a forceful exhalation, and a ripple of movement, transferring the spermatophore along the groove in the hectocotylized arm (pumping).

these at light microscope level, and reviews earlier work. Octopuses die within days if the branchial glands are removed, but can survive removal of one gland, an operation that induces hypertrophy of the remaining gland. Taki concluded that the branchial glands were endocrine organs, providing a hormone having widespread metabolic effects. More recent experiments and studies at electron microscope level suggest an alternative explanation; the glands synthesize haemocyanin and the animals die because their removal destroys the oxygen transport system (Section 3.2.14).

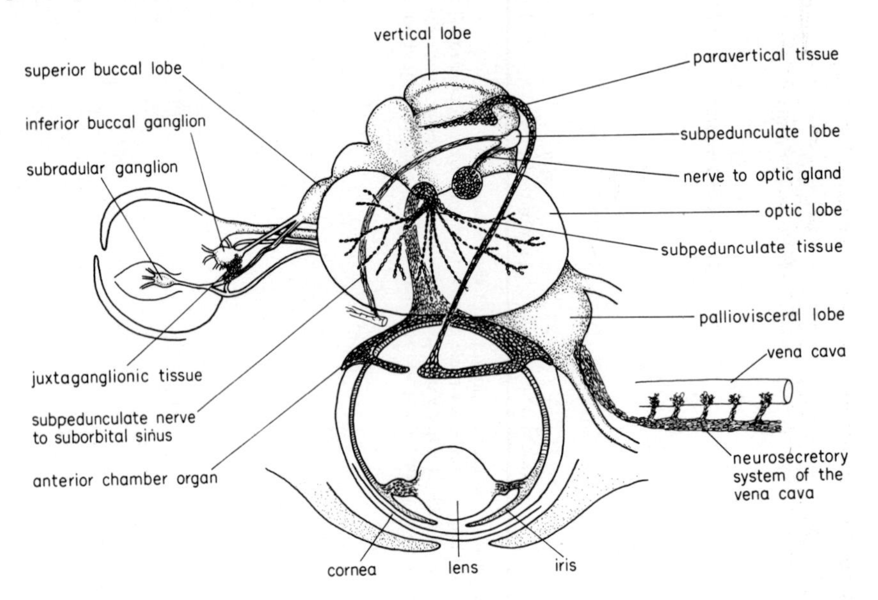

Fig. 6.16 Neurovenous organs that have, or may have endocrine functions in *Octopus*. The left eye and optic lobe have been deflected downwards and are shown in horizontal section; semi-diagrammatic (after Young, 1970b).

(6) In addition to these more or less obviously secretory tissues there are further 'paraneural' structures, ranged around the central nervous system (Fig. 6.16). These resemble the NSV and juxtaganglionic system in that they are all closely associated with the CNS and all consist of small cells, similar to neurones, with nuclei 4–6 μm in diameter. The cells give rise to fine nerve fibres, about 1 μm in diameter. The cells and their processes tend to form tubes, on a collagenous framework, radiating from their origins close to nerve ganglia. Two series can be distinguished.

These are:

(a) The subpedunculate tissue (Thore, 1939) which arises within the optic lobe as a series of strands lying along blood vessels deep inside the lobe. The tissue emerges close to the optic stalk and runs downwards and across the floor of the orbit.

(b) The paravertical tissue (Boycott and Young, 1956a). Masses of small cells lying along the sides of the vertical lobe of the supraoesophageal brain, probably, but not certainly, nervous (no nerve fibres have been seen, unlike the other paraneural tissues), with a strand that penetrates the cranium to follow a dorso-lateral course across the orbit to the 'anterior chamber organ'. This is a ring of paraneural tissue linked to both the paravertical and the subpedunculate strands, lying at the back of the space between lens and cornea. Young (1970b, 1971) has reported on the anatomy of the juxtaganglionic and other paraneural tissues.

The possible functions of these tissues are reviewed below.

6.2.1 *The neurosecretory system of the anterior vena cava*

This very extensive tissue is undoubtedly neurosecretory. The endings of its many nerve cells penetrate the walls of the vena cava and are filled with a variety of large vesicles. Martin (1968) recognizes four types and also reports large cells, embedded in the NSV system, with inclusions of 0.3–2.0 μm diameter, too large perhaps for neurosecretion but possibly endocrine secretions also due for discharge into the bloodstream. The peripheral NSV system described by Alexandrowicz (1964, 1965) and the neurosecretory cells reported from the visceral lobe by Bonichon (1967) and Martin (1968) are almost certainly parts of the same system. The neurosecretory cells are present and active in animals of all ages and at all times of year (Laubier-Bonichon, 1973). Martin (1968) notes that the cells in the palliovisceral lobe of the NSV system were notably short of dense vesicles in an animal that had fallen from its aquarium and spent some hours on the floor. Laubier-Bonichon (1973) kept octopuses in the laboratory for periods of several weeks and found that there was an initial depletion of stainable neurosecretory material followed (after 40 days) by eventual replacement, starting in the cell bodies and moving down the axons towards the blood vessels in the visceral lobe.

In an alternative approach, Blanchi (1969), Blanchi and De Prisco

(1971) and Berry and Cottrell (1970) made extracts of the peripheral part of the NSV system and were able to show that these had marked effects upon the heartbeat. They used the isolated heart preparation, from *Octopus* or *Eledone*, and obtained almost identical results: extracts produced a prolonged (10–20 min) acceleration and increase in amplitude of the heartbeat. The effects of adrenaline and 5HT were, in contrast, transient, lasting for less than two minutes. Extraction procedure such as immersion in distilled water (Blanchi, 1969) or boiling (Berry and Cottrell, 1970) were more effective than simply grinding the material up in seawater, which suggests that it is necessary to rupture cell membranes to release the active principle. Blanchi and De Prisco (1971) have established that the material responsible for changing the heartbeat is dialysable, with a molecular weight of about 1400. It is soluble in water, slightly soluble in alcohol and insoluble in chloroform or acetone.

Berry and Cottrell (1970) also attempted to accelerate the heart of *Eledone* by stimulating the NSV trunks. In one specimen out of eleven tested they appeared to be able to mimic the effects of extracts in this way.

We do not know what function NSV system has in the living animal. It seems too elaborate for its role to be limited to cardioregulation and, although this may well be a function of some fraction of it, the failure of the electrical stimulation experiments is worrying; cardiovascular responses should normally be rapid, and one would expect neural control of the discharge of secretion. Martin's (1968) observation on an animal subjected to osmotic stress is in keeping with Young's (1970b) suggestion that the NSV and other paraneural tissues are all in some manner concerned with the regulation of body fluids. Young cites the anterior chamber organ as almost certainly having this function (since the fluid in the anterior chamber is known to be different from blood or seawater, Amoore, Rodgers and Young, 1959 see also Section 3.3.12). He also points out that comparative studies show the juxtaganglionic tissue to be particularly well developed in the pelagic octopods, where body fluid regulation may play a part in buoyancy control (Denton, 1961, 1974). But such an interpretation can hardly explain the depletion and subsequent replacement of neurosecretory material in aquarium-kept animals observed by Laubier-Bonichon (1973) at Banyuls, which has an open circulation with seawater that is hardly likely to differ significantly from that in the sea outside.

6.2.2 *Neurosecretory cells in the subpedunculate lobe*

Two nerves leave this lobe and terminate outside the CNS. One serves the optic gland, as described above in Section 6.1.1. The other runs forward and down onto the floor of the orbit and out along the pharyngo-ophthalmic vein (POV) (Young 1970b); Froesch (1974) has described endings in the walls of the vein full of neurosecretory granules. Froesch and Mangold (1976) followed this observation with a demonstration that extracts of the POV will produce marked increases in the mean pressure, pulse amplitude and beat frequency of the systemic ventricle *in vitro*. Quite what the POV hormone does *in vivo* is less certain. Neurosecretory granules are plentiful in young animals, few or absent in mature octopuses (Froesch 1974). Mature individuals and octopuses with their subpedunculate lobes removed show no gross abnormalities in blood pressure, pulse, or heartbeat frequency (Wells, unpublished). The subpedunculate nerve to the POV is absent in squids (Young 1970b).

6.2.3 *Other paraneural tissues, and neurosecretory cells within the central nervous system*

The only justification for discussing these together is that we have no evidence as to what they do. In the case of the NSV and POV systems we have a few experimental results, albeit inconclusive. Here we have none. The paravertical and subpedunculate tissues may be concerned with regulating the activity of the anterior chamber organ which in turn probably regulates the fluid contents of the cavity between lens and cornea; the juxtaganglionic tissue is perhaps concerned with the regulation of the body fluids generally (Young, 1971, see above). The neurosecretory cells in the superior buccal lobe may be involved in the control of secretion of 5HT and tyramine by the posterior salivary glands (Bonichon, 1968; Martin and Barlow, 1971). But these are guesses.

The neurosecretory cells in the olfactory lobe are presumably concerned in some way with reproduction since they appear to discharge their contents into the intercellular spaces when a female lays eggs (Bonichon, 1967).

6.2.4 *The posterior salivary glands*

The posterior salivary glands produce a wide range of pharmacologically active substances, including tyramine, 5HT, octopamine, histamine and acetylcholine as well as proteolytic enzymes and cephalotoxin (Section 4.2.1, and 4.3.1, Hartman *et al.* 1960). Some of these substances are also found in the blood. Sereni (1930) used the behaviour of the chromatophores to show that tyramine and histamine circulating in the blood almost certainly originate in the salivary glands which are thus both endocrine and exocrine. Tyramine and histamine are found in greater quantities in the blood of *O. macropus* and *Eledone moschata* than in *O. vulgaris* which is typically paler at rest than the other species. Blood from *O. macropus* or *Eledone* injected into *O. vulgaris* caused darkening, while *O. vulgaris* blood caused the chromatophores to contract when injected into *O. macropus* or *Eledone.* After removal of the posterior salivary glands, *O. macropus* and *Eledone* became very pale, but their normal shade could be restored by injections of tyramine or histamine. These effects are apparently caused by the action of the drugs at a central nervous rather than at a peripheral level, since they were not found in freshly de-enervated areas of skin (Sereni, 1930).

Bacq and Ghiretti (1953) developed a system for perfusing the salivary glands through their normal aortic blood supply and were able to collect their secretion from the salivary ducts. Stimulation of the salivary nerves caused an increase in secretion (after a short delay) and the liberation of acetylcholine as well as tyramine, octopamine and 5HT into the perfusion fluid. Acetylcholine was also released in small quantities in the absence of stimulation. Acetylcholine and 5HT are known to have effects upon the heartbeat and peripheral circulation (Section 3.2.11).

Plate 1 Follicle cells surrounding a developing oocyte in *Octopus*. The inset, top right, shows a section through a developing oocyte at the stage when the follicular envelope folds into the egg (cf. Fig. 5.4, p. 89); this section has been stained with sudan black and shows the lipid droplets which are prominent at the onset of yolk protein synthesis. The electron micrograph covers parts of two actively secreting follicle cells and the processes of one or more of the cells forming the thin outer envelope (see Fig. 5.4). O (at base) – oocyte; C – chorion; G – golgi complex; N – nucleus, with NP – nuclear pores and a prominent NO – nucleolus; L – remains of lipid droplets which largely disappear as yolk protein synthesis accelerates. Arrows show vesicles of material packed by the golgi, apparently migrating towards the finger-like processes which penetrate the chorion. (Photographs, S. K. L. Buckley.)

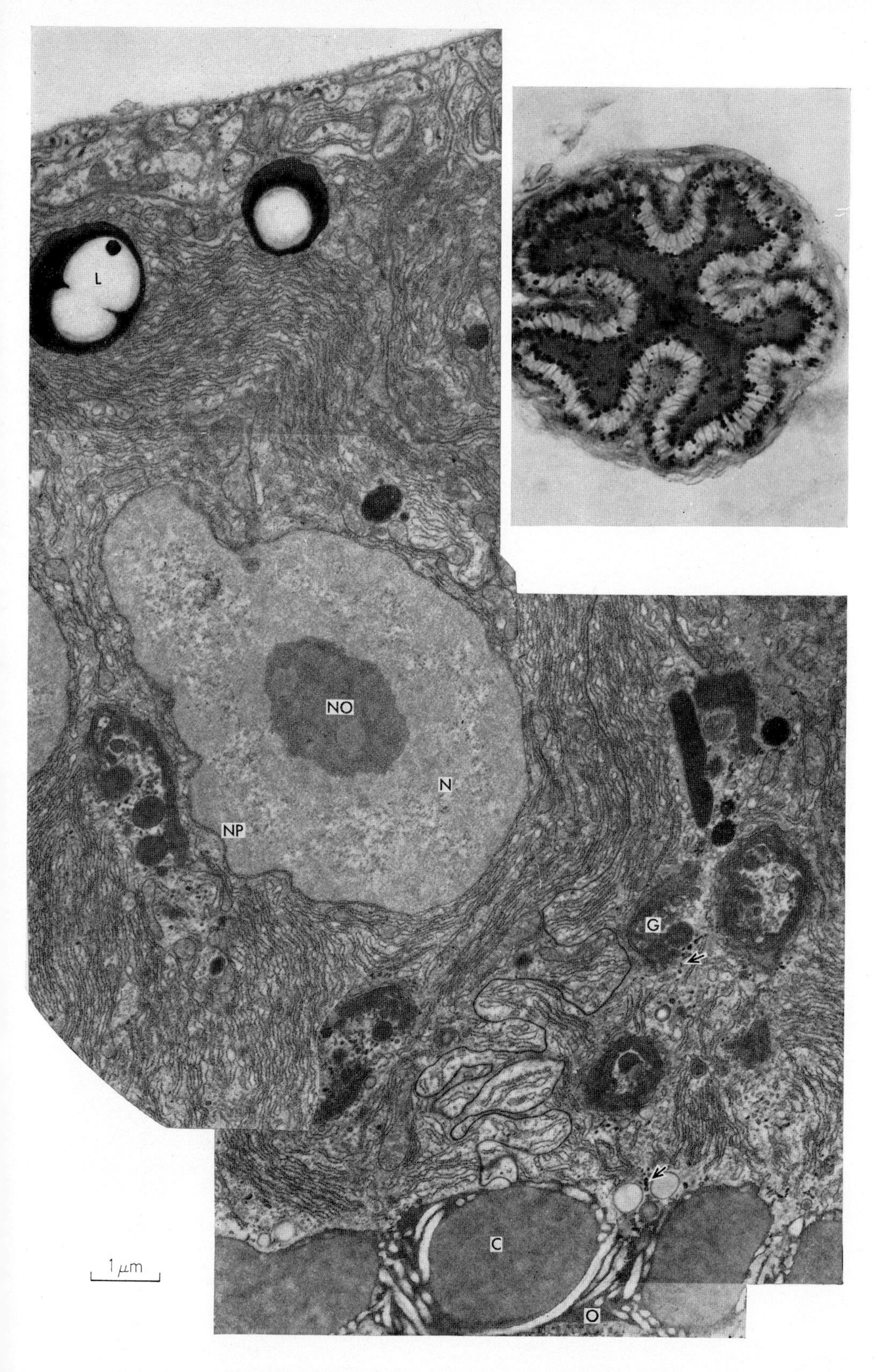

Plate 1 (for caption see facing page)

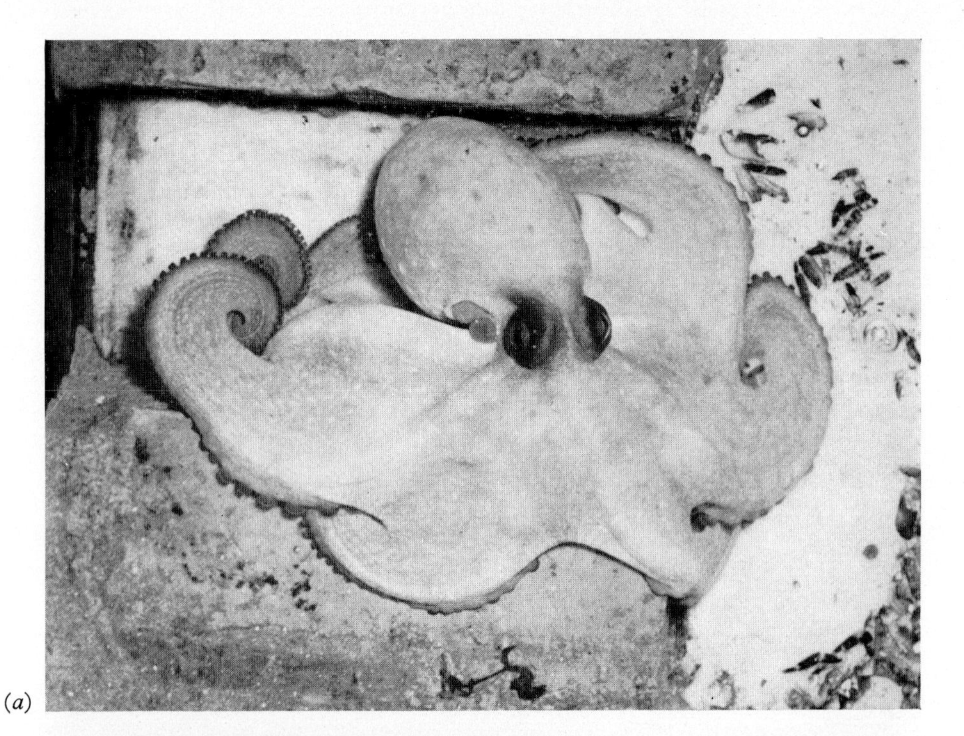

(*a*)

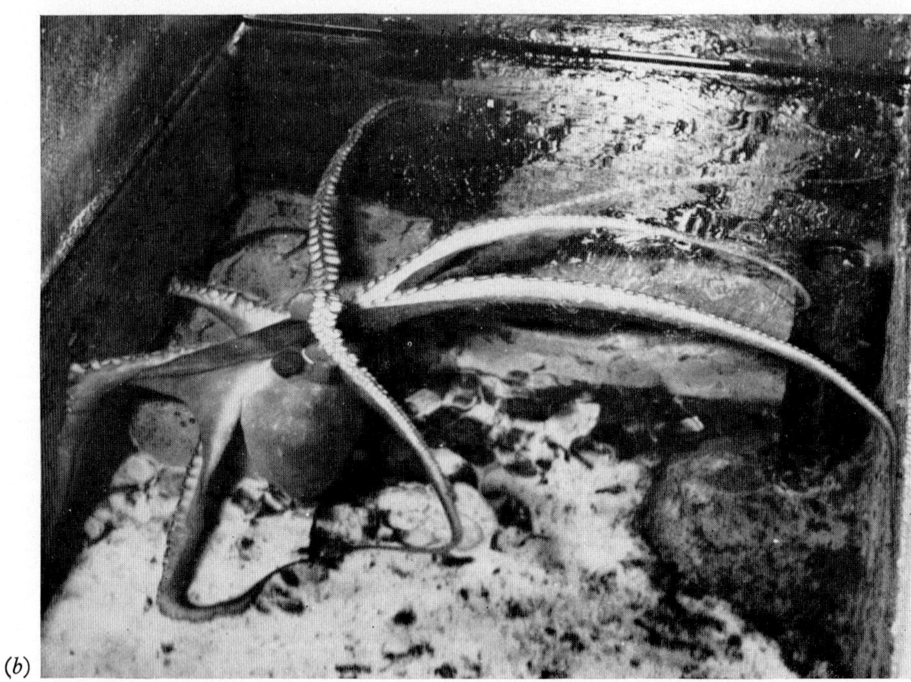

(*b*)

Plate 2 (*a*) The dymantic response of a startled octopus, a display that may serve to deter would-be predators – see p. 152; part (*b*) in the Frontispiece shows the same response in a smaller animal.

(*b*) An octopus blinded by section of the optic nerves sits with the arms outstretched along the sides and surface of the aquarium – see p. 220. (Photographs M.J.W.)

CHAPTER SEVEN

An inventory of the sense organs

7.1 Eyes

Anyone watching an octopus for the first time is bound to notice that he is being observed. The animal has eyes that stare back. It responds to movement, cowering if anything large approaches it, or leaning forward in an alert and interested manner to examine small happenings in its visual field. Anyone who keeps the creatures in a laboratory, or any diver who approaches them in the sea soon discovers that they can see just about as well as we can. The eyes are very prominent and so clearly important that they have tended to distract attention from the rest of the animal's sensory equipment. We know more about the structure and physiology and the use that an octopus makes of its eyes than about all the rest of the sense organs put together. Because we are ourselves so dependent on vision we find it relatively easy to manipulate the stimuli and devise experiments to investigate visual systems. We can think of ways to test hypotheses about the visual analysing system so that there is a considerable literature on *Octopus*' responses to figures of differing shapes, sizes and shades. There is no equivalent body of information about, for example, the animal's chemotactile sense, which is probably quite as important to it.

Information about the brain's analysis of its visual inputs, and the effect of brain lesions on performance in visual learning experiments will be reviewed in the chapters that follow. Here we are concerned with the eyes as instruments without, for the present, considering what happens to the information that they provide.

7.1.1 *The retina*

The octopus eye has the same major components as a vertebrate eye. There is a cornea, an iris and a lens. The lens is suspended by a ciliary body and it casts an image on a hemispherical retina. The eye can

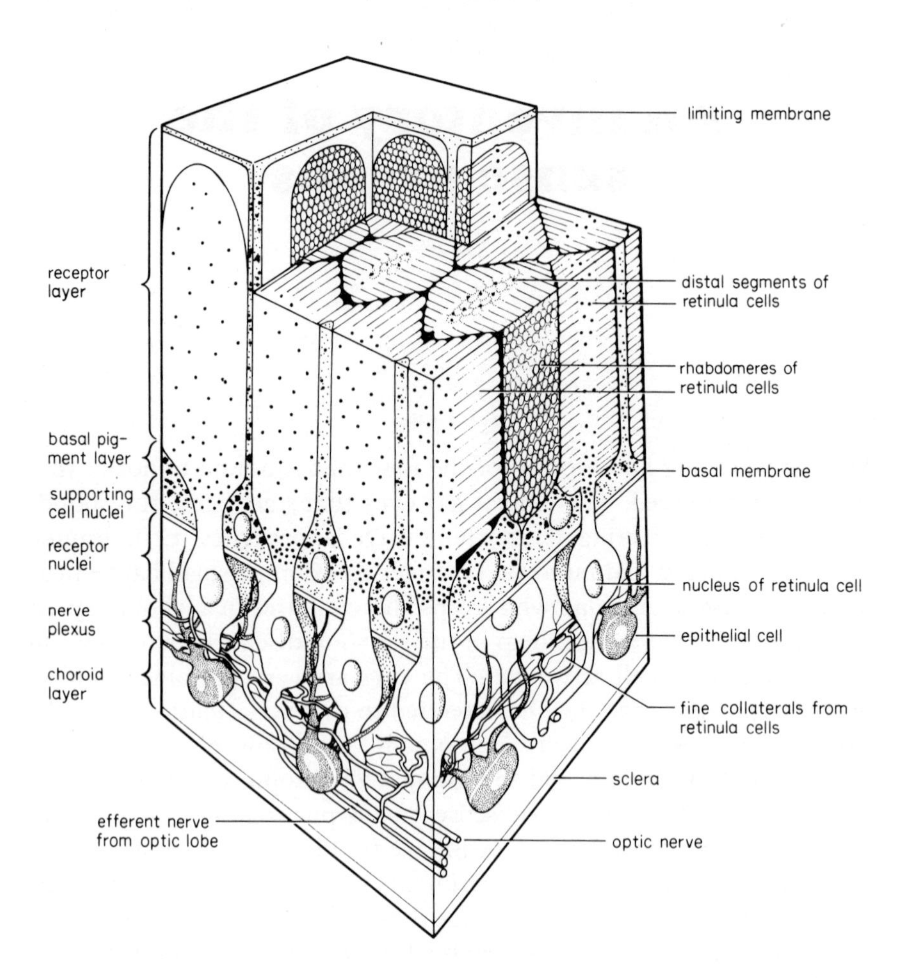

Fig. 7.1 The retina of *Octopus.* For this diagram the vertical scale is much contracted. In life the distal segments of the retinula cells would be of the order of 5μm across and 150μm in length. The supporting cells have processes running up between the rhabdomes, and they either secrete, or themselves expand to form, the limiting membrane, in which pigment granules are found in the light-adapted eye. The nerve plexus includes centrifugal fibres from the optic lobes as well as the axons and the fine collaterals from the retinula cells. The layer of epithelial cells has processes running in the plexus and upwards among the cell bodies of the retinula cells. Details of the choroid, which includes blood vessels, connective tissue, and muscles, are not shown. No blood vessels penetrate beyond the basal membrane. Finally, there is a thin cartilaginous sclera pierced only by blood vessels and the bundles of optic nerves (after Wells, 1966b; diagram compiled from figures and data in Moody and Parriss, 1961; Young, 1962a, 1963).

compensate for long- and short-term changes in light intensity and it can focus on near or distant objects. The whole structure is suspended in a socket by muscles that can move it around to scan the animal's surroundings.

The octopus eye is unlike the vertebrate eye in two important respects; the retina is not inverted and it is structurally far less complex. There are no bipolar and no ganglion cells, the equivalent structures being buried in the outer layers of the optic lobes (Section 8.1.8).

Retinal structure is summarized in Fig. 7.1. The retinula cells have long thin outer segments pointing inwards towards the lens, and nucleated inner segments on the far side of a basement membrane that also carries the cell bodies of glial 'supporting cells' between the photosensitive elements. Blood capillaries run among the inner segments (Yamamoto, Tasaki, Sugawara and Tonosaki, 1965; Young, 1971). Each retinula cell gives rise to an axon. The axons join together in bunches, cross a loosely packed choroid space and penetrate the cartilaginous sclera. Collaterals arising close to the origins of the axons form a neuropil which is joined by the fine endings of efferents running out in the optic nerves from the optic lobe (Young, 1962a, 1971).

The outer segment of each retinula cell consists of a central core with two series of finger-like processes ('rhabdomeres') sticking out at right angles on opposite sides of this axis. As the outer segments pack together, the central axes tend to form a rectangular array, so that at light microscope level the apparent unit is not the retinula cell and its pair of rhabdomeres, but the larger complex formed from the cores and inward-facing rhabdomeres of four retinula cells, a 'rhabdome'.

The retinula cells differ in diameter and length, both locally and regionally. On average, the rhabdomes are narrowest and longest along an equatorial strip, getting broader and shorter dorsally and ventrally. At maximum density there are about 55×10^3 rhabdomes to the mm^2 (Fig. 7.2, Young, 1963b).

7.1.2 *Light and dark adaptation*

Between the outer segments of the retinula cells are the narrow processes of the supporting cells. These, and the cores of the outer segments, contain pigment granules that move up and down as the eye adapts to changes in light intensity. Movement is quite rapid, a

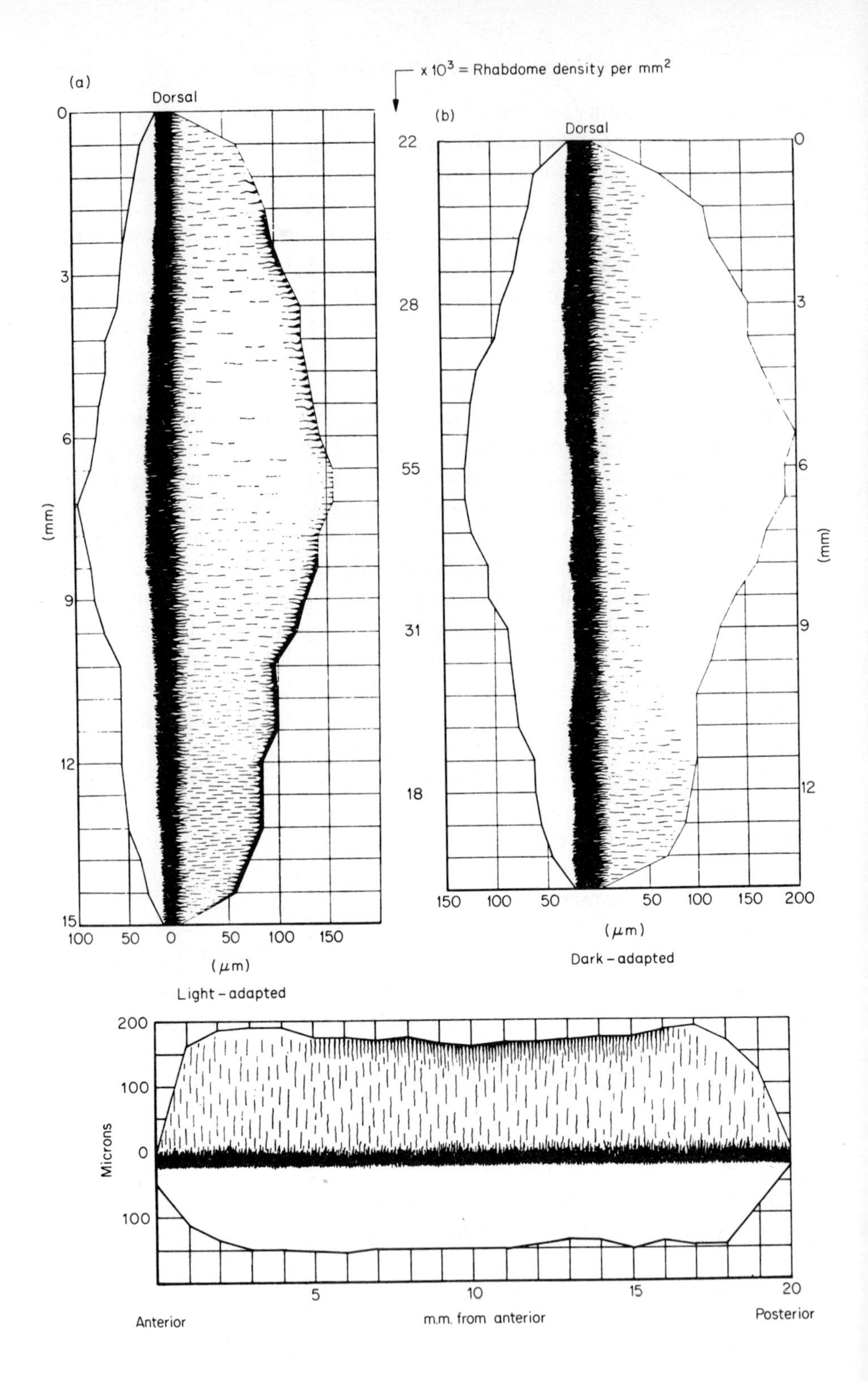
x 10^3 = Rhabdome density per mm^2
(a)
Dorsal
0
3
6
9
12
15
(mm)
100 50 0 50 100 150
(μm)
Light - adapted
(b)
Dorsal
22
28
55
31
18
0
3
6
9
12
(mm)
150 100 50 50 100 150 200
(μm)
Dark - adapted
200
100
0
100
Microns
5
10
15
20
Anterior
m.m. from anterior
Posterior

matter of minutes; changes of shorter duration are dealt with by alterations in the aperture of the iris and, if a very bright light is shone suddenly into the eye, by contracting the muscles in the skin around the eyeball, shutting the eye in altogether.

There are regional differences in the behaviour of the screening pigment. The pigment in the equatorial strip clears most rapidly in dark adaptation and, in general, remains close to the basement membrane, so that the tips of the rhabdomes are always relatively pigment-free in this part of the eye. The anterior and posterior ends of the strip tend to be clear at all times. In contrast the ventral part of the retina shows a very large response to bright light, with pigment migrating (through more slowly than in the central area) to the extreme periphery (Fig. 7.2).

Pigment movements are accompanied by changes in the length of the outer segments, which contract in bright light (Fig. 7.2). Both sorts of change take place in isolated eyes, so they are presumably not under nervous control, unlike the iris, which remains permanently wide open if the optic nerves are cut (Young, 1963b, 1971).

7.1.3 *Visual pigments*

The outer segments contain a rhodopsin, with maximum absorption at 475 nm in the blue-green part of the spectrum. The rhodopsin changes to metarhodopsin on illumination, but there is no further breakdown and no bleaching under physiological conditions (Brown and Brown, 1958). The visual pigment is thought to be attached in an oriented manner to the walls of the tubules of the rhabdomeres, since these act as dichroic analysers of polarized light, with maximum absorption when the electric vector is parallel to the tubules. This condition can be related to the capacity to distinguish the plane of polarization, which is demonstrable at a behavioural level (Moody and Parriss, 1961; Moody 1962) and in the octopus retinogram (Tasaki and Karita, 1966; see also Ito, Karita, Tsukahara and Tasaki, 1973).

Fig. 7.2 Lengths of rhabdomes and basal segments in the retinae of two octopuses fixed (a) after 10 h of illumination, (b) after 10 h in dark. Measurements made on transverse sections at the mid points of the eyes. The ordinates give distances from the dorsal margin of the retina and the abscissae lengths in microns of the rhabdomes (to the right) and the nervous layer of the retina (to the left). The positions of the pigment granules are approximately shown. (c) Lengths of rhabdomes as seen in a horizontal section taken at the equator of the light-adapted eye of an octopus of 150 g (from Young, 1963b).

A second pigment, retinochrome, with an absorption maximum at 490 nm, is found deep in the retina, at the base of the outer segments. This is more sensitive to dim light than the relatively plentiful rhodopsin and it does bleach out under physiological conditions. The rhodopsin is clearly the primary visual pigment and the function of this second material, with an absorption maximum so close to the first is problematic. It is believed to 'play a supplementary part, perhaps acting on a direct supply of active retinal for the rhabdomes' (Hara, Hara and Takeuchi, 1967).

7.1.4 *Electrical responses from the retina*

If an electrode is placed on the inner surface of the retina a response to light can be measured. The response is negative with respect to a reference electrode in seawater outside the eye and is sustained for as long as the eye is lit. A corresponding positive potential can be observed if an electrode is placed at the back of the retina. The origin of the current appears to be inner segments of the retinula cells; Hagins, Zonana and Adams (1962) used microelectrodes to probe the retinae of squids and found that a light spot focussed at any point along the outer segments always carried with it the region of minimum potential, while the region of maximum potential remained steady at the level of the retinula cell bodies. 'The fields observed were those to be expected if the positive membrane current arose in the cell bodies of the illuminated group of receptors, flowed through the intercellular spaces and entered the outer segments'.

Tasaki, Oikawa and Norton (1963), also working with excised retinae, found that the positive and negative potentials of the octopus electroretinogram are to some extent independent. The response to a light spot of increasing intensity or area reaches a maximum at the level of the inner segments, while continuing to increase in the outer segments. The spread of the electrical response to a limited lit area is about twice as great on the lens side as on the back of the retina (Tasaki and Norton, 1963). A further difference is shown in Fig. 7.3; the electrode at the back of the eye picks up oscillating potentials that begin about 15 ms after a light flash and continue for some 40 ms (Boycott, Lettvin, Maturana and Wall, 1965; see also Ito, Karita, Tsukahara, and Tasaki, 1973). The origin of these oscillations is unclear. They cannot be due to the receptors (since they are not found inside the eye) and must arise in some way through interaction between the optic nerves. They are considerably reduced if the animal is intact (see below).

Boycott *et al.* (1965) implanted wires into the eyes and optic lobes of freely moving unanaesthetized octopuses and were then able to study the effect of an intact efferent innervation of the retina. With

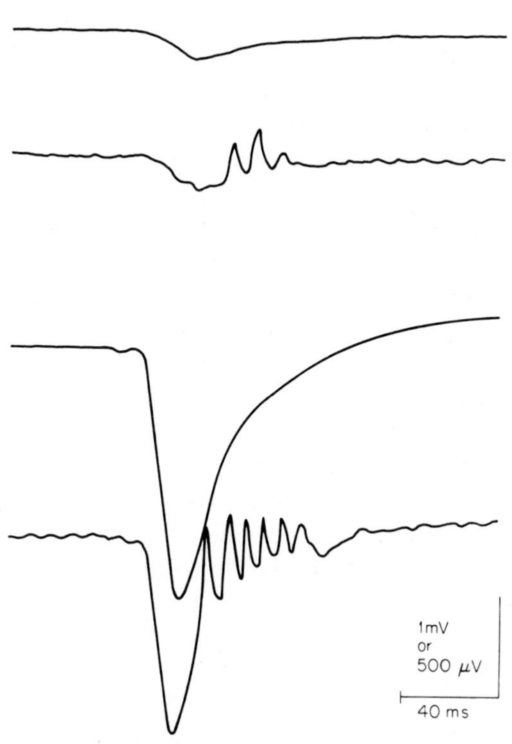

Fig. 7.3 *Octopus* electro-retinograms. Upper pair: response to dim brief flash. The upper trace is the ERG recorded within the eye. In the lower trace the monopolar recording electrode was placed immediately behind the eye on the optic nerves and shows the brief burst of oscillations in addition to the smooth ERG. Lower pair: response to intense flash. ERG within the eye recorded on the upper trace and ERG and oscillations behind the eye recorded on the lower trace. Voltage 1 mV for electrode within the eye, and 500 μV for electrode behind the eye (after Boycott, Lettvin, Maturana and Wall, 1965).

the optic nerves cut, electroretinograms (ERGs) like those from isolated retinae were recorded. With the optic nerves intact, the ERG could be increased by as much as 100% by touching the animal or stamping on the laboratory floor. The increase was shown in the absence of any detectable motor response; the iris, for example, did not expand or contract. Since it correlated with increased activity in the efferent components of the optic nerves it seems that the animal when alerted can increase the sensitivity of its visual receptors directly as well as, perhaps, indirectly by altering the aperture of the iris.

The ERG has been used to test for the intraocular discrimination of light polarization plane, postulated on morphological and behavioural grounds by Moody and Parriss (1961). Tasaki and Karita (1966) examined the retina and recorded from the outer segments, using a glass electrode that, presumably, was not fine enough to record from single retinal cells, because they found little or no change in response when a polaroid screen covering the light source was rotated. Prior exposure to a polarized light source did, however, alter the condition of the area under observation, so that subsequent rotation of the test source produced a rise and fall in potential. Tasaki and Karita were careful to retain the orientation of their retinae and found that selective adaptation of the population of retinal cells from which they were recording could be achieved by lights at 0–180° *or* 90–270° but not at angles at 45° to these. The effective angles, horizontal and vertical with respect to the position of the retina in life are, of course, parallel with one or other of the predominant orientations of the tubules of the rhabdomeres. Similar results were obtained with squid eyes by Ito *et al.* (1973). ERG studies and their relation to the problem of colour vision in octopuses are discussed in Section 8.1.16 below.

7.1.5 *Electrical responses in the optic nerves*

With a considerable length of optic nerve from retina to optic lobe, it ought to be easy to record from single units. In fact, it is rather difficult, for a number of reasons. One has already been mentioned in relation to control of the peripheral circulation (Section 3.2.4); octopuses can control the dilation of their arteries and shut down the bloodflow to damaged areas. By the time the orbit has been penetrated to expose the optic nerves, circulation to the retina has stopped. The only hope seems to be to open up as widely and rapidly as possible and direct a flow of aerated seawater onto the back of the retina and around the nerves. A second problem then becomes apparent; the nerves possess a sheath that is impenetrable by conventional microelectrodes. This can be solved by softening the sheath with a jet of magnesium-free seawater, directed at the point to be penetrated. Having evolved this technique, Lettvin and Maturana (1965) were able to record from groups of several hundred nerve fibres coming from the retina. They found two main types of response; fibres giving a large, adapting response to the onset of light and a smaller group that responded to light off, after a delay. In two instances they found single units that appeared to respond to a light/dark edge rather than to light, an analysis at the retinal level that could represent one function of the network of collaterals that underlies the inner segments.

The efferent nerves are larger than the afferents and it was easier to record single units here. They responded to a wide variety of stimuli, such as tapping the side of the tank or switching on a light shining into the contralateral eye. Some responded only to combinations of stimuli or changed their response when the same stimulus was repeated.

Both afferents and efferents tend to fire 'spontaneously' so that there is a continual background of activity in the intact animal (Lettvin and Maturana, 1965).

7.1.6 *The optic chiasma, and the 'deep retina'*

Something like 2×10^7 optic nerves leave the retina of an animal of 350 g (Young, 1963b) and the number increases with size since the packing density of the rhabdomes remains about the same as the octopus grows (Packard, 1969). These nerves project onto the optic lobes, with a vertical inversion. There is no anterior-posterior weaving of the nerve bundles.

Young (1971) has suggested that an inverted projection of the visual map is necessary if a topological relation between this and any projection from the gravity receptors is to be preserved. But this is unconvincing since, so far as we know, the animal does not need to take its own position with respect to gravity into account when it assesses the orientation or the movement of things that it sees. Instead the animal's visual analysing mechanism is organized on the assumption that the retina remains in a constant orientation with respect to gravity, a matter that depends directly on the integrity of the statocysts (Section 9.2.1). In this situation, the movement of objects relative to the octopus could be interpreted and responses organized just as readily with an inverted image. It should be noted, also, that *Nautilus* has no chiasma (Young, 1965c). *Nautilus* has a primitive, lensless eye which responds to changes in light intensity. The animals respond, for example, to the shadow of a diver swimming over them. They must, moreover, be able to form some sort of visual image, since they will move directly towards shadows if they are released in the sea (Cousteau and Diolé, 1973). Bidder (personal communication) reports that *Nautilus*, active at night in their aquaria, will follow the light of a torch shone through the glass. Both performances suggest that the animal is well able to relate the position of things seen with its own position in space in the absence of any sort of chiasma.

The fact remains that all modern squids and octopuses have a vertical chiasma. The magnitude of the morphogenetic problems that must have been solved in the course of evolving such a complex structure surely indicates that it can hardly have arisen by chance. But its adaptive significance is unknown.

Compared with the vertebrate retina, the retina of *Octopus* is very simple. There are no equivalents of the amacrine, bipolar or ganglion cells in the cephalopod; peripheral processing of the visual input must be much simpler.

As soon as the optic nerves are followed into the optic lobes, however, it becomes clear that the apparent relative simplicity of the cephalopod system is an illusion. It is a matter of stacking; the amacrines, bipolars and ganglion cells are all there, but stuck onto the outer layer of the optic lobe rather than onto the back of the retina. Fig. 8.4, p. 192 taken from Young, 1962b, summarizes the anatomical situation.

There is no direct evidence about the function of any of the anatomically distinguishable cell types in the optic lobes. Certain features of the arrangement, notably the shapes and orientations of the dendritic fields of the bipolar cells, can be related to the manner in which the animals appear to analyse visual images in behavioural experiments, and these are dealt with below, in Chapter 8. By analogy with the visual systems of vertebrates and arthropods, it seems probable that the visual analysing system in the optic lobes of *Octopus* will ultimately prove a fruitful ground for electrophysiologists. But as yet it is a practically untouched field, preliminary reports suggesting only that it will prove interesting, very complicated and technically rather difficult (Boycott, Lettvin, Maturana and Wall, 1965; Lettvin and Maturana, 1965).

7.1.7 *Focussing the eyes*

The refractive index of the fluid on the two sides of the cornea being very similar, images of things seen must be brought to focus on the retina by the lens. The lens is spherical, as in fish, and has a very short focal length of about 2.5 × the lens radius, projecting onto a retina that is nearly hemispherical (a little longer fore and aft than vertically). Such a lens will only focus an image at full aperture if the refractive index varies from about 1.53 at the centre to 1.33, approaching that of seawater, at the periphery (Pumphrey 1961). This seems to be achieved by differences in the proportions of the α-, β-, γ- and δ-fractions of the crystalline proteins in successive layers of the lens (Bon, Dohrn and Batink, 1967).

Early reports of cephalopod eyes noted the general resemblance to mammalian eyes and assumed that they were normally focussed for distant objects when at rest, in which event the muscles of the ciliary body would have to move the lens forward to view things close to the octopus. Beer (1897) disposed of this assumption by showing that stimulation of the muscles suspending the lens pulled this back towards the retina, not forwards and away from it, and Heine (1907) followed this up by showing that the lens moves both backwards *and* forwards in the living animals. Reviewing these, and other accounts (Hess, 1909, for example, disagreed and states that the eye is focussed on infinity at rest) Alexandrowicz (1927) has confirmed Heine's obervations; the ciliary muscles pull the lens inwards, while muscles on either side of the sclera raise the internal pressure of the eyeball and push it out, a double focussing system that has no parallel among vertebrates.

The odd thing about this system is that it would seem, on the face of it, to be quite unnecessary. The focal length of the lens is so short, and the relative length of the receptors so long, that objects from a few centimetres to infinity should be in sharp focus all the time. It may be, as Alexandrowicz himself suggests (1928c) that some of the musculature is concerned not so much with focussing the eye at rest, as with avoiding distortion when the animal accelerates; the eyes of cephalopods are very large and the stresses imposed on them when the animal jets itself backwards or forwards must be considerable. This would apply particularly to squids (Alexandrowicz worked on *Sepia*) but octopuses move relatively slowly and we have somehow to account for the observations made by Beer, Heine and Hess (see above) all of whom claim that the lens moves when the animal is at rest and observing the observer and his opthalmoscope.

The intrinsic and extrinsic muscles of the eyes are served by nerves arising within the lateral pedal lobes on the sides of the suboesophageal part of the brain (Young, 1971).

7.1.8 *The roles of the iris diaphragm*

The iris is a fast-working diaphragm that can respond to changes in light intensity more rapidly than the pigment migrating among the rhabdomes (Fig. 7.4). It contains brown and yellow chromatophores, which absorb the light, and iridophores which reflect it, and provide a white background to the chromatophores (Froesch, 1973b). The pupil, surrounded by a sphincter muscle (and presumably by a dilator as well; though this has not been traced morphologically, see Froesch) is a

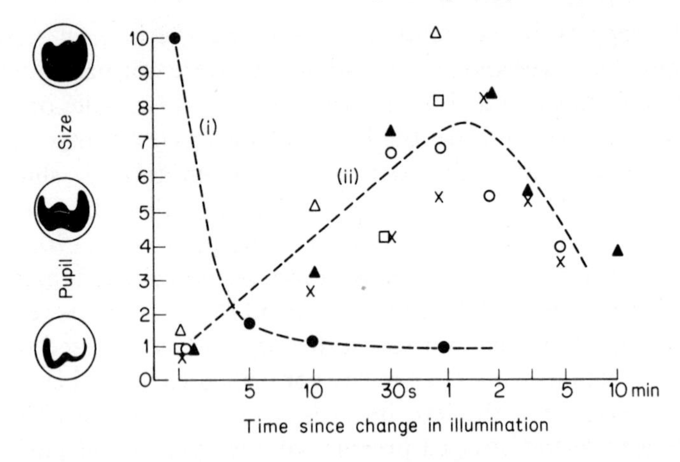

Fig. 7.4 Adaptive response of a cephalopod iris. Pupil size change in the cuttlefish *Sepia.* Time course of the response (i) to sudden increase, (ii) to sudden decrease in illumination of the eye by 3 log-units. The response of the partially dark-adapted eye (i) is immediate. The response of the light-adapted eye (ii) is slow and reverses. (Separate symbols indicate different experimental animals.) (Muntz, unpublished, from Packard, 1972).

horizontal oblong that can be expanded to a square or closed to a narrow slit. The response of the chromatophores and pupil are under nervous control, independent on the two sides of the body, through nerves arising in the anterior chromatophore and lateral pedal lobes respectively (Froesch, 1973b; Young, 1971).

Although the aperture of the iris is broadly predictable if the light intensity and adaptive state of the eye is known, the response is not simply reflex (van Weel and Thore, 1936). The emotional state, and the colour pattern that the animal happens to be showing (Packard and Sanders, 1969, 1971) are also clearly correlated with pupil size. In general, the pupil enlarges if the animal is startled, an effect that is particularly well shown in the 'dymantic' display (ex Greek δειμαντικος (adj.) 'a frightening (thing)', Plate 2) that the octopus makes when it is disturbed away from home by something larger than itself. Pupillary dilation also occurs in social situations, when the animals are fighting or mating and is sometimes seen when the octopus is alerted by the movement of potential food objects in its tank. More often than not the enlargement is brief and the effect passes in a second or less so that its cause and purpose are not easy to determine. Certainly some of the colour patterns shown by octopuses have a signalling function, and it is possible that the size of the pupil is also important in this respect (Packard, 1972).

7.1.9 *Astigmatism and the resolution of horizontal and vertical extents*

Because of the shape of the iris, the octopus eye will be astigmatic in bright lights. With a slit pupil the image should be blurred in the horizontal plane. In fact the evidence is that at the light intensities likely to matter to it, *Octopus* can resolve horizontal differences rather better than vertical (Section 8.1.7). Sutherland (1961) found that they were able to learn to recognize differences in the horizontal extent of rectangles more readily than vertical differences, under conditions when the pupil was expanded to about two-thirds of its full aperture. In experiments on visual acuity, to be examined below, vertical stripes were more readily resolved than horizontal (Sutherland, 1963b). These facts can be matched to Young's (1960a, 1962b) observations on the shapes of the dendritic fields of the bipolar cells in the optic lobes. As the optic nerves enter they encounter a tangential array of broadly oval fields, with their long axes mainly horizontal and vertical. Fields with a horizontal axis appear to be slightly more numerous than the verticals. The further significance of these arrangements in relation to the visual analysing mechanism is considered in Chapter 8.

7.1.10 *Visual acuity*

Two sorts of experiments have been made to see if the octopus eye is able to distinguish detail as well as one might expect from the dense packing of the rhabdomes and the enormous numbers of optic nerves that project onto the optic lobes.

Sutherland (1963b) trained octopuses to discriminate between plaques with vertical or horizontal stripes and between striped plaques and grey plaques matched for brightness. The animals learned to distinguish between grey and 3 mm wide stripes, at a distance of 60 cm, averaging about 70% correct responses in the first 200 trials when trained with the stripes vertical and 65% when they were horizontal – further evidence that lengths are more readily assessed in the horizontal plane (Section 8.1.7). When training was continued with progressively narrower stripes, discrimination declined progressively. 1.5 mm can be distinguished from grey, but 0.75 mm probably cannot, at least under the conditions of these experiments (see below). Further groups of animals were trained to distinguish between stripes 1.5 and 3 mm wide, a discrimination that they learned readily (again performance was better when the stripes were vertical, Table 7.1).

The animals were also subjected to a series of unrewarded transfer

Table 7.1 Visual acuity tested by training animals to distinguish between striped and grey figures. There were 20 trials per day. The shapes were moved for one second in every five (from Sutherland, 1963c).

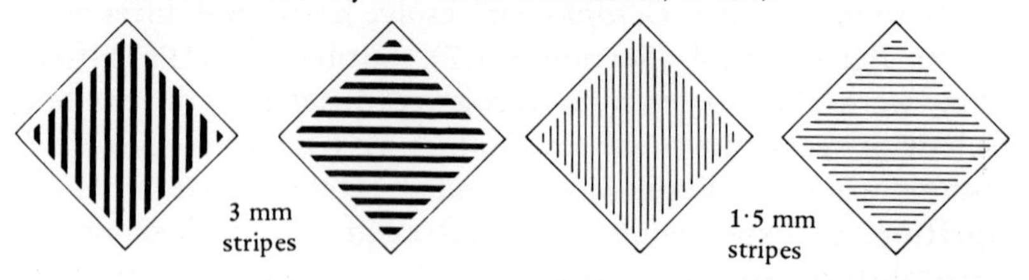

Group	*Days*	*Training*	*Per cent correct* Stripes Vertical	*Per cent correct* Stripes Horizontal
1	1–20	Grey (G) versus 3 mm stripes	70	65
$n = 8$	21–25	(Transfer tests); score in training trials interspersed with these	80	73
	26–27	Further training G versus 3 mm	90	81
	28–38	G versus 1.5 mm	64	69
	39–43	G versus 0.75 mm	54	59
	44	G versus 3 mm	74	61
	45–46	(Transfer tests); training score	76	74
2	1–20	1.5 mm versus 3 mm	71	64
$n = 8$	21–25	(Transfer tests); training score	80	70
	26–27	1.5 mm versus 3 mm	74	56
3	1–6	1.5 mm versus 3 mm	81	61
$n = 8$	7–9.5	(Transfer tests); training score	78	64
4	1–4	1.5mm versus 3 mm	68	53
$n = 8$				

tests with grey and striped plaques, including stripes both wider and narrower than those used in training and plaques rotated through 90°. The results confirmed the results of training with progressively narrower stripes; at about 0.75 mm wide stripes begin to be indistinguishable from a uniform grey background; the failure occurs earlier if the stripes are shown horizontally, with 1.5 mm stripes treated as grey in this orientation.

The difficulty of trying to assess acuity from this type of experiment is, of course, that the angle subtended by a stripe depends on how far it is from the octopus when the animal makes the decision to attack or leave it alone. Sutherland notes that decisions about 3 mm stripes were made regularly at the full 60 cm length of his training tanks. Confronted by stripes narrower than this, the animals were likely to approach the plaques before leaping at them or retreating to their homes. Taking these distances into account, Sutherland (1963c) concluded that *Octopus* could reliably distinguish stripes subtending an angle of 17 minutes of arc.

Packard (1969) used a quite different technique to estimate visual acuity. He placed small octopuses in a stationary dish within a rotating vertically striped drum. The animals make compensatory eye and body movements to hold the visual image steady (see also Section 7.2, below, on statocysts) and their ability to resolve stripes moving around them can be assessed from this.

Table 7.2 is a summary of the results that he obtained. It is quite clear that the animals were able to see stripes subtending angles of 49 minutes of arc and greater. Below that, responses were erratic, with no clear tendency to follow the stripes except, perhaps, by the largest animals in the 29′ situation.

Because of the magnification of the air-glass-water lens of the jar, an angle of 49′ in air corresponds to an angle of about 65′ as seen by the octopus. This is four times the angle estimated by Sutherland (1963c). The apparent discrepancy is attributable to the relative sizes of the animals. Sutherland's octopuses, in the 250–500 g range, were ten or twenty times the size of Packard's largest specimens. The packing density of the rhabdomes does not alter appreciably as the animals grow (it is always about 7 μm between centres) so the number of rhabdomes covered by the image of a stripe of any given size will increase with eye size. In Packard's experiments with very small animals, the image of a stripe 2.5 mm wide at 17 cm (49′ in air, 65′ to the octopus) would range from 17 μm (eye diameter 2.2 mm) to 54 μm

Table 7.2 Responses made in optomotor experiments (after Packard, 1969).

		Animal weight (n = 4 in each group)		
		0.47± 0.03 g	*2.7±0.4 g*	*17.0±5.0 g*
	Angle subtended	*Score maximum = 24*		
1 Plain white	–	0	5	15
2 'Noisy' grey, with spots and stain patterns	–	1	9	23
3 Vertical stripes	9°	17	22	24
4 Vertical stripes	49′	9	11	18
5 Vertical stripes	27′	−1	2	9
6 Vertical stripes	16′	5	4	−3
7 Vertical stripes	9′	3	−6	−1

Individuals were observed for 2 minutes divided into 6 periods of 20 seconds. If a compensatory movement was made in a 20 second period, one point was scored (so maximum score is 4 × 6 = 24). In experiments 3–7 control runs were made with the same stripe widths, with the stripes horizontal. Each entry here has the control score deducted from the score made with the vertical stripes (from Packard, 1969).

(7 mm) across the retina. The threshold size at which such images produce a response seems to be about 21 μm, or the width of 3 rhabdomes. A stripe subtending an angle of 17′, the value found by Sutherland, above, would yield an image covering about 4 rhabdomes in the much larger eyes of his experimental animals (Packard, 1969). It seems that visual acuity must improve progressively as an octopus grows. The eyes of animals of the size (±300 g) used in most of the visual training experiments to be described in Chapters 8 and 10, can safely be assumed to be capable of resolving objects with a diameter of somewhat less than half a centimetre at a range of one metre. The human eye, in air, is about thirty times as good.

7.1.11 *The epistellar body*

Young (1936) noted this small yellowish organ at the back of the stellate ganglion and suggested that it might represent the cell bodies of the third order giant fibres of squids, now given over to neurosecretion. At light microscope level, the cells look like neurones with short axons terminating in a central lumen. Subsequent work has shown that the body originates independently of the ganglion so that it cannot be the remnant of the third order giant fibres (Sacarrão, 1956, 1965). Ultrastructural examination shows clearly that it is a photoreceptor and not a gland at all. The cells of the epistellar body are structurally similar to the receptor cells of the retina, and they contain large quantities of rhodopsin (Nishioka, Hagadorn and Bern, 1962; Bern, 1967). Intracellular recordings show generator and action potentials in response to light. Response to light flashes, from a tungsten microscope lamp, adapted with repetition (Mauro and Baumann, 1968).

The function of this photoreceptor, which is inside the mantle and must be subject to lighting that fluctuates with the animals respiratory movements, is entirely unknown. The epistellar bodies are very large in some deepwater octopods (Young, personal communication).

7.2 Statocysts

The eyes of cephalopods have many structural similarities to the eyes of vertebrates. This is perhaps not very surprising, since there are only three ways of constructing an eye capable of forming sharp images of objects seen at a variable distance. If the animal cannot scan, like a television camera, rebuilding a picture from successive impressions, the only alternatives left are some sort of fixed focus compound eye, with a great many facets each responsible for a fraction of the total field, or a camera eye with an adjustable lens bringing the image to focus on a single continuous receptor surface. Vertebrates and cephalopods have similar visual requirements and have evolved very similar eyes by convergence.

They also require very similar information about their positions in space. A swift-moving animal needs to know the direction of gravitational pull, and the angular accelerations that result when it changes course. Most arthropods obtain the necessary information from hair plates recording angles between the elements of the jointed skeleton, and from gauges detecting strains in the cuticle. Among the arthropods, the crustacea alone have gravity receptors that depend, like our own, on the shear forces generated by a relatively heavy lump of material supported on hair cells.

With no hard external skeleton, the cephalopods were obliged to adopt the crustacean/vertebrate solution to the gravity problem. In the absence of cuticular strain gauges there is inevitably also convergence with the vertebrates in the manner of determining angular accelerations; cephalopods and vertebrates both have systems that depend upon the deflection of flaps when the body moves relative to fluids inside it.

The structure of the *Octopus* statocyst has been described by Young (1960b, 1971), and the outline given below is based on his work with details of the fine structure added by Barber (1965, 1966a, b, 1968) and by Budelmann, Barber and West (1973). The system includes gravity and angular acceleration detectors, and a further system of hair cells lining the capsule of the statocyst, the function of which is at present unknown.

7.2.1 *The gross structure of the statocyst*

Cephalopods have a pair of statocysts embedded in cartilage below and to either side of the brain. In octopods, each statocyst hangs from the roof of a cavity, surrounded by and filled with clear fluids, the perilymph and endolymph, that are ionically similar to the blood. Fixation produces a considerable precipitate of unidentified organic materials (Amoore, Rodgers and Young, 1959). A web of fibrous strands and blood vessels connect the statocyst to the walls of the cartilaginous chamber. Among decapods and in the developmental stages of *Octopus,* the perilymphatic space is absent.

The general structure of an *Octopus* statocyst is shown in Fig. 7.5. The walls of the suspended bag consist of two single-layered epithelia, inside and outside, sandwiching strands of connective tissue and striated muscle. In places, the walls are stiffened by cartilage, and there is a cartilaginous plate, the anticrista, projecting into the endolymphatic cavity. The statocyst cartilages support two fields of hair cells. One field 'the macula' is oval and capped by a calcareous cone, the otolith. The other runs in a strip around the capsule, the crista. The hair cells of the crista support a series of nine flaps arranged in three sets at right angles. With the octopus sitting on a horizontal surface, the lateral and transverse flaps lie horizontally, the third set in the vertical longitudinal plane. From their gross morphology, one would expect the macula to be a gravity receptor, and the flaps of the crista to deflect in response to angular accelerations. Physiological observations, discussed below, show that these assumptions are essentially correct.

In addition to the cells concentrated along the crista and in the macula, there are hair cells scattered all over the inner surface of the statocyst, contributing axons to a network of nerves that is particularly rich around the bulging posterior sac (Fig. 7.5).

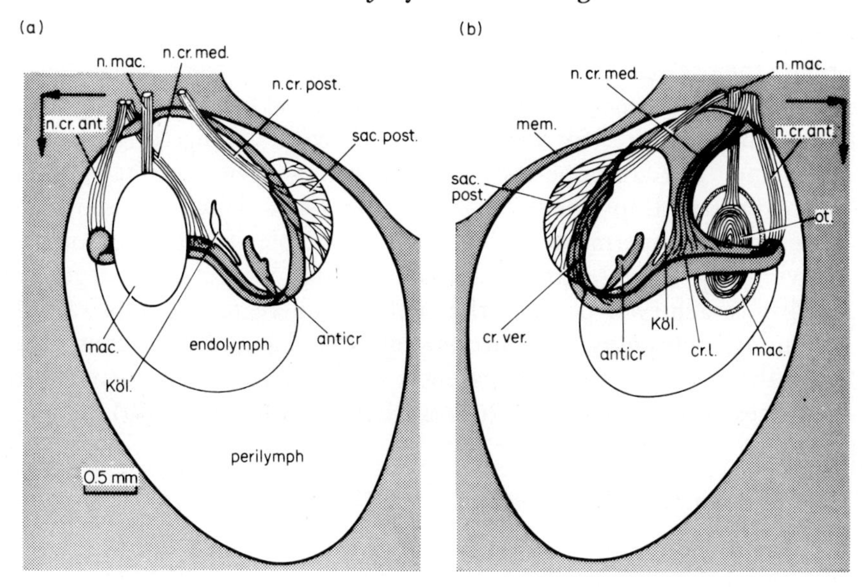

Fig. 7.5 The right statocyst, seen (a) from the midline and (b) from the right hand side (arrows show anterior and ventral). antricr. anticrista; cr.l. and cr.ver. lateral and vertical sections of the crista; Köl. Kölliker's canal connecting endolymph and perilymph; mac. macula; n.cr. ant., n.cr.med and n.cr.post. nerves to the anterior, median (= lateral) and posterior sections of the crista; n.mac. macular nerve; ot. otolith on macula; sac. post. posterior sac. Shadowing shows cartilaginous stiffening (from Young, 1971).

7.2.2 *Behavioural observations*

Removal of one statocyst has very little effect on the movements of an octopus walking or swimming around its tank. The animals tend to lean a little to the unoperated side, and there is a tendency to circle in a wide arc with this side on the inside when swimming. If both statocysts are removed the animal becomes noticeably unsteady, though it can still walk right way up and, as we shall see later (Section 9.2.2, below), it can still make visually oriented detours to get food. The arms are perhaps spread more widely than usual, as if the animal was obliged to keep a firm grip on the bottom, and the head rocks backwards and forwards and from side to side as it moves along. It can still right itself if placed upside-down on the floor of its tank.

In part, orientation must be visual. Blinded animals, which normally orient themselves as successfully as those with their optic nerves intact show the same effects of statocyst removal but in an exaggerated form. They require, for example, about twice as long to right themselves and they are hopelessly disoriented on vertical surfaces, where the body may flop sideways or over the head, which the animal no longer holds uppermost as is almost invariable in intact octopuses.

Swimming is more affected than walking, where some degree of stability is achieved because the suckers adhere to the substrate. In swimming, octopuses without statocysts zig-zag and corkscrew erratically through the water, apparently quite unable to compensate for the forces developed by jets from the funnel (Boycott, 1960).

7.2.3 *Orientation of the eyes*

At rest, the slit pupil of the eye of an intact octopus runs horizontally, more or less regardless of the posture of the animal (Fig. 7.6). Removal of one of the statocysts has little effect on this, but removal of both does; pupillary (and therefore retinal) orientation then depends upon how the animal happens to be sitting (Wells, 1960a).

The fact that the slit pupil tends to remain horizontal has been used as a basis for a number of experiments on the function of the macular gravity receptors. Budelmann (1970) used the apparatus shown in Fig. 7.7. The octopus, with its arms confined in a sack, or strapped to a board, was slowly rotated around an axis defined by the centre of the pupils of its two eyes. There were no visual cues to orientation. The eyes cannot, of course, continue to roll through 180°, but they try to do so. The results of a large number of experiments are summarized in Fig. 7.8. Removal of one statocyst reduces the compensatory counter-rolling and removal of both eliminates it altogether (Fig. 7.8A, B, C). Similar results are obtained if the animal is rotated around its longitudinal axis. In this case, the eye movements are in the vertical plane, with the eye on the down-going side moving upwards relative to the other (Budelmann, 1970).

The maculae of the two statocysts lie at 90° to one another, with the hair cell fields set at 45° to the vertical longitudinal plane of the body. If, as might be guessed by analogy with other statolith-operated sense organs, the stimulating force is shear, tending to deflect the kinocilia, differing results should be obtained with unilaterally operated animals, depending upon the direction of rotation relative to the plane

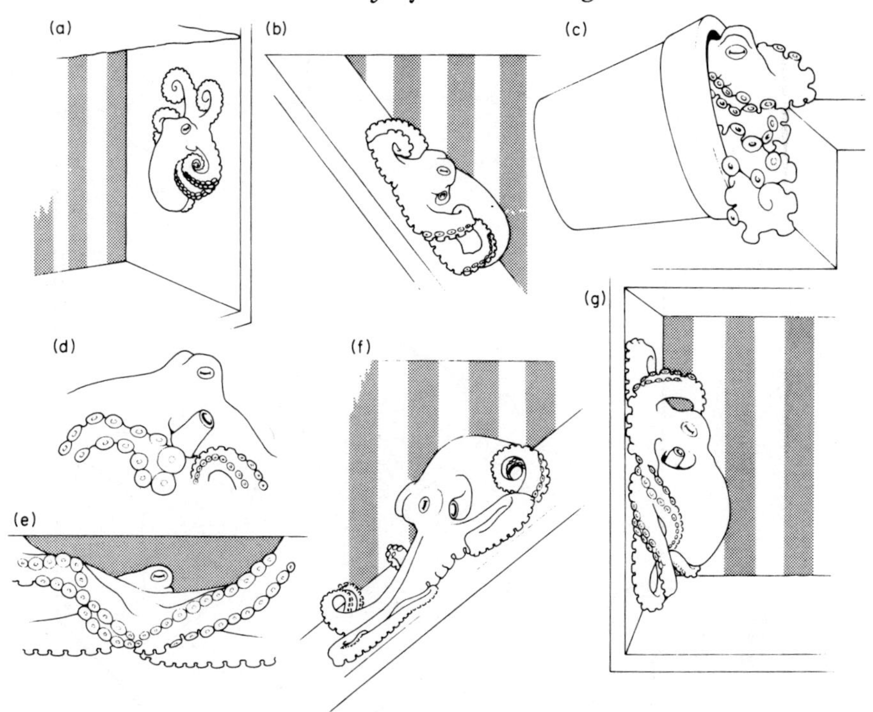

Fig. 7.6 Orientation of the eyes before and after bilateral statocyst removal. In unoperated animals the slit-like pupil normally remains horizontal or very nearly so (d–e), whatever the position of the octopus. After removal of both statocysts this ceases to be true, and the orientation of the retina, as indicated by the position of the pupil, thereafter depends upon the position in which the animal is sitting (f–g). Pictures traced from photographs (c, d) and (e) are of comparatively large (500 g) octopuses, the rest of small (15–25 g) animals in an aquarium set up in front of a vertically striped background; in (b) and (f) the aquarium, with the animal sitting on the bottom, has been tipped through 45° (from Wells, 1960a).

of the hair cell field. Budelmann (1970) rotated animals around axes parallel with, at right angles and at 45° to the plane of the remaining macula. His results are summarized in Fig. 7.8B, E, and F. The plane of rotation matters; compensatory eye movements fail if the macula field is rotated around an axis that does not change the direction of shear. In the experiments summarized in Fig. 7.8E, the magnitude of the shear force would alter, but (apart from its sign) its direction would not. In experiments shown in Fig. 7.8F, conversely, the direction of shear would change but its magnitude would not and these animals did make clear compensatory eye movements. This result was checked with

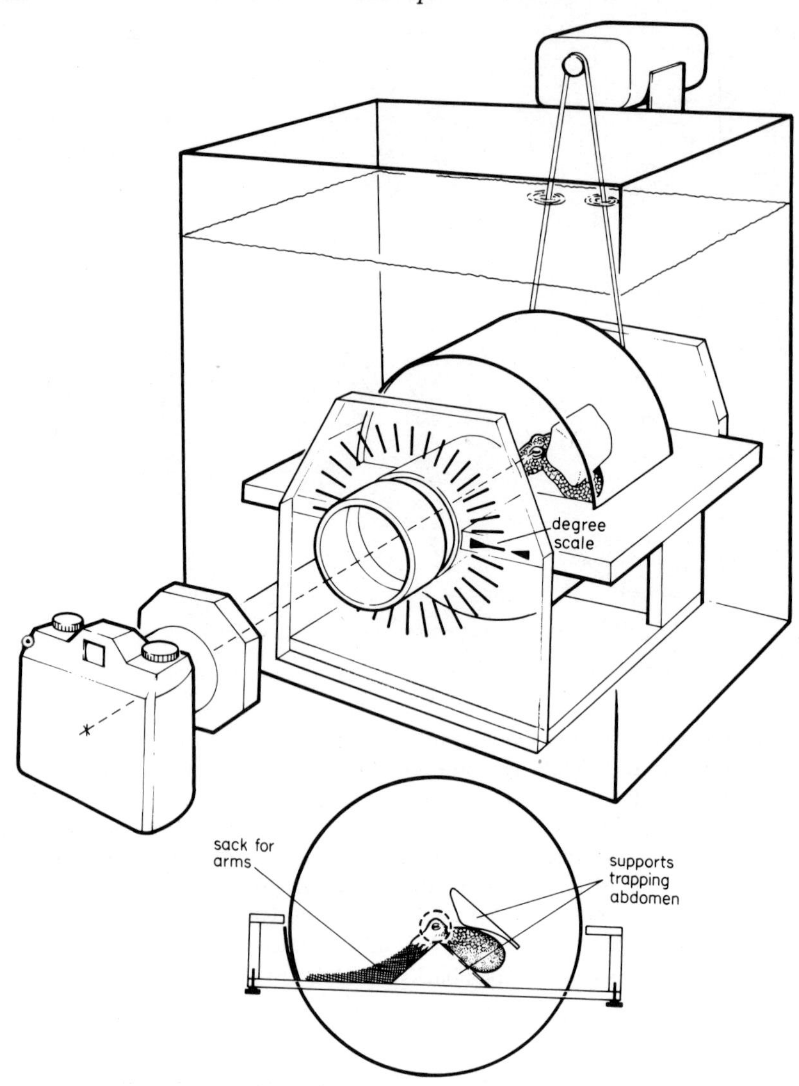

Fig. 7.7 Apparatus for studying the effect of statocyst removal on compensatory eye rotation (after Budelmann, 1970).

reference to vertical eye movements, and with the same result; shear direction matters, the magnitude of the shear force does not.

In crustaceans and vertebrates, the gravity receptors respond to the magnitude as well as to the direction of the shear force (Budelmann, 1970; Schöne and Budelmann, 1970) so the conclusion from the work on *Octopus* statocysts was entirely unexpected. A further check was

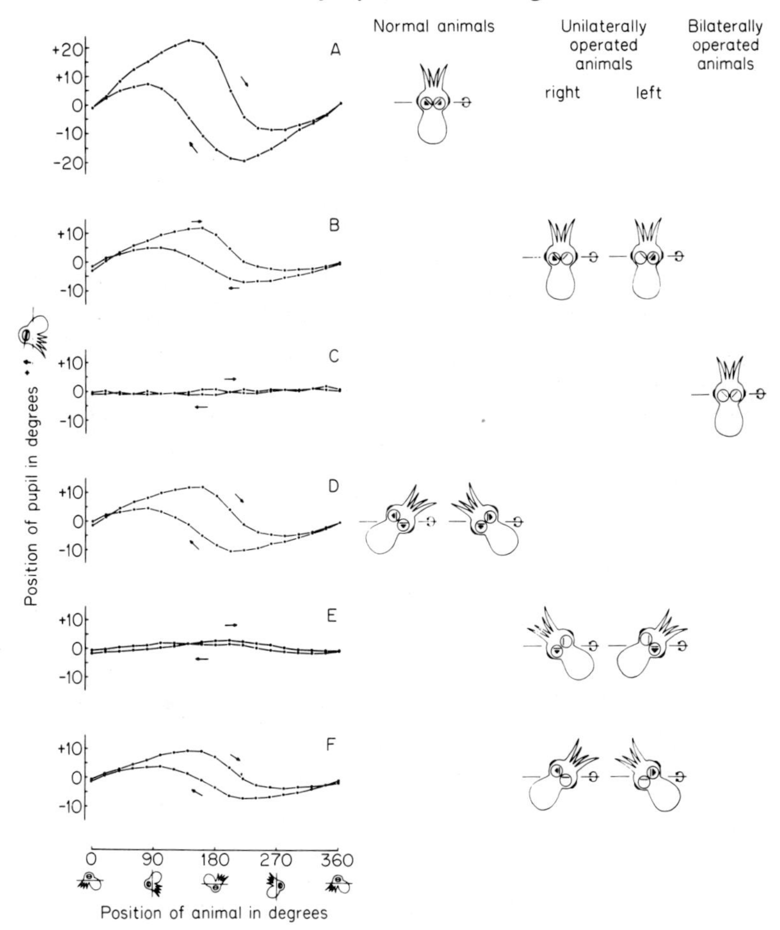

Fig. 7.8 Compensatory eye movements in octopuses as a function of body position relative to gravity during rotation around the animal's transverse and diagonal body axes. The degree of counter-rolling is measured as the angular deviation of the pupil slit from an animal-fixed reference line. Diagrams on the right show the position of the animal from above with the position of the macula and otolith relative to the axis of rotation. Arrows on the curves show the direction in which the animal is being tilted on the apparatus shown in Fig. 7.7 (from Budelmann, 1976).

made by repeating the eye-rotation experiments in a centrifuge, where the magnitude of the gravitational force could be increased independently of its direction. As Fig. 7.9 shows, an increase from 1 to 1.5 g has no effect whatever on eye rotation. Somehow, the system in *Octopus* is organized to eliminate the effect of changing the magnitude

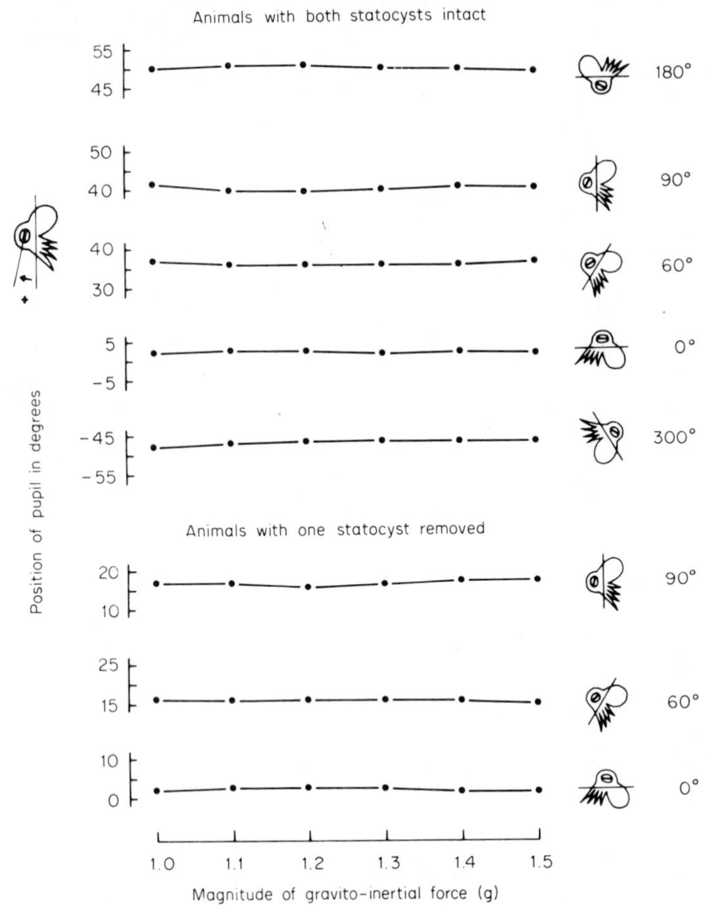

Fig. 7.9 Compensatory counter-rolling of the eyes of octopus as a function of the magnitude of the gravito-inertial force at the body positions indicated on the right hand side. Each line consists of single measures on one animal. Ordinate, angle between pupil slit and animal-fixed reference line; abscissa, magnitude of gravito-inertial force (from Schöne and Budelmann, 1970).

of the shear force, while preserving information about its change of direction. The development of a system with this property is perhaps related to the very considerable accelerations that cephalopods can produce by mantle contraction. An octopus that is making a jet-propelled dash at its prey, or towards the safety of its home in the rocks needs to preserve the orientation of its eyes, since its whole system of form recognition depends upon this (see Chapter 8).

Counter-rolling and vertical displacement of the eyes both depend upon changes in the direction of shear in the macula. The two sorts of movement should therefore, always occur together. In fact, they occur together only in animals with one statolith removed; in normal octopuses counter-rolling occurs alone in response to rotation around a transverse, and vertical displacement alone in response to rotation around a longitudinal axis. The central organization of an appropriate response clearly takes into account the information derived from both statocysts (Budelmann, 1970).

7.2.4 *Electrophysiology of the macula*

The calcareous otolith (Fig. 7.5) lies on a pad of about 500 hair cells, each of which sprouts a group of some 120 kinocilia, arranged at right angles to the field centre (Barber, 1966a, 1968). The hair cells are primary receptors, apparently contributing one axon each to the plexus of nerve cells that underlies the macula. Elaboration at this level could account for the existence of more than 9000 axons in the macular nerve, but it is more probable that a high proportion of these are efferents running to the sense organ from the CNS (Budelmann, Barber and West, 1973).

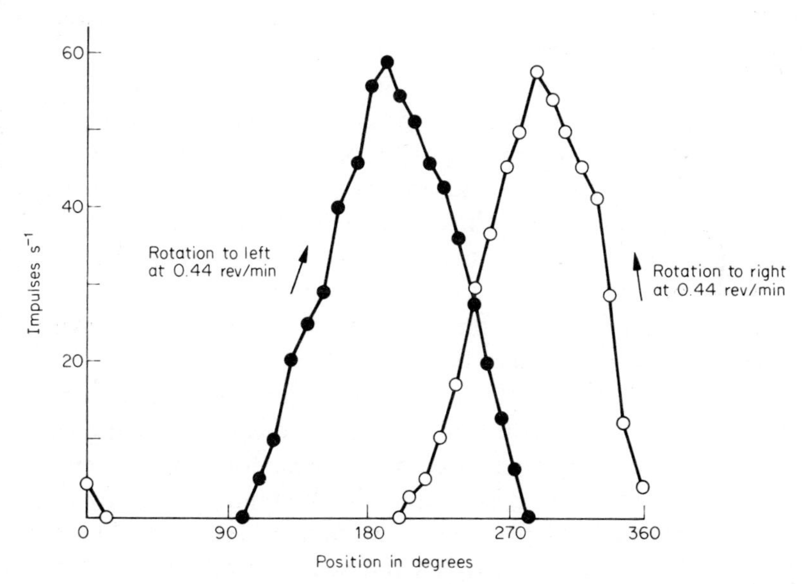

Fig. 7.10 Rates of firing in a few fibre preparation, recording from the macular nerve of the right statocyst, while the octopus is rotated around its longitudinal axis (from Budelmann, 1976).

Recordings from groups of fibres in the macular nerve show impulse rates that rise to a maximum and then fall as the animal is rotated around its horizontal axes (Fig. 7.10). Maximum rates arise from groups of kinocilia orientated at right angles to the direction of shear and bent in the direction of their basal feet. Individual hair cells are active over a limited range (about 180°), adapt rapidly and are very sensitive to vibration. So far, no trace of any efferent effects has been found. The individual receptors behave like the hair cells of crustaceans and vertebrates; the elimination of shear magnitude as a property determining compensatory eye movements must be achieved centrally since it is plainly not a property of the individual receptors (Budelmann, 1976).

7.2.5 *The crista, and responses to rotation*

If its aquarium is rotated an octopus will attempt to keep the image of the outside world stable on its retina by moving its eyes and head, and when this fails by walking against the direction of movement (Section 7.1.10). If it is rotated at about 1 revolution per second, the animal attempts to keep pace with the non-rotating environment for a few revolutions, and then spins with its tank. If rotation is now stopped suddenly, about half of the animals tested make weak but unmistakable post-rotationary movements, with short lateral sweeps of the head (to the right after clockwise rotation) and after-nystagmus flicks back. If the animals are blinded by cutting their optic tracts, the same compensatory eye and head movements are made at the beginning of rotation, and the post-rotationary flicks are even better developed. These responses cease if both statocysts are destroyed (Dijkgraaf, 1961).

Blinded octopuses with one statocyst removed and the other damaged by section of all but the posterior (= vertical) cristal nerve still showed post-rotationary head flicks. Out of 5 such animals tested, Dijkgraaf (1960) found the direction of horizontal rotation to be irrelevant in two, while three showed signs of distinguishing between clockwise and counterclockwise spins. Thus, for example, an animal with the left posterior cristal nerve intact (and everything else disconnected) showed stronger compensatory movements when rotated counterclockwise than clockwise; the same animal's post-rotationary movements were most marked after clockwise rotation. The two experiments would, of course, result in the same (clockwise) direction of endolymph displacement. Recordings from the cristal nerves (see below) confirm that each section of the crista is indeed more responsive to deflection in one direction than the other.

7.2.6 *The structure of the crista*

The strip of hair cells that forms the crista includes sensory elements of varying size, arrayed along a ridge, with a distinct single or double row of very large hair cells at the crest. Fig. 7.11 is a reconstruction based on light microscope studies; the hair cell processes are embedded in an amorphous ground substance to form a series of flat plates, the cupulae. Each section of the crista bears three such plates, sticking

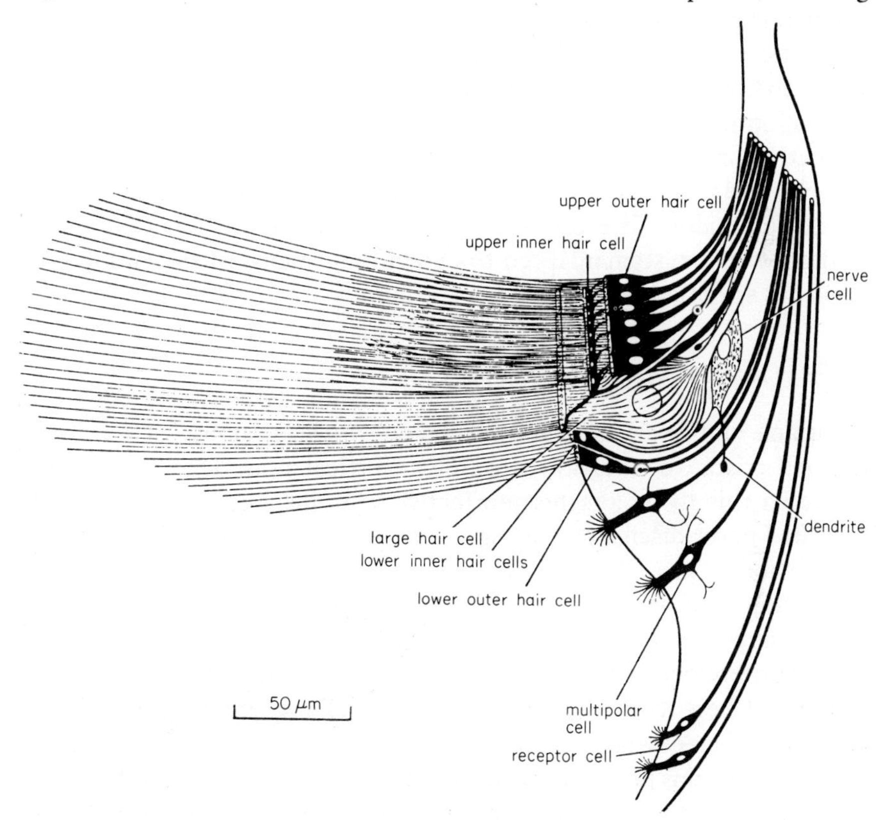

Fig. 7.11 A section of the crista, part of a plate with a single row of large hair cells. The axons from the multipolar and receptor cells shown here join a plexus of further nerves, not shown, underlying the crista and forming a network around the axons of the hair cells; at least some of the nerves in the plexus appear to be efferent (from Young, 1960b).

out at right angles into the endolymph. The large hair cells lie in a double row in the first plate of the transverse crista, single in the second, and then alternately, ending with a final double row in the last plate of the vertical crista, a very distinct anatomical feature for which there is as yet no functional explanation (Young, 1960b).

Further work at EM level has suggested that the enormous length of the large hair cell processes may be illusory. The kinocilia could be about 10 μm, not 400 μm long with the rest of their apparent length made up of fibrous material embedded in the cupula (Budelmann, Barber and West, 1973). Studies at this level have also shown that the orientation of the kinocilia is very consistent, so that the basal feet of the two hundred or so arising from any one large hair cell are all facing in the same direction. This direction, moreover, is always at right angles to the cristal ridge. Evidence from vertebrates suggests that deflection of kinocilia in the direction of their basal feet depolarizes, and deflection in the opposite direction hyperpolarizes the receptor cells (Flock, 1965; Wersall, Flock and Lundquist, 1965). The orientation of the cilia along the crista is thus a further indication that one is dealing with a system adapted to detect movements of the endolymph in planes specified by the flaps of the cupula.

7.2.7 *Electrophysiology of the crista*

Maturana and Sperling (1963) reported on the response of the longitudinal crista to rotation around a vertical axis. One might have expected this to be a rather ineffective stimulus, but they successfully recorded nerve discharges during acceleration and after-discharges when the turntable was stopped. Their preparations ($n = 13$), which showed no resting discharge, were sensitive to low frequency vibration (but not to airborne sounds) and able to differentiate between clockwise and counterclockwise accelerations; the crista from the left statocyst responded best to the onset of counterclockwise rotation, and to the cessation of clockwise movement.

Following this preliminary study Budelmann and Wolff (1973, and Wolff and Budelmann, 1977) carried out an extensive series of experiments, which included recording from all three of the cristal nerves. A sample of their results is given in Fig. 7.12 which shows the (few-fibre) response of the right side longitudinal crista to the onset and cessation of rotation around the animals longitudinal axis. The system clearly differentiates between rotations to the left and to the right, and the

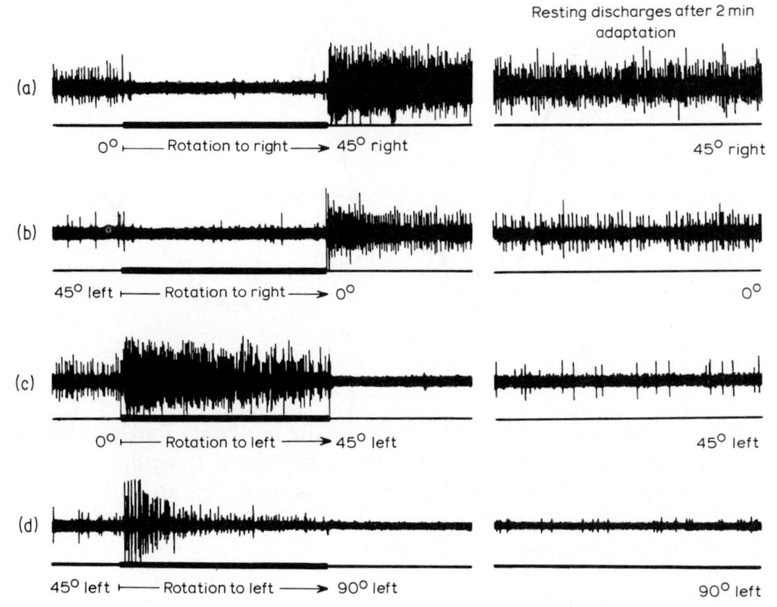

Fig. 7.12 Discharge pattern (few fibre responses) of the nerve of the right crista longitudinalis of *Octopus vulgaris* after intracranial section of all right statocyst nerves. Horizontal broad traces indicate the positional changes of 45° with an angular acceleration stimulus at the beginning and a deceleration stimulus at the end of each constant speed rotation at 3.33 rev/min (20°/s). Each horizontal axis rotation was around the animal's longitudinal axis and was carried out after two minutes' adaptation to the initial resting position. To the right: resting discharges after two minutes' adaptation to the various final resting positions (stop positions) are shown (from Budelmann and Wolff, 1973).

results obtained depend upon the position with respect to gravity at the start of the run. The records also show that there is a steady resting discharge (cf. Maturana and Sperling, 1963) which changes with the attitude of the preparation. Similar results were obtained from the transverse crista with rotation around the earth-horizontal transverse axis, and the vertical crista with rotation around the vertical axis (Wolff and Budelmann, 1977).

Responses to gravity can be distinguished from responses to angular accelerations if the preparations are moved very slowly. At 2.67 deg s^{-1} no response to acceleration was found but the resting discharge remained and changed as the statocyst was rotated through either of the earth-horizontal axes (Fig. 7.13) (Budelmann, 1977; Budelmann and Wolff, 1973).

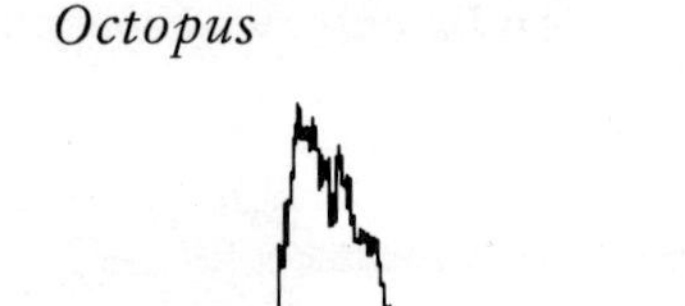

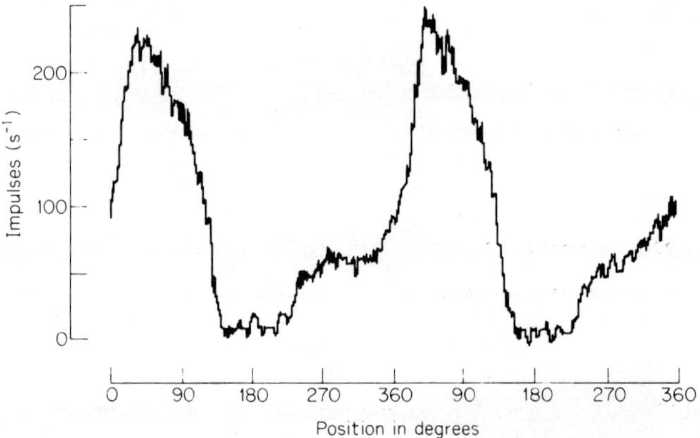

Fig. 7.13 Few fibre responses of the nerve of the left crista longitudinalis after intracranial section of all left statocyst nerves. Recording during two successive earth-horizontal axis rotations to the left around the animal's longitudinal axis. Constant speed rotation at 0.44 rev/min (2.67°/s) (from Budelmann and Wolff, 1973).

7.2.8 *The anticrista*

The anticrista is a thin flexible plate with a rough spiny surface, extending into the cavity of the statocyst in a position where it must alter the flow of endolymph over the vertical crista. It is cartilaginous with a larger cell population than the other cartilages of the statocyst. It varies in shape from one individual octopus to the next.

The purpose of this structure is unknown. Young (1960b) has suggested a protective function, shielding the vertical crista when the octopus makes a sudden rush forward at its prey. But the animal also propels itself backwards by means of the jet, and there is no corresponding protection for the transverse crista, along the opposite wall of the sac.

7.3 Receptors in the skin and suckers

An octopus may react to a light touch anywhere on its body surface. It will respond to pieces of fish dropped onto its back by reaching to pick them up and it is likely that the whole body surface is covered with both mechano- and chemoreceptors. The sense organs responsible have, however, only been studied systematically in the region of the suckers, since these are used to examine objects touched and seem to be more sensitive than the rest of the surface.

The suckers, and especially the rims of the suckers, carry huge numbers of primary receptors (Rossi and Graziadei, 1958). A single sucker of 3 mm diameter (about half way along the arm of an animal of 250 g) carries several tens of thousands of these cells; the total compliment for the eight arms, allowing something for receptors in the skin outside the suckers, has been estimated at 2.4×10^8 (Graziadei, 1971). Graziadei and Gagne (1973) have reviewed light and EM studies, and Graziadei (1965a) has given an account of what is known about the central connexions of the receptors that they have described. There are at least four morphologically distinct types, all with cell bodies embedded in the columnar epithelium of the sucker (Fig. 7.14). Type 1 is round and buried beneath the surface. Type 2 is long and narrow with a ciliated tip, reaching between the epithelial cells. Type 3 is multipolar, irregular and deeply buried, with a narrow neck protruding towards the surface. Type 4 is fusiform, like 2, but terminates in a globe-shaped pore armoured with stiff immobile cilia. Type 2 receptors are about ten times as frequent as types 1, 3 and 4. Types 2 and 3 have been described at EM level (Graziadei 1964, Graziadei and Gagne 1976).

There appear to be two morphological variants of the type 2 receptor. One of these forms clusters of flask-shaped cells, each group associated with a ciliated non-sensory accessory cell; clusters are found

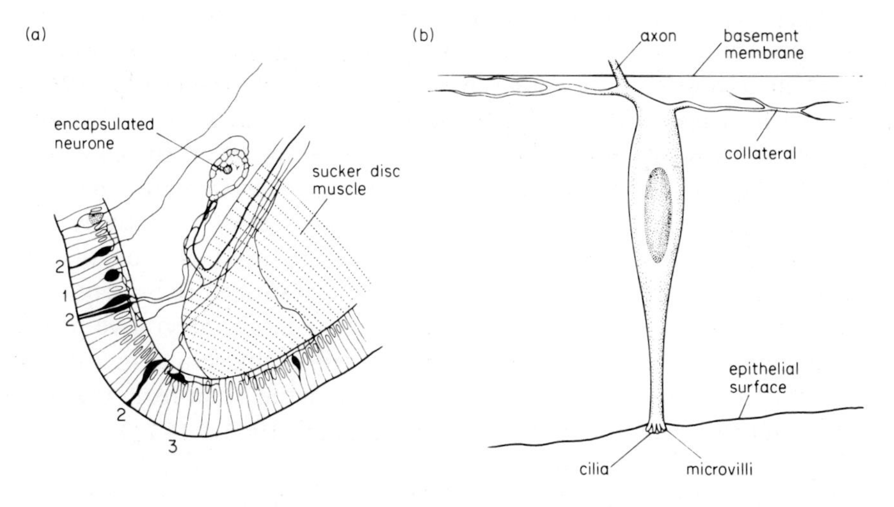

Fig. 7.14 (a) The position of sense organs and an encapsulated neurone in the rim of a sucker. Sense organs types 1 and 3 are presumed mechanoreceptors while the more numerous type 2 which sometimes occur in clusters, open to the exterior and are presumed to be chemoreceptors. (b) shows more detail of a type 2 receptor. (b) after Graziadei, 1964).

only at the edge of the sucker where they will be exposed at all times, whether or not the sucker is gripping a surface. The other form occurs as single cells, many of them distributed across the infundibulum; the cuticle covering of the disc of the sucker is perforated over each receptor. From their appearance and distribution Graziadei (1964) believes that type 2 receptors must be chemosensory.

Type 3 receptors have radiating dendrites at the base, and a long thin neck, which runs to the surface. The neck encloses a canal lined with microvilli and filled with an electron-dense material; the canal lining is reminiscent of photoreceptive elements in the eye and epistellar body. Most of the type 3 receptors are found in the epithelium of the flat downward-facing disc of the sucker. In this position it seems most unlikely that they are indeed photoreceptive and Graziadei and Gagne (1976) conclude that (like types 1 and 4) they are most probably mechanoreceptors, responding to distortion. Type 3 receptors differentiate later in development, and are replaced more slowly in regenerating suckers than the fusiform ciliated cells that make up types 2 and 4 (Graziadei and Gagne, 1976).

7.3.1 *Central connexions*

Degeneration experiments show that at least some of the axons from primary receptors in the suckers travel direct to the arm nerve cord ganglia. Others run to the subacetabular ganglia. These probably include axons from all four types of skin receptor (Graziadei, 1965a).

In the rims of the suckers, the very numerous type 2 (presumed chemosensory) receptors have axons that terminate in the connective tissue between sucker disc muscle and sucker epithelium. Here they form synapses with the cell bodies and processes of a relatively small number (hundreds, compared with ten thousand or so) of large encapsulated neurones (Fig. 7.14). The encapsulated cells relay to the arm ganglia. This is evidently the first stage in an integrative process (further discussed in Section 10.2.9) that eventually results in a sensory input of less than 30 000 neurones running to the brain from the thirty million or so receptors in each arm (Graziadei, 1965c, 1971).

7.3.2 *Electrophysiology*

Recordings from the arm nerves can be made if the animal is first immobilized by removing the supraoesophageal part of the brain. This stops it from walking about, but it will still live for several weeks. The arms no longer collaborate with one another (Chapter 10) and it is possible to trap the animal, in a box, with one arm extended through a hole in the side, pegged down near the base but otherwise free-moving on the recording table. Such a preparation retains the blood supply, without which the CNS soon fails, and it is possible to expose the axial nerve cord and record from units within this, as soon as the animal relaxes the vasoconstriction that at first seals off any damaged area (see also Section 7.1.5, above, where the same problem is found in recording from the optic nerves).

Using animals treated in this way, Rowell (1966) was able to record from the peripheral nerves, from interneurones in the arm cords and from the longitudinal tracts running along these to the brain.

Peripheral nerves from the suckers were cut at the point of entry into the arm ganglia. The nerve distal to the cut was drawn into a glass suction electrode. Recordings made in this way were always dominated by cells responding to mechanical distortion of the rim of the sucker concerned. Single units were rapidly adapting, firing 20–40 times in response to a stimulus, which could range from gentle pressure to a jab with a needle. The form or direction of the stimulus did not appear to matter, once it was above threshold. The units were not particularly sensitive – they did not respond, for example, to a drop of water falling onto the rim of the sucker from a height of 1 cm. They were not spontaneously active, and they were readily fatigued. After six or eight repetitions within as many seconds, a unit would cease to respond altogether, and remain inactive for a minute or more.

No responses to chemical stimulation (dilute acetic acid) were found. This is surprising since by far the largest cells in the rims of the suckers are the encapsulated neurones described by Graziadei (1965c) and these appear to be relays for the input from presumed chemoreceptors.

Responses from neurones in the arm nerve cords are discussed in Chapter 10. The huge number of neurones in the arm nerve cords carry out elaborate analyses of the inputs from the primary receptors, so that the information finally relayed to the brain is very different from the messages recorded at the level of the peripheral nerves (Rowell, 1966).

7.3.3 *Behavioural studies*

Recordings from the peripheral nerves, discussed above, would imply that the mechanoreceptors and chemoreceptors in the suckers are rather insensitive. Behavioural observations suggest otherwise. It is almost impossible to touch the sucker of an intact octopus without eliciting some sort of reaction. Ten Cate (1928), following the earlier work of von Uexküll (1894), used stimulants that included silk threads and fine bristles to study reflex responses to stimulation of the arms and suckers. These responses, which include withdrawal and subsequent extension of the sucker touched towards the origin of the stimulus, are described in Section 10.2.8 and could be evoked by gentle stroking as well as by more violent poking or pinching.

The apparent disagreement between the electrophysiological and behavioural observations must be a matter of the sampling method used in the former. Rowell himself (1966) found interneurones in the arm nerve cords that were far more sensitive to mechanical disturbance than any of the receptors that he was able to record from at the peripheral nerve level.

In chemotacile training experiments, octopuses proved capable of learning to discriminate between solutions of hydrochloric acid, sugar and quinine in seawater at concentrations well below those detected by the human tongue. They could distinguish between seawaters of differing concentration (half, normal and double strength; Wells, 1963b) and they were able to detect potassium chloride added to seawater in concentrations at least as dilute as 10^{-5} M (Wells, Freeman and Ashburner, 1965). In these experiments, considered in more detail in Section 9.1.4, the capacity to discriminate was tested by dipping spongy objects into the solutions concerned and presenting these to the arms of the octopus. The results depended, presumably, on chemoreceptors in the suckers.

7.4 Receptors in the muscles

Many cephalopod muscles have stellate neurones among their fibres. Alexandrowicz (1960b) described such cells from the mantle of *Eledone* (Fig. 7.15) and Graziadei (1965b) subsequently demonstrated their presence at a number of sites in the arms and suckers of *Octopus.* They are, as Graziadei points out, rather difficult to stain and probably more widespread than their reported distribution would suggest.

These branched neurones are presumed to be proprioceptors, responding to distortion of the webs of muscle that they embrace. Very little is known about their function. Gray (1960) recorded from the stellate nerves in the region of Alexandrowicz's receptors but

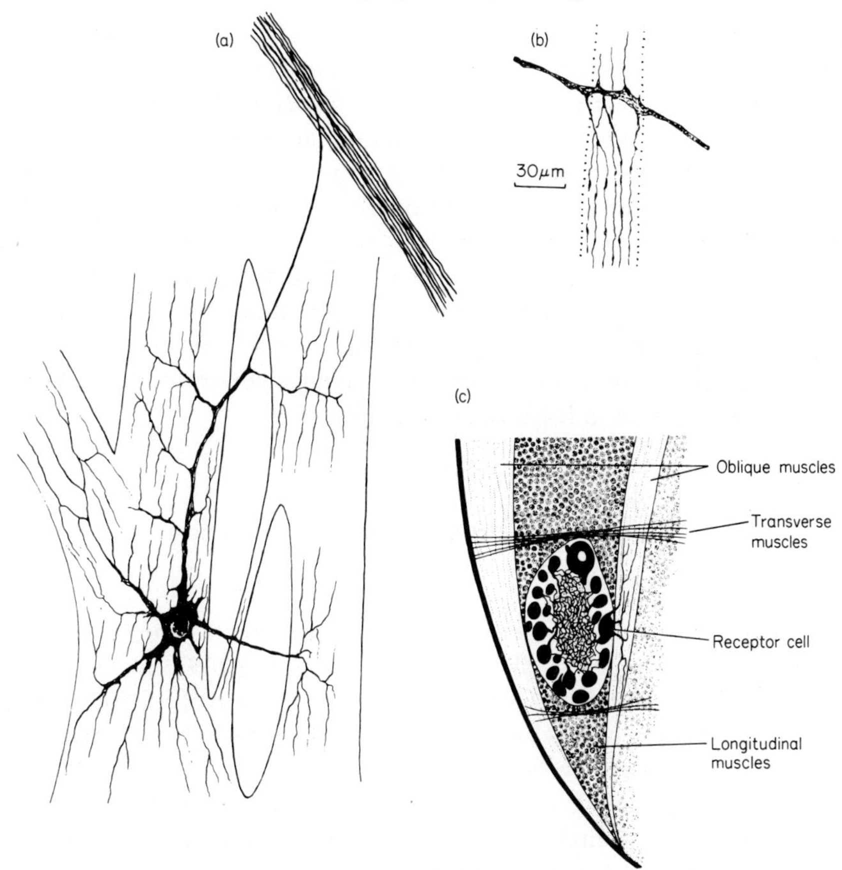

Fig. 7.15 Receptors in the muscles of octopuses. (a) a cell from the mantle musculature of *Eledone*; (b) fine branches of a similar cell in a small strip of muscle (outlined by dots); (c) shows a receptor cell in the intramuscular nerve cord in the arm of *Octopus*, with dendrites extending into an oblique muscle. The monopolar elements in the cord are presumably motor neurons. For the relation between the intramuscular cords and the other structures in the arm, see Fig. 10.9, ((a) and (b) after Alexandrowicz, 1960b and (c) after Graziadei, 1965b).

found no trace of proprioceptive feedback in normal respiration (Section 3.1.3). Rowell (1966) recorded from about forty neurones in the arm cords without discovering any that were relaying positional information.

In contrast with these entirely negative findings, behavioural observations certainly suggest that octopuses make use of muscle stretch information. If an arm is grasped and pulled, it pulls back. If an object is presented to the arms the muscles in the arms visibly tighten to take the weight. Both sorts of responses could, of course, be based on peripheral mechanoreceptors, as could the operation of the interneurones which Rowell (1966) found to discriminate between voluntary and forced movements of the arms. But it seems rather elaborate to postulate this when neurones of an appropriate structure are clearly sited in places where one would expect them to perform as stretch receptors. Most probably the information from proprioceptors is used very locally, perhaps at the level of the intramuscular nerve cords in the arms (see Fig. 7.15b). Certainly, it does not seem to be available to the highest levels of the brain; there is no evidence whatever for a proprioceptive component in learning (see Chapter 9).

7.5 Other sense organs, of unproven function

7.5.1 *The 'olfactory pit'*

An 'olfactory' pit or tubercle is found on the side of the head, posterior to the eyes, in many cephalopods (Watkinson 1909). In *O. vulgaris* it is visible as a small (2 mm diameter in an animal of 500 g) clear patch in the skin just below the point at which the edge of the mantle joins the neck between head and abdomen. Beneath the skin there are many large ciliated cells with processes that run to the periphery between the epithelial cells (Woodhams and Messenger, 1974). The region is served by a small nerve of a few hundred fibres running to the olfactory lobe on the optic stalk of the brain (Boycott and Young, 1956a; Messenger, 1967a).

There is no proof that either structure has anything to do with olfaction. The name seems to have arisen with Hancock (1852) who assumed that the pit was in some way equivalent to the osphradium in gastropods; it is obviously well-sited to sample the stream of water entering the mantle. But Wells (1963b) removed it from blinded octopuses and found no difference in their behaviour as a result; sardine blood in the water caused them to move around their tanks, indian ink did not. Messenger (1967a) repeated this experiment with *Sepia*, pipetting fish juice onto the olfactory organ (or the region where it had been before extirpation) with exactly the same result. Either this organ is not a chemoreceptor, which seems unlikely in view of the EM evidence of Woodhams and Messenger (1974), or it is tuned to pick up some more subtle stimulus.

7.5.2. *Further sense organs*

In addition to the above, primary receptors have been described from the lips (of *Sepia*; Graziadei, 1965a) and gut (Alexandrowicz, 1928a), from the internal surface of the cornea (Alexandrowicz, 1927) and from the gill hearts (Alexandrowicz, 1960a). Most have been examined only at light microscope level, and seem little different from the array of chemo- and mechanoreceptors in the suckers.

The function of these receptors can be guessed from their forms and locations, and it is clear that the list is by no means complete. It shows only where people have bothered to look.

We can be reasonably certain from what we know of their behaviour that octopuses do not possess any further major detector of external events. The animals do not, for example, have an equivalent of the lateral line system of fish and it is unlikely that they enjoy mechanisms paralleling the electrical or sonar devices that enable some vertebrates to find their prey. Enough octopuses have been kept in aquaria under close observation to make the emergence of any further external sensory systems an unlikely event. No such categorical statements can be made about internal sensory devices and here we are still truly in ignorance. We do not know how an octopus regulates its blood pressure, osmotic or ionic condition, egg laying or spermatophore production. We do not know how its strange system of digestion by instalments is monitored and controlled. Receptors have been described from many of the relevant structures but the questions of which do what are all unanswered.

CHAPTER
EIGHT

What an octopus sees

8.1 Untrained responses and the results of training to discriminate by sight

8.1.1 *Growing up and learning*

Octopus vulgaris hatches at a length of 2–3 mm and at once swims upwards to join the drifting clouds of plankton. The surface waters of the sea form its habitat for the first 6–12 weeks of its life, before it settles onto rocky ground on the seabed. We do not know whether recognition of its prey in the plankton is innate, though this seems likely by analogy with *Sepia*. Cuttlefish apparently start life with a visual system that ensures attacks on mysid crustaceans and little else during the first few days after hatching. As the animals age this restriction eases progressively, until it is replaced or overlaid by learning based on individual experience (Section 4.1.3).

This seems a very reasonable way to organize an animal that must change its habits as it grows. Shrimps, isopods and amphipods may form a useful source of food when the animal weighs a few grams but are unlikely to be of much interest to an octopus several hundred times more bulky. Fish, most of which must be avoided at all costs by newly settled octopuses, may no longer represent a threat as the animals grow towards an armspan of a metre or more and a weight of several kilograms. Although the need for continuous relearning must be lessened by the parallel growth of many of the foods that an octopus eats – a crab is a crab and rather unlike other objects, irrespective of its size – the detailed appearance of desirable prey species will keep changing as the octopus grows through and beyond the range of sizes in which these animals can be found. Vision is acute enough to permit fine distinctions to be made and, as we shall see below, it is quite certain that octopuses can learn to discriminate very rapidly.

It also seems inescapable that the octopus must learn to find its way about its habitat, largely by sight since it is a swift-moving creature that swims as well as crawling. Individuals certainly return to particular homes in the rocks (Chapter 4) and they must relearn the whereabouts of 'home' whenever they outgrow the cavities that they have adopted, and move to set up a base elsewhere.

In the laboratory an octopus can be trained to distinguish between many pairs of figures, including some that appear geometrically quite similar to us. It can distinguish at least one quality of light (its polarization plane, Section 8.1.17 below) that is invisible to ourselves, and it can generalize from its individual experience to predict the consequences of interference with unfamiliar objects. All this makes it a very fine experimental animal, and a considerable range of experiments have been made in attempts to deduce the manner in which the octopus central nervous stystem handles its visual input. The findings have led to hypotheses about the likely nature of visual analysing systems in general, so that the experiments with octopuses have helped us to make successful predictions about visual learning in animals as diverse as goldfish and small children.

8.1.2 *Untrained preferences*

Any attempt to analyse the system by which *Octopus* classifies the things that it sees must start with stimuli that we can ourselves define; there are too many differences between a crab and a fish to make the analysis of the means by which these are distinguished a very practicable proposition. Though there are dangers in the approach through 'simple' stimuli (as physiologists since the days of Lettvin *et al.*'s 'What the frogs eye tells the frog's brain' (1959) will be well aware) it has been customary to investigate the visual analysing systems of octopuses and other animals using simple geometric shapes. A standard technique has been to compare the number of trials that the animals require to learn to discriminate between the members of a series of pairs of figures thus discovering the comparative difficulty of a series of problems. Alternatively the animals can be trained to distinguish between the members of a single pair of shapes and the similarity to these of further figures assessed from the reactions of the octopuses in subsequent unrewarded 'transfer' tests. Details of such techniques are given below. The main difficulty in evaluating the results has been that *Octopus* clearly generalizes on the basis of its own past experience. It may also have innate preferences; we cannot distinguish between these and preferences arising out of common experience in the sea before capture. Either way, we have to know about untrained preferences before we can usefully assess the results of training experiments.

A number of investigations have been made. The preferences of untrained octopuses can be tested by showing them objects and observing how often these are attacked. Alternatively the animal can be trained to discriminate between the members of pairs of objects by rewarding it when it attacks one (the 'positive' object) and punishing it when it attacks the other (the 'negative'). If it learns much more readily when trained in one direction than in the other, then it is reasonable to infer that the positive object in the 'easy' direction is the more attractive of the two to be distinguished.

In the first category (unrewarded tests with unfamiliar objects) Young (1958b) found that circles were attacked more often than rectangles of the same area, white figures were attacked more than black and vertical (10 × 2 cm) rectangles more than horizontal rectangles of the same size when both were moved up and down. He showed 22 octopuses each figure 5 times. Only the black/white preference was sufficiently marked to be significant at the 5% level.

Subsequent work has shown that the black/white preference is at least in part dependent upon contrast with the background. Young carried out his 1958 experiment in grey asbestos tanks. In a more extensive series of tests, made ten years later, he obtained the same result – white is preferable to black – in grey tanks, and a distinct preference for black when the animals (some of them the same individuals) were tested in tanks with a creamy-white background (Young, 1968).

Messenger and Sanders (1972) confirmed Young's result in white tanks, and also re-investigated the preference for vertical over horizontal rectangles. They found a consistent preference for the vertical shape (moved, as before, up and down) irrespective of contrast with the background. Superimposed on this, however, there was an apparent interaction between orientation and brightness; black vertical rectangles were attacked more often than white, while the reverse was true of the horizontals (Table 8.1).

In training, the untrained preference for vertical tends to persist, and has been shown to contain at least two components. Other things being equal, octopuses apparently find shapes that move up and down more attractive than those that move from side to side. The effect is particularly well-marked in the case of vertical rectangles because it seems that octopuses also prefer shapes that move along their long axes (Table 8.2), or in the direction of their points (Sutherland, 1960c; Sutherland and Carr, 1963; Sutherland and Muntz, 1959).

Table 8.1 Preference tests (from Messenger and Sanders, 1972).
Total number of unrewarded attacks made by thirty-six octopuses on eight shapes each presented twice.

	Shape preference							
	▮	▬	▥	▤	▯	▭	●	○
No. of attacks	51	17	43	28	32	28	55	37

Maximum possible number of attacks on each shape is 72.

Mean number of unrewarded attacks made by thirty-six octopuses on the four classes of shape

	Shape					
	Verticals		*Horizontals*	*Black*		*White*
Mean	3.50		2.03	3.42		2.69
SEM	0.28		0.27	0.24		0.25
t		3.80			2.05	
P		<0.001			<0.05	
(*df* = 70)						

Maximum possible number of attacks for each animal with each category of shape is 6.

Table 8.2 Attacks on rectangles shown with their long axes vertical or horizontal and moved as indicated (after Sutherland and Carr, 1963).

No. of attacks made on different shapes				
Up down		*Side to side*		*No. of trials*
612		538		720
V	H	V	H	
335	277	240	298	360

Horizontal and vertical rectangles have been used very extensively as the shapes to be discriminated in training experiments, including much of the work on the effects of brain lesions on learning (Chapters 11 and 12). In these experiments, the shapes have nearly always been moved up and down. Under these conditions the experimental animals, predictably, learn more readily when the vertical rectangle is 'positive', a state of affairs that creates problems in the interpretation of the results of training since animals that attack more (as they are liable to do when trained in the preferred direction) will tend to acquire more experience per trial than their counterparts.

Other visual preferences have been less studied. Very large shapes are clearly frightening – the animals pale and withdraw deeply into their homes – and perhaps for this reason Boycott and Young trained their animals with the smaller shape as positive in most of their earlier experiments on visual discrimination (see Boycott and Young, 1956b). But the animals soon learn that shapes are not dangerous and can mean food and it is then found that among figures in the range 2–10 cm diameter (that is up to about the size of the octopus' abdomen) the larger of two shapes is generally the more attractive (Rhodes, 1963; Sutherland and Carr, 1963). The evidence is thus that octopuses have no preferred target size when they come into the laboratory. They seem, too, to have no predilictions in favour of shapes with simple or complex outlines; a cross *vs* square discrimination is learned equally readily whichever of the figures is positive (Sutherland, 1962b).

8.1.3 *Training to make visual discriminations*

Octopuses have been trained in a great many visual discrimination experiments. The results all confirm the impression that one gets from observations on the behaviour of octopuses in aquaria and in the sea; these animals can recognize shapes and sizes and they rapidly learn to discriminate between new figures.

The overwhelming majority of the training experiments made to date have taken advantage of the octopus' habit of selecting a 'home' to which it retires and from which it keeps a watch on its surroundings. In aquaria, a pot, box or a pile of bricks form a suitable home and can be used to ensure that the animal views any stimulus from a fixed starting point. The usual technique has been to show the animal an object – commonly a black or white shape cut from plastic and presented on the end of a transparent rod – at a distance of a metre or so from the home. The object is moved from side to side or up and down 2–4 times a second to attract the attention of the animal, which can choose to attack or ignore it. A healthy octopus, after a few days in captivity, will normally slowly approach and grasp any moving object that it sees, provided only that the target is smaller than itself.

If the animal is then fed, the object can be jerked away and attacks upon it will continue and become more rapid on future occasions. Repeated presentations and feedings lead to increasingly vigorous attacks, so that after a few trials, the typical approach is a jet-propelled leap from the home onto the target as soon as this is lowered into the water.

Alternatively, the octopus can be punished for attacking, by giving it

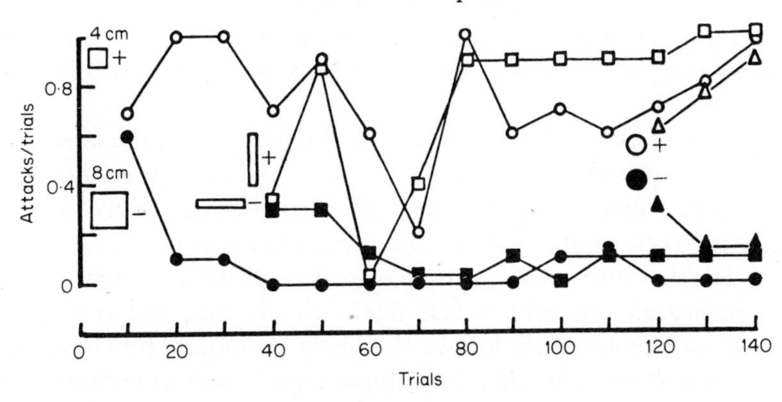

Fig. 8.1 Record of a training experiment in which an octopus was required to distinguish between three different pairs of figures. + means that the animal was fed and – that it was shocked when it attacked. There were 8 trials per day, those with the different figures intermixed (from Young, 1958b).

small (6–9 V A.C.) electric shocks, applied by touching the animal with a pair of electrodes on the end of a probe. Shocked in this manner if it attacks, the animal soon comes to leave an object, even a crab, alone when it sees it at the far end of its tank. The fact that the octopus has seen the object is obvious; it will lean forward from the entrance to its home, bob its head up and down and it may change colour before retreating more deeply into its hole.

Discrimination between the members of a pair of figures can be built up by rewarding the animal when it attacks one and shocking it when it attacks the other. Fig. 8.1 shows the progress of an experiment in which an octopus was trained to distinguish between the members of 3 pairs of figures, differing in size, orientation, or contrast.

In experiments of this sort it has generally been the practice to present the test objects through a slot in the lid of the animal's tank so that the octopus and the experimenter remain out of sight of one another until the octopus hits the target presented, or until some predetermined period, commonly about 15 s, has elapsed. The theory behind this is that the octopus may pick up cues from the attitude of the experimenter, a possibility considered because it has been known to occur in mammalian experiments. In practice this sort of interaction seems unlikely. Mammals can reasonably be expected to pick up cues from mammals because their innate signalling devices can be very similar, but it seems a bit far-fetched to suppose that the octopus can learn to recognize subtle differences in stance or facial expression in his human teacher. However, the matter has never been specifically investigated.

Much more important is the effect of the octopus on the observer. Octopuses are quite large and they can move and change colour with great rapidity. Trainers unfamiliar with the animals (many of the experiments to be quoted have used undergraduate labour) may themselves be startled into movements that in turn distract the octopuses. The sophisticated observer is less likely to do this, but is suspect because he or she generally knows the likely (or hoped for) outcome of the trial and may be able to influence the result consciously or unconsciously by moving the target in an irregular manner. To avoid this possibility movement has been mechanized in a few experiments (Sutherland, 1958a; Rhodes, 1963), a precaution generally omitted since it is tedious to transport the apparatus along a row of tanks and it makes it difficult to wrest the figures from the octopus. An alternative is to keep the shapes stationary. Sutherland (1963a) compared learning with stationary and moving shapes. Stationary shapes were set up behind a screen, which was removed at the beginning of each trial. Attacks on shapes presented in this manner were slower and at first less frequent than those on the same shapes, moved about in the 'standard' manner, so that discrimination developed more slowly. But the standard of performance attained after 200 trials, about 70% correct responses on a ⌶ *vs* + discrimination (shapes l and m in Fig. 8.8), was not different for the two groups.

Manual presentation of moving test shapes remains a theoretical defect of the method in most of the experiments to be reviewed below. Just how serious this is in practice is difficult to assess. It could be important, given a biased observer (a student anxious to show that the animal can learn, for example) in the early stages of a training experiment or in cases where the animal is required to make a discrimination at the limits of its ability. It is certainly *not* important in the case of easy discriminations that the animal has learned thoroughly, because attacks are here commonly too quick to allow for any variation in the manner of moving the test objects. One could add that the relatively small number of experiments made with movement mechanized or with stationary objects have yielded results that are not detectably different from those obtained using the cruder hand-held method of presenting the figures to be discriminated. It is probably adequate that the observer attempts to 'mechanize' himself as far as possible and that he remains out of sight of the octopus until it has attacked the target. Doing the experiments in this way discards a great deal of potentially valuable information that could be gained from close observation of the animals as they vacillate between attack and retreat. But it largely disposes of criticisms based on the possibility of man-octopus interactions when, as has often been the case, the people actually doing the training are kept in ignorance of the 'expected' result.

8.1.4 *Analysis of the visual input*

The similarity of any two figures can be assessed from the number of trials that an animal takes to learn to discriminate at a specified standard of accuracy, or from the accuracy with which it performs after a fixed amount of training. It is also possible to assess similarity from transfer experiments. Animals trained to discriminate between two figures are shown further shapes that they have not seen before and the similarity of these to the originals assessed from the frequency with which they are attacked.

Sutherland (1957a, b *et seq.*) made a systematic study of the visual analysing system of *Octopus,* using flat black or white perspex figures, moved about on the end of transparent plastic rods. His results have led to a series of hypotheses by himself and others (see Section 8.1.6 *et seq.* below) about the nature of the mechanisms in the brain of *Octopus* responsible for the analysis of the visual image. Not all species of animals pay attention to the same features of shapes that they are shown, and some interesting comparisons of the performance of *Octopus* with that of fish and mammals are made in Sutherland (1962a and 1969).

8.1.5 *Recognition of the orientation of figures*

Octopuses can learn to discriminate between the members of many pairs of figures including some that differ only in orientation (Boycott and Young, 1956b). Rectangles of the same size are readily distinguished when one is shown horizontally and the other vertically. If one of the two is offered at 45° instead of at 0° or 90°, the discrimination is more difficult, but still well within the animals' capacity; groups trained in this way made 71% (| vs / or \) and 65.5% (– vs / or \) correct responses over the first 60 trials of training (Table 8.3). It was therefore somewhat surprising to find that the same two rectangles,

Table 8.3 Scores made in the first 60 trials of discrimination training (from Sutherland, 1957b)

Shapes	*Shape dimensions (cm)*	*n*	*% Correct*	
| versus —	10 × 2	6	81	*
| versus / or \	10 × 2	8	71	
— versus / or \	10 × 2	8	65.5	*
⊤ versus ⊥	10 × 2 (with tail 4 × 2)	6	59	
⊢ versus ⊣	10 × 2 (with tail 4 × 2)	7	56	*
/ versus \	10 × 2 (with tail 4 × 2)	6	50	

* Difference significant at $p = <0.05$.

shown at right angles to one another, but obliquely, were not distinguished at all. These two shapes appear to be identical, so far as the octopus is concerned (Table 8.3; Sutherland 1957a, b, 1958a). The orientation of the image of an oblique rectangle on the retina will, of course, depend on which eye the octopus happens to be using at the time; it normally attacks monocularly, but there is no evidence of a 'preferred' eye, generally or individually (Section 10.2.1) so that both eyes are likely to be used in the course of any training experiment. Confusion between obliques (and other left-right mirror images) could arise because the animal learns to recognize apparently different shapes with the two eyes. That this is not the explanation of their failure has been demonstrated by blinding the animals in one eye; the animals perform just as badly as before (Messenger, 1968; Messenger and Sanders, 1971). *Octopus* in any case can learn to discriminate between pairs of figures such as ⊢ and ⊣ (Table 8.3) or [and] (Sutherland 1960a). The failure to discriminate between / and \ must reflect some feature of the mechanism by which octopuses analyse the properties of figures that they see. It can only mean that in the shape-classifying system of *Octopus*, the horizontal and vertical axes have a status that is not shared by oblique axes.

8.1.6 *A possible mechanism*

The results described could arise if the animal were classifying shapes in terms of their horizontal and vertical extents. Sutherland (1957a)

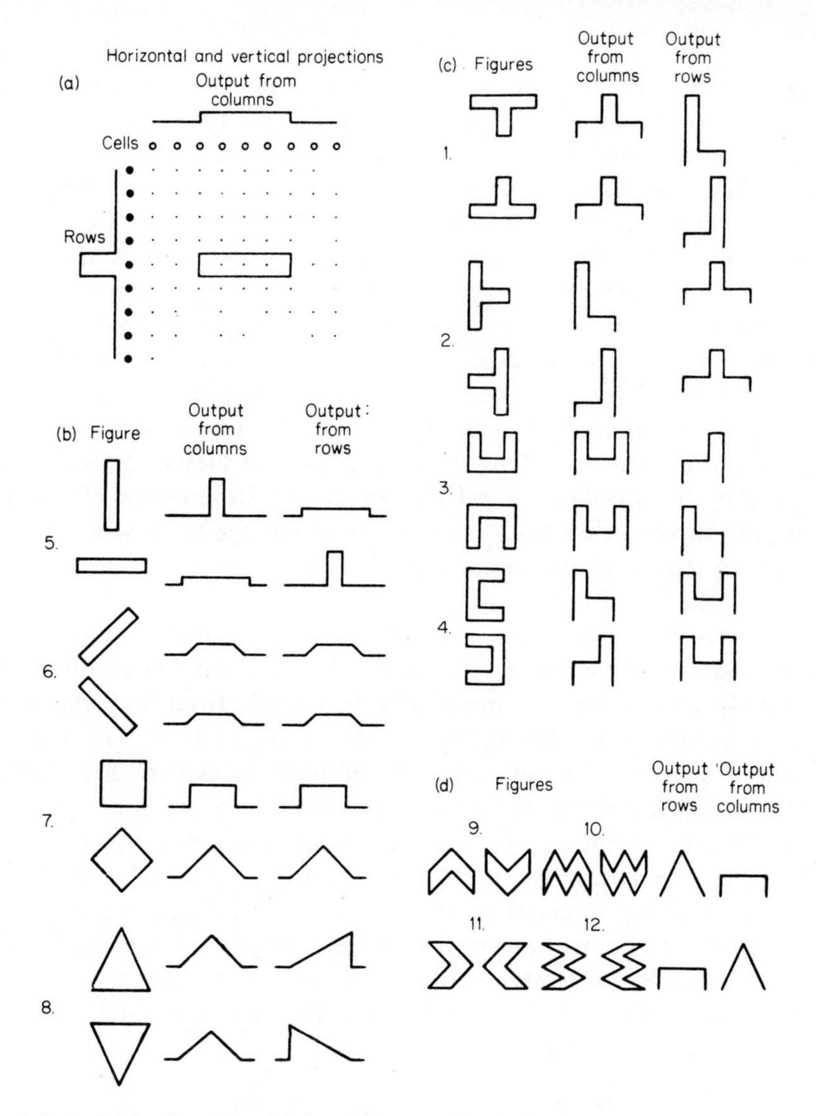

Fig. 8.2 Sutherland's hypothesis of shape discrimination. (a) the small dots represent the retinal array with a horizontal rectangle projected onto it. Open circles at the top represent cells specific to each column, filled circles at side of cells specific to each row. The cells for the rows and columns are supposed to be connected to distinct outputs (Sutherland, 1957a). (b) the outputs produced by various rectangles, square, diamond and triangles (Sutherland, 1958a). (c) mirror images (Sutherland, 1960a). (d) figures that should be confused (Sutherland, 1960c). Numbers indicate pairs of figures to which reference is made in the text (from Young, 1961).

suggested that the retinal array is projected onto the optic lobes to form a second rectilinear projection and that the first stage in analysis carried out by the CNS is to summarize the distribution of stimulation along the rows and columns of this internal projection. Any shape can thus be represented in terms of two projections at right angles (Fig. 8.2). Pairs of shapes that yield very different projections should be easy to distinguish while those that produce similar projections should be difficult. Oblique rectangles produce the same projections (Fig. 8.2) and are not distinguished at all.

Sutherland never explained how the next stage in the analysis was supposed to proceed. Having abstracted the projections, the CNS would presumably somehow be obliged to scan these; no mechanism based on the absolute or relative degree of stimulation of specific neurones in the 'rows' and 'columns' is acceptable because the animal is able to recognize the same shape at different sizes and irrespective of where the image falls on the retina.

The model, nevertheless, has considerable predictive value. It is possible to estimate the relative difficulty of a number of discriminations by examining the projections that result from breaking down shapes in this way. Among the shapes considered in Fig. 8.2, for example, pair 5, with very different projections, is easy, 3 and 4 are difficult, while pairs 6 and 9 are impossible. A square and a diamond (pair 7) are recognized as distinct shapes, while the triangle (8) is treated as resembling the diamond in transfer tests. Individual variation, particularly in relation to the more difficult discriminations, is quite high, as one might expect in animals taken from a natural population. Table 8.4 summarizes a typical series of results to show the range of performance that can be expected in a moderately difficult series of discriminations.

8.1.7 *The horizontal and vertical axes*

Experiments with mirror images of shapes show that up-down inversions are easier to distinguish than left-right inversions of the same figures. Table 8.3 includes results obtained with ⊤ shapes; the animals scored 59% correct responses in 60 trials with ⊤ and ⊥, 56% with ⊢ and ⊣. In a considerably longer experiment with ⊔ shapes (Fig. 8.3) the up-down inversions were again more readily distinguished.

This was to be expected. It is true of all the vertebrates that have been tested (see Sutherland, 1960a, 1962a) and it is what one would

Table 8.4 The range of performance shown by individual octopuses in two moderately difficult discriminations. The animals had two days positive training to attack the + shape; scores given here were made in the course of the 80 training trials (at 10 per day) that followed (from Sutherland, 1959b).

Training shapes +	*Training shapes* −	*% correct responses* *Individual animals*	*Group average*
○	□	59	
○	□	55	
○	◇	66	
□	○	82	
□	○	70	69
◇	○	86	
○	◇	73	
□	○	66	
◇	○	64	
○	△	76	
○	▽	71	
○	◁	61	66
△	○	60	
▽	○	63	

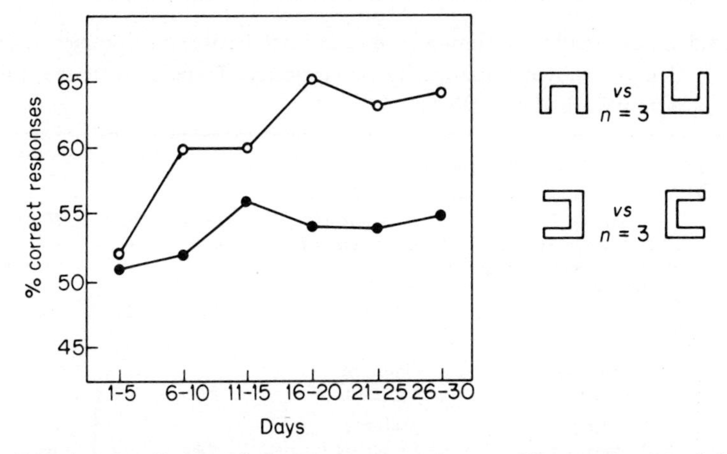

Fig. 8.3 Training to distinguish between mirror images. There were 20 trials per day (after Sutherland, 1960a).

expect from the habits and habitat of any bottom-living organism. Left-right mirror images represent the same object seen from in front and behind, so it is biologically sound for the animal to treat the two in the same way. There is no posture that the animal is likely to adopt in the course of its everyday life that will invert images in the vertical plane.

Considered in terms of the shapes' projections, this means that differences in the horizontal projections are likely to be of more importance to the animal than differences in the vertical (Fig. 8.2). This can be correlated with an apparently greater accuracy of assessment of the horizóntal extents of figures, and a tendency to pay attention to these differences in preference to differences in vertical extent.

Sutherland (1961) carried out a series of training experiments in which animals were required to learn to recognize the difference between 20 × 2 and 10 × 2 cm, or 17 × 2 and 13 × 2 cm rectangles. Eight octopuses were trained to make each of these discriminations, four being trained with the rectangles shown vertically, and four horizontally. Since octopuses show a preference for elongate shapes moving along their long axes (Table 8.2), half of the animals in each group were trained using shapes moved up and down, the other half using shapes waved from side to side. After the first 320 training trials, a series of transfer tests was run (see Table 8.7) and the group (I) trained on the 'easy' discrimination moved onto progressively more difficult pairs of rectangles. The scores made are summarized in Table 8.5.

Table 8.5 Scores made in training to discriminate between rectangles of differing length, shown horizontally or vertically. Transfer tests see Table 8.7 (from Sutherland, 1961).

Days	*Treatment*		*Correct responses (%)*			
	Group I *20 × 2 versus 10 × 2*	*Group II* *17 × 2 versus 13 × 2*	*I–H* *n=2*	*I–V* *n=2*	*II–H* *n=2*	*II–V* *n=2*
1–16	Training 20 versus 10	Training 17 versus 13	76	64	56	53
17–19	Transfer tests Retraining 20 versus 10	Training 17 versus 13	86	67		
20	Training 19 versus 11	Training 17 versus 13	78	69	67	55
21	Training 18 versus 12	Training 17 versus 13	84	55		
22–30	Training 17 versus 13	Training 17 versus 13	68	57	61	51
31–36	Retraining 20 versus 10	Training 20 versus 10	74	61	70	63
37–39	Transfer tests Retraining 20 versus 10	Transfer tests Retraining 20 versus 10	76	63	74	64

The main finding from this series of experiments, which is quite clear despite the small number ($n = 4$) of animals in each of the subgroups, is that octopuses learn to recognize a difference in horizontal extent far more readily than a difference at right angles to this.

A further indication that horizontal extents are relatively important is found when transfer tests are run after training to recognize figures that differ in both the horizontal and vertical dimensions. Octopuses trained to distinguish between 10 × 10 and 4 × 4 squares and subsequently shown horizontal or vertical 10 × 4 rectangles, treat the rectangles as equivalent to squares having the same horizontal extent; a horizontal 10 × 4 rectangle evidently resembles a 10 × 10 square, while a vertical rectangle of the same size is equivalent to the 4 × 4 – the vertical extents of the figures seem to be relatively unimportant (Mackintosh, Mackintosh and Sutherland, 1963; Sutherland, 1960b, 1964).

8.1.8 *The anatomy of the optic lobes*

An attraction of the rectilinear projection hypothesis is that it is compatible with the anatomy of the visual system. The structural arrangements in the outer layers of the optic lobes – the 'deep retina' which would be the site of Sutherland's internal projection – are quite consistent with the theory that the animal somehow sums stimulation along horizontal and vertical axes. The dendritic fields in the plexiform layer spread out tangentially across the radially arranged inputs from the optic nerves (Figs. 8.4 and 8.5). Each such field could respond proportionately to the length of the strip of an image falling across it, or there could be some arrangement by which fields respond only to strips of exactly the right lengths. Horizontally oriented fields are apparently a little more numerous than their vertical counterparts (Table 8.6) and this is again consistent with the experimental findings; horizontal extents are assessed more accurately than vertical, while transfer tests indicate that the animals pay particular attention to differences between figures that would change the pattern of stimulation along the horizontal axis.

The outputs from cells in the plexiform layer feed towards the medulla of the optic lobe, where they make contacts with further neurones that presumably abstract features that depend upon a series of horizontal and vertical extents. These, and cells yet further down the chain of connexions (Figs. 8.4 and 8.5) could be responding to

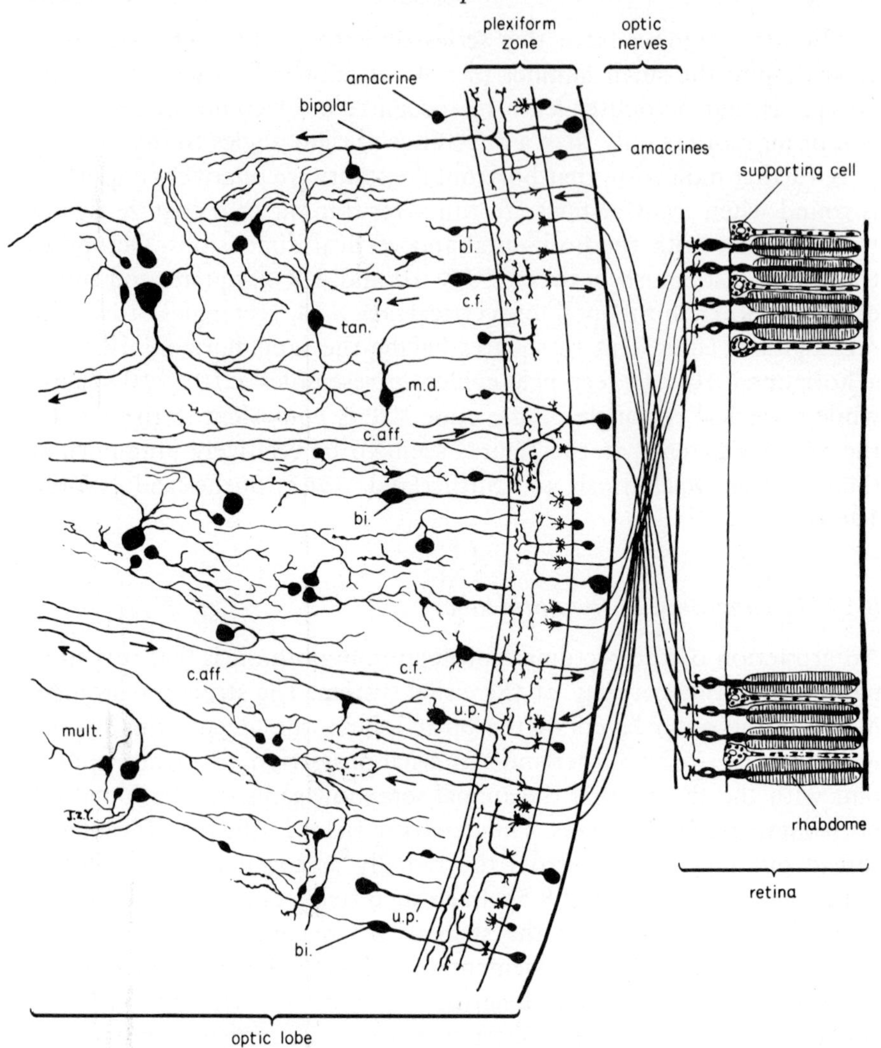

Fig. 8.4 Connexions in the outer part of the optic lobe. Nerve cell types: bi., bipolar; c.f., centrifugal (with axon to the retina); c.aff., afferent to the plexiform layer; m.d., cell with centrifugal dendrites; mult., multipolar; tan., tangential; u.p., unipolar. Details of retinal structure are considered in Section 7.1.2 (from Young, 1962b).

whole geometric shapes, or to such features as points and angles between sides. Young (1965e) has called them 'classifying' cells; they form an important feature of models of learning discussed in Chapter 12.

The anatomical arrangements outlined above could only function in the manner suggested if the projection on to the 'deep retina' somehow

Table 8.6 Number and lengths of fibres in an area of 22 500 μm^2 in the plexiform layers of the optic lobe of an octopus. There are more vertically running fibres than horizontal fibres and relatively few obliques. The horizontal fibres tend to be slightly longer, which could indicate a preponderance of horizontal axes with vertical offshoots (from Young, 1962b).

	Degrees from horizontal					
Layer of plexus	75–105	105–135	45–75	135–165	15–45	165–180 0–15
2nd tangential						
number	39	8	13	6	5	12
length (μm)	1455	393	462	204	179	531
mean length (μm)	37	49	36	34	35	50
3rd tangential						
number	24	18	11	16	11	19
length (μm)	625	447	373	490	445	664
mean length (μm)	26	25	34	31	41	35
4th tangential						
number	26	14	14	11	13	16
length (μm)	760	450	321	357	359	415
mean length (μm)	29	32	23	32	29	26
total						
number	89	40	38	33	29	48
mean length (μm)	32	32	30	32	34	34

preserved the orientation of objects in the outside world. This means that the retina must remain in a constant orientation with respect to gravity.

The existence of a mechanism to ensure this was demonstrated by Wells (1960a) who noted that the slit pupil of the eye remained horizontal, more or less regardless of how the animal happened to be sitting. This orientation was lost if the statocysts were removed (Fig. 7.6). Destruction of the statocysts was followed by a loss of the ability to discriminate between horizontal and vertical rectangles, which could be restored if one was careful to match the orientation of the figures to the orientation of the retina at each trial. The proper functioning of the visual analysing system thus depends upon a mechanism that locks the retina to an artifical horizon (Fig. 8.15, p. 215).

8.1.9 *Absolute extents*

Transfer tests, run after training to discriminate between rectangles of differing dimensions show that such discriminations are made on a basis of extent, not area. Animals trained to discriminate between

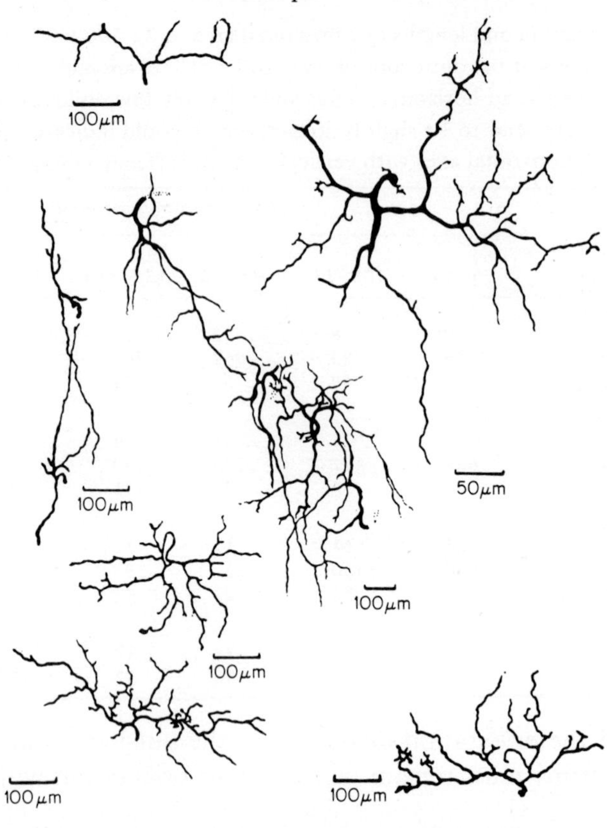

Fig. 8.5 Further details of nerve cell arrangements in the optic lobes. (a) traces the shapes of typical dendritic fields of bipolar cells collecting inputs in the plexiform layer (from Young, 1961a). (b) opposite is a diagrammatic reconstruction of the layout of these tangential fields beneath the surface of the optic lobe. d.v. and d.h. are fields with their long axes vertical and horizontal. m.c.1. and m.c.2., cells in the medulla of the lobe with more than one dendritic field. mult., multipolar cell in the medulla; aff., afferent neurone (from Young, 1960a).

20 × 2 and 10 × 2 rectangles (Table 8.5) would be expected to transfer to 10 × 4 and 20 × 1 rectangles if area were important, and they should recognize the shapes used in training irrespective of their orientation. In fact, they continue to attack rectangles of the correct length, regardless of area, and they are scarcely able to recognize the shapes used in training when these are rotated through 90° (Table 8.7).

Further tests (Sutherland, 1961) in which the proportions of rectangles were equated gave less straightforward results. A 20 × 2 rectangles is longer and proportionately narrower than a 10 × 2. If

(b)

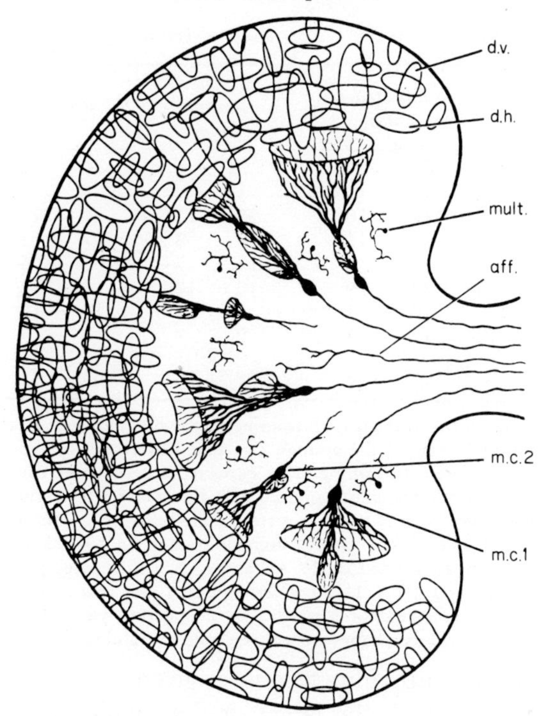

discrimination were based on horizontal and vertical projections, in the manner suggested in Fig. 8.2, 20 × 4 and 10 × 1 rectangles should, respectively, be treated as resembling the 10 × 2 and 20 × 2 shapes used in training, since the training and transfer shapes then yield the same projections. If, on the other hand, it is the absolute horizontal or vertical extent of the figures that matters to the octopus, then the 10 × 1 and 20 × 4 rectangles should be treated as equivalent to the 10 × 2 and 20 × 2 shapes used in training, despite the differences in projections that these shapes yield.

As Table 8.7 shows, the animals appear confused by the substitution. Any tendency to assess shapes in terms of their projections should yield 'perverse' scores when the results are considered in terms of the absolute extent of the figures used in training, as is done in the table. The scores made were very much poorer than those made in training, indeed some animals failed to discriminate at all, but insofar as they indicate any consistent tendency, it is to treat the rectangles in terms of their absolute extents rather than their projections.

Further evidence that octopuses can learn to recognize figures in

Table 8.7 Attacks made on transfer test shapes (from Sutherland, 1961).

Group		*Equated areas*		*Equated propor-tions*		*Rotated*		*Expected scores*	
		+	—	+	—	+	—	+	—
Group I	H	33	10	28	14	14	11	31	2
Days 17–19	V	21	11	19	8	17	13	20	6
Group I	H	24	2	14	7	5	6	22	2
Days 37–39	V	22	9	10	10	6	2	12	2
Group II	H	18	1	9	11	16	12	21	2
Days 37–39	V	12	3	9	9	5	5	14	3
Total		130	36	79	59	63	49	120	17

Transfer tests made with the animals with training scores summarized in Table 8.5. Responses were scored as plus and minus if the animals continued to attack or avoid rectangles of the same *length* as those offered in training. The 'expected scores' show the average number of attacks made per 10 training trials interspersed with the transfer tests; they indicate what the transfer scores might have been if transfer were on the basis of length alone. 'Equated area' rectangles were 20 × 1 and 10 × 4 cm for animals trained on 10 × 2 and 20 × 2 rectangles (and similarly for Group, II, trained on 17 × 2 and 13 × 2). 'Equated proportions' rectangles were 20 × 4 (10 × 2) and 10 × 1 (20 × 2). Equated area and equated proportions tests should have produced perverse scores if area or proportion were the basis on which the shapes were recognized. Rotated shapes were those used in training, shown at 90°.

terms of their absolute dimensions is given by Rhodes (1963). He trained animals to attack the larger or the smaller of two discs shown simultaneously. In transfer tests, pairs of discs were again presented, one of each pair having the same diameter as the positive training shape, the other being larger (when the animal had previously been trained to attack the larger) or smaller (when training had been to attack the smaller of the two). In 5 tests with each of 6 animals, the disc with the correct diameter was attacked 22 times, while the member of the pair with the correct relative size (the larger or the smaller of the two shown together) was taken only 8 times. As Rhodes points out, this was not the result that one would expect from similar experiments with mammals.

8.1.10 *'Open' and 'closed' forms, and reduplicated patterns*

The tendency to attend to absolute dimensions rather than to the proportions of shapes implies that analysis in terms of horizontal and vertical projections is not invariant, or at least not the only manner in which the central nervous system of octopus handles its visual input.

Early experiments with rotated shapes indicated that these were not recognized, as would be expected on the 'projections' hypothesis.

Table 8.8 Results of transfer tests after 400 trials of training to distinguish from $n = 8$ (from Sutherland, 1960c).

Transfer test	Pair of shapes used in transfer tests	No. of trials with each shape	Attacks on original shapes during day's retraining		Attacks on transfer shapes		Transfer index
			+	−	+	−	
1	Original shapes with movement at 90° to originals	70	41	2	43	3	1.0
2	Shapes rotated by 90° from orientation of originals	70	33	1	29	5	0.75
3	Shapes 4/9th area of originals	70	35	4	39	6	1.1
4	Outline shapes	140	62	5	57	39	0.3

Results with rectangles are shown in Table 8.7. Squares and diamonds are not equated; transfer tests indicate that a diamond resembles a triangle rather than a square, as might be expected (Sutherland, 1958b).

But rotated figures do not always pass unrecognized. Each member of the pair *n* and *o* in Fig. 8.8, for example, is clearly recognized as distinct even after rotation through 90° (Table 8.8), while a cross rotated through 45° is attacked just as often as an upright cross after training to distinguish this from a square (Fig. 8.6a; Sutherland, 1960c, 1962b). There must be some additional discriminatory mechanism at work.

Transfer tests indicate that octopuses tend to classify shapes along a dimension that seems to depend upon the relation of outline to area. Fig. 8.6 summarizes the results of two series of transfer tests run after training to discriminate between a square and a cross, or a (different) cross and an H shape of similar area. To us, the figures fall into a series that is perhaps best described as 'open' to 'closed'; an open shape is one that has a large perimeter in relation to its area, or a shape containing several re-entrants. A closed shape has a small outline/area ratio and few or no re-entrants (Sutherland, 1962a, b, 1963b).

The differences between the projections of these shapes would not allow us to predict the order in which *Octopus* ranks them. Sutherland 1962b has done his best to stretch the theory, suggesting that the distinguishing feature of 'open' as opposed to 'closed' forms could be the number and relative magnitude of peaks on the projections. On this view the quite different ranks of horizontal and vertical rectangles

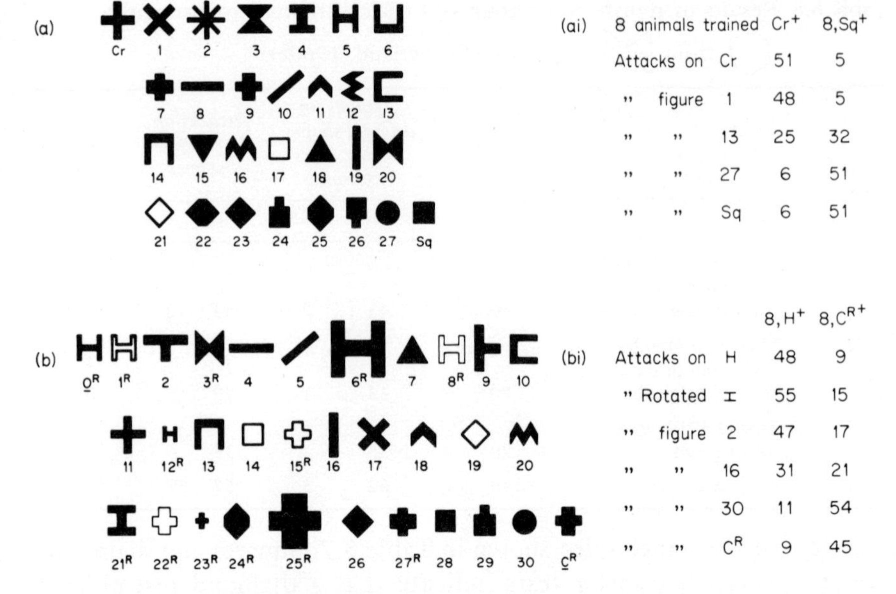

(ai)

8 animals trained			Cr⁺	8,Sq⁺
Attacks on		Cr	51	5
"	figure	1	48	5
"	"	13	25	32
"	"	27	6	51
"	"	Sq	6	51

(bi)

			8,H⁺	8,C^{R+}
Attacks on		H	48	9
" Rotated		I	55	15
"	figure	2	47	17
"	"	16	31	21
"	"	30	11	54
"	"	C^R	9	45

Fig. 8.6 The order in which octopuses rank shapes in transfer tests. Animals were trained to discriminate between a cross and a square in (a) or an H and a different cross in (b). They were then shown each of the other shapes ten times in unrewarded tests interspersed with further training on the original shapes. In (b), a suffix R shows shapes that were presented in various rotations. Tables (ai) and (bi) show examples of the actual number of attacks made in training and transfer (after Sutherland, 1962b, 1963a).

(or ⧗ and ⋈ shapes) is accountable since the animals are known to pay more attention to horizontal than to vertical projections. But it fails to predict the position of ⊔ shapes, which yield sharp peaks of large relative magnitude (and which should therefore be considered very 'open' forms) and it cannot, for example, account for the position of an oblique rectangle, which is clearly a closed form yielding no peaks at all, mid-way along the scale in one series of tests, and towards the 'open' end in the other (Fig. 8.6).

A final blow to the proposition that octopuses analyse visual images only in terms of their horizontal and vertical projections is given by the results of training experiments made with reduplicated patterns. The patterns shown in Fig. 8.7 yield identical projections but are nevertheless readily distinguished; eight octopuses averaged 80% correct responses over the first 160 trials of training to discriminate between these (Sutherland, Mackintosh and Mackintosh, 1963).

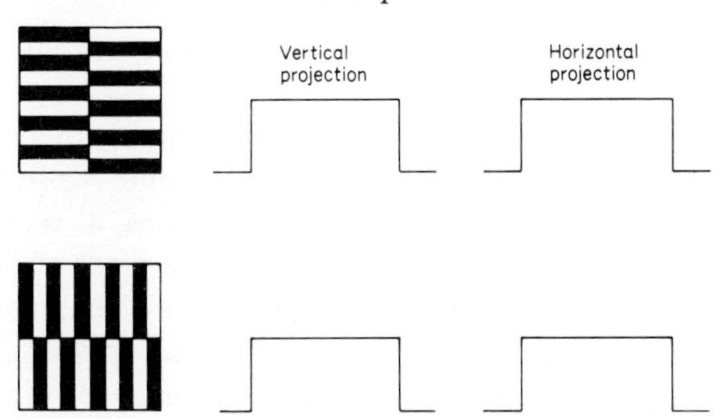

Fig. 8.7 Reduplicated patterns. 8 octopuses averaged 80% correct responses over the first 160 trials of training to distinguish between these. The vertical and horizontal projections of the shapes are identical (after Sutherland, Mackintosh and Mackintosh, 1963).

8.1.11 *Alternative theories*

The demonstration that the recognition of shapes by *Octopus* was likely to be dependent in some manner on the distribution of stimulation along axes that remained horizontal and vertical implied a system different from and simpler than that of mammals which, it seems, must also take information from gravity receptors into account when octopuses assess what they see. The rapidity with which octopuses learn and Sutherland's numerous experiments meant that by 1960 more data on visual discrimination was available for *Octopus* than for any other animal (Deutsch, 1960). This and the relative simplicity of the cephalopod system encouraged the outbreak of a minor rash of theories of cephalopod shape recognition. Besides Sutherland, who proposed at least three systems (1957b, 1960c, 1963b; it depends on how you score successive modifications), Deutsch (1960) and Dodwell (1961) have both evolved models which, like Sutherland's, take for granted the stability of the horizontal and vertical axes. Subsequent work (reviewed below) has shown that none of these theories (except perhaps Sutherland's last, which is not readily tested) can account for all the known facts of visual discrimination in *Octopus.* But each yields some correct predictions and may therefore have some relation to the way(s) in which *Octopus* handles its visual input. The sequence is, in any case, of historical interest as a commentary on the nature of experimental psychology.

Sutherland's (1957b) theory has already been outlined, together with its first additional specification, that horizontal distances (and therefore differences in projection) are more accurately assessed than vertical.

The ancestor of this and the several subsequent theories to be considered was probably Deutch's (1955) scheme to describe visual analysis in vertebrates. Deutsch's model was a rectilinear array, projected from the retina. The assumption was made that only boundaries constituted a sufficient stimulus to evoke discharge in the retinal nerves. The units of the array were interlinked laterally, and also fed into a final common cable (f.c.c.). The occurrence of a boundary in the retinal area represented by any unit caused this to discharge into the f.c.c. and to excite its neighbours. Excitation was then supposed to spread in all directions until the wave encountered a further set of units stimulated by a boundary, whereupon a further signal was discharged down the f.c.c. Any shape is thus translated into a pattern of discharge in time. A horizontal rectangle, for example, causes an initial discharge followed by a pause, then two further discharges, one as the spreading waves of excitation from the long sides strike the opposite sides, then a later, weaker discharge as the discharges from the short sides finally reach the side of the rectangle. A circle causes a single discharge, followed by a further single sharp volley. And so on.

Some of the predictions that would arise from this simple mechanism are that circles and squares, and mirror images will be confused as indeed they are in *Octopus* and a number of other animals. But so should horizontal and vertical rectangles, and this is plainly not the case.

To overcome the insufficiencies of Deutsch's theory, Dodwell (1957a) adopted the linear array/projection hypothesis but supposed that the connexions between units were in two layers. Each layer connected in one direction only, so that the grid resolved itself into two linear arrays, each line of units terminating in an output to a final common cable summarizing the output from the whole parallel array. Dodwell further postulated that the other end of each linear strip was linked to its neighbours, and that as soon as one of these terminal units fired, all fired. In an array consisting of a series of vertical strips, the effect of this would be a descending curtain of excitation running simultaneously down all the strips and out along the f.c.c. The effect of any image (again, boundaries only) superimposed on this would be to alter the transmission characteristics of the grid units affected, such that any stimulated unit passed on excitation more slowly than usual. A vertical boundary superimposed on a vertical strip would result in the long delayed discharge of a single impulse or volley into the f.c.c., while a horizontal, or oblique line would only slightly delay the discharge of many linear strips. The second layer was to have precisely similar properties, but the parallels were arranged horizontally. The output from the system would be a sequence of signals down each of the two final common pathways. A square, for example, would be indicated by a discharge in the undelayed chains peripheral to the figure, followed by a volley signalling the two horizontal lines, followed after a longer interval by signals indicating the vertical sides, the same code from each of the two linear arrays. Other figures would be indicated in the forms shown in Fig. 8.8. In a further account (Dodwell, 1957b), published shortly after Sutherland's first series of experiments (1957b), it was pointed out that Sutherland's scheme merely postponed the problem of shape recognition because it nowhere suggested how the horizontal and vertical projections were handled at what had to be a further stage in analysis.

In 1960 Deutsch produced a fresh 'rectilinear array' scheme, specifically for *Octopus*, following a study of Cajal's (1917) account of the anatomy of the plexiform layer in the optic lobes. The scheme proposed again that the only effective stimulus was a boundary, thus differing fundamentally from Sutherland's which worked by assessing extent. The processes of the cells in the plexiform layer tend to run, as we have seen (Section 8.1.8), horizontally and vertically, at right angles to the incoming projection from the retina. Deutsch postulated that the chance of any one of the plexiform elements firing would depend upon the proximity of active retinal inputs; two boundaries close together would cause more activity than two far apart. He supposed, moreover, that only those elements of the plexiform layer that run vertically behave in this manner, and that vertical boundaries somehow inhibit output altogether. All plexiform units discharge into a final common cable, so that the signal derived from the projection of the image of a shape will be the summated discharges of all the units excited. A maximum discharge would, presumably, result from an array of parallel horizontal lines, a minimal from a similar array of verticals.

Deutsch's system worked well for a number of pairs of shapes tested by Sutherland up to that date. It could account for size to size transfers; doubling the size of a rectangle doubles all the vertical distances, but balances this by doubling the number of units likely to be excited, halving the size has the opposite effect, but the total signal

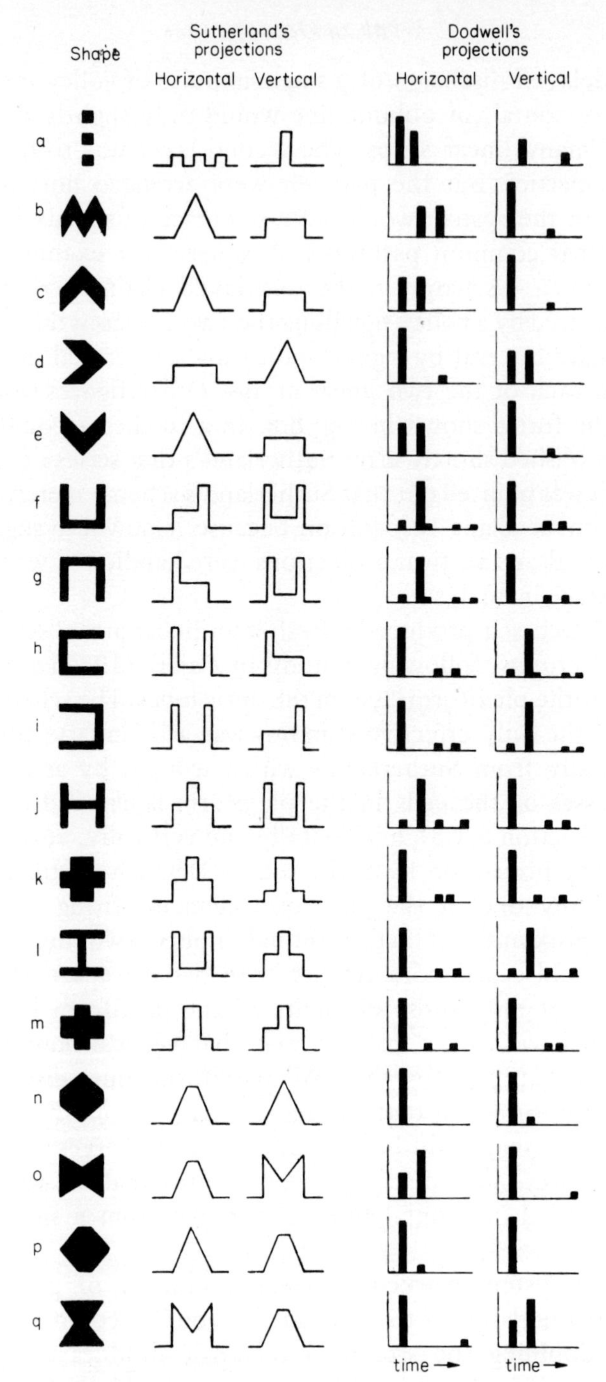

Fig. 8.8 Shapes, and their projections in pathways signalling horizontal and vertical extents (from Dodwell, 1961).

will remain about the same. | is readily distinguished from –, and / is more like – than | because the corners of the oblique rectangle will be placed where there are short vertical distances between boundaries. V and W shapes produce the same output but < and ⋚ (which are more readily distinguishable) do not, since the number of close boundaries in the vertical plane is greater in ⋚ .

There followed a period of argument about the relative merits of the three systems. Sutherland 1960d criticized the other two, and this was followed by pairs of papers by Deutsch and Sutherland (1960) and Dodwell and Sutherland (1961) in which the protagonists demonstrated the virtues of their own proposals, criticized each others' and claimed to have been misunderstood or misrepresented. The discussion revealed that none of the theories could account for all the results obtained even up to 1961.

8.1.12 *A critical test of the rectilinear theories*

In 1970, Muntz returned to an examination of the three sets of hypotheses, using the shapes shown in Fig. 8.9. The shapes used in training, A1 and A2, were of the same area and length of outline. They also yield the same horizontal and vertical projections so that the fact that the animals learned to discriminate between A1 and A2 clearly eliminated both Sutherland's original (1957b) theory and his subsequent outline/area (1962a, b) propositions in relation to open and closed forms (Section 8.1.10).

A1 and A2 would also give identifical outputs if processed according to Deutsch (1960), or Dodwell (1961).

The └┬┐ *vs* ┌┴┐ result (the animals discriminated at 63% correct, t = 5.54 and p = <0.05, with 9 d.f.) is thus incompatible with all but Deutsch's original (1955) theory, which was not specifically tailored to the octopus. Muntz then continued with transfer tests, using the figures B1 and B4 shown in Fig. 8.9. In these tests B1 clearly resembled A1, its left-hand mirror image, but B4 was *not* taken in place of A2, the up-down mirror image, nearly as often as in place of A1. This eliminates the remaining theory, Deutsch 1955.

Muntz's (1970) results also appear to be incompatible with any form of piecemeal analysis based on recognition of the assemblage of components specifying a shape. Sutherland (1963b) finally abandoned his 'projections' model for the visual analysing system and suggested instead that the retinal (or plexiform) units that respond to shapes are

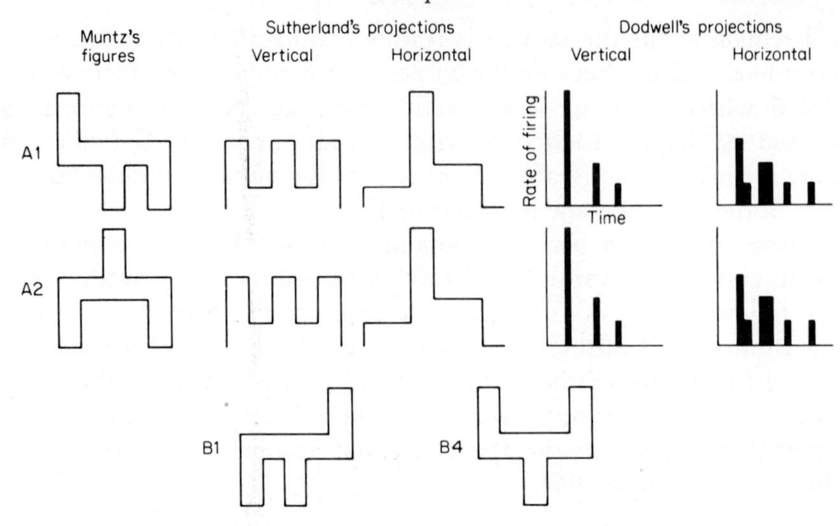

Fig. 8.9 Muntz's (1970) test of rectilinear theories. Shapes A1 and A2 yield the same projections, but are readily distinguished. In plotting Dodwell's projections, it is assumed that elements ▭ and ▯ yield outputs of the same intensity but that ▯ gives a delay slightly less than twice that of ▭ . Deutsch's (1960) hypothesis would predict signals in the form of a single complex shower of impulses indicating the proximity of the boundaries scanned vertically, plus a further signal of some sort (never specified) to indicate the distribution of activity in the vertical plane. No attempt has been made to represent these, but they would presumably be the same for figures A1 and A2. Figures B1 and B4 were used in transfer tests (see text).

fired through dendritic fields that include an excitable strip with an inhibitory surround, in the manner that Hubel and Wiesel (1959) have found to be characteristic of many cells in the visual cortex of cats. Shapes, or components of shapes would be coded in terms of the intensity of output from one or more such units. Considered very generally, such a system would discriminate between open and closed forms and it could explain the rather ready discrimination of reduplicated patterns. But it is difficult to see how it would manage to distinguish between A1, A2 and the transfer shapes in Fig. 8.9.

8.1.13 *Back, almost to the beginning*

None of the models outlined above can explain all of the phemonema of shape recognition in *Octopus* (or any other animal). That does not mean that they are useless, for the theories have yielded a high proportion of successful predictions; more often than not the three rectilinear theories have agreed about what should be easy and what difficult to discriminate. This surely indicates that there is some element of reality in such models; the octopus visual analysing system must be based on a rectilinear up-down, left-right geometry. The facts of the anatomy, and the existence of a mechanism to ensure that the retina remains in a constant orientation with respect to gravity are consistent with this finding from the psychologists 'black box' experiments. This is very convenient, because we are ourselves conditioned to work in terms of rectilinear maps and projections. We find it easier to think about a visual system that is constructed in this way and the happy accident that *Octopus* carries the relevant anatomy in a form that we can recognize is an unexpected boon. It may eventually mean that this is the first complex pattern-recognition system to be understood in any animal. The electrophysiology that will be necessary to check behavioural experiments may be difficult but the structures concerned are at least potentially accessable.

In the meantime it becomes a question of how best to approach the problem. Sutherland (1968) in an essay on pattern recognition in animals and man has outlined twelve capacities that should be present in any satisfactory model of visual pattern recognition. These range from the capacity to recognize shapes regardless of their size, through the capacity to 'fill in' outlines, to such features as a tendency to improve in performance with practice and to reject redundancy in random patterns of dots.

To a zoologist, this sort of global approach would seem to be a gross mistake. It is surely in the last degree unlikely that men, rats, goldfish and octopuses (let alone worms and insects) all process their visual input in the same way. These attempts to evolve a model that will have all the capacities found in man let alone the rest of the animals is an ambitious project which could consume a great deal of time and resources. It could, as Sutherland (1968) plainly intended, become the basis on which to evolve a computer program that might eventually mimic our own capacities. But 'this would still not prove that our brains used the same logic as the computer program. It would, however be an outstanding scientific achievement' (Sutherland 1968). It would seem better to try first to understand the mechanism of one simple system with some, but probably not all of the same properties as our own. The octopus visual analysing system is evidently limited in its complexity by constraints – notably the very restricted use of information from the statocysts, and evident tendency to rectilinear geometry – that make it unusually accessable to analysis. It seems as good a place to start as any.

8.1.14 *Selective attention and switching in the correct analyser*

One of the lessons from the attempt to evolve a universal model for the visual analysing system of *Octopus* is that the animal almost certainly employs more than one analyser. Shapes like and (Fig. 8.2) should have been indistinguishable according to Sutherland's original theory and he was obliged to postulate (1959a) that the animals were able to distinguish between figures moving in the direction of their points, and that some other factor (outline/area, or the acuity of the points) was also taken into account. The extensive series of transfer tests indicating a tendency to range shapes along the 'open to closed form' dimension and the ease with which octopuses identify reduplicated patterns all suggested that the animals may employ different analysers for different problems. Indeed, individuals may choose to solve a discriminatory problem in different ways, depending upon the context and their own individual previous experience.

Tests of these views have been made. Sutherland, Mackintosh and Mackintosh (1963) trained two groups of octopuses to distinguish between a square and a parallelogram that differed only slightly from the square. One group was presented with this difficult discrimination from the start, the other began with an acute-angled parallelogram, quite unlike the square, and moved towards the difficult problem by easy stages. The results of the two are summarized in Fig. 8.10. The group that approached the difficult discrimination through training on progressively squarer parallelograms made this discrimination much more reliably after the same number of trials. *Ex hypothesis*, this group had learned to employ the 'correct' analyser consistently, whereas the group trained on the difficult pair of shapes were still experimenting with alternative means of solving the problem; since the shapes are in any case so readily confused, these animals would have got the wrong answer from time to time even when using the correct analyser, a state of affairs unlikely to lead to its consistent employment.

In a further series of training experiments, animals were pre-trained to make a size or a shape discrimination, and then moved on to a second discrimination in which both sets of cue were available. Transfer tests at the end of training revealed that the animals trained to classify shapes in terms of size learned significantly less about the shape of the figures to be distinguished than their shape-trained controls

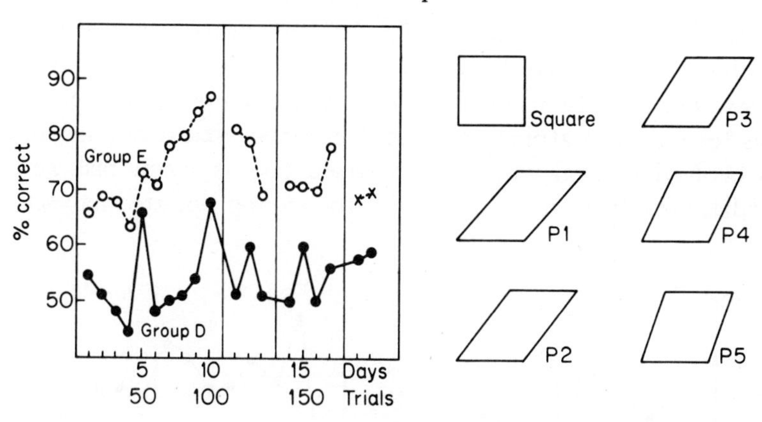

Fig. 8.10 Group E (n = 8) was trained to discriminate between the square and P1 for 10 days then given one day each with P2, P3 and P4 (phase 2 of the experiment) before going on to square vs P5 in phase 3. Group D (n = 8) was trained on the more difficult square vs P5 throughout. In phase 4, D was given unrewarded transfer tests (scored x – x) with P1. In these experiments, the shapes to be discriminated were shown simultaneously and unmoving; the animals had some pretraining to attack (after Mackintosh, 1965 and Sutherland, Mackintosh and Mackintosh, 1963).

(Sutherland, Mackintosh and Mackintosh, 1965). In a further experiment it was found that prolonged training on an orientation discrimination actually hindered subsequent learning of a shape discrimination (Mackintosh and Mackintosh, 1964b).

Where there are two or more possible means of distinguishing between shapes even octopuses without prior training may learn different things about them. Messenger and Sanders (1972) trained animals in a situation where both the brightness and the orientation of rectangles were relevant. Animals with both cues available learned more rapidly than octopuses with only one or the other. Transfer tests run at the end of training showed that of the 29 subjects trained in the two-cue situation, 22 had learned to pay particular attention to brightness while 6 were responding mainly to orientation – the remaining animal gave no clear indication of the basis on which it was making the discrimination. Presumably these individual differences reflect individual experiences in the sea before capture.

8.1.15 *Reversal of training*

A further situation which ought to reveal a tendency to persist with the use of a particular analyser occurs when training is reversed.

Reversal, where the animal is obliged to switch its responses between two objects that it has learned to discriminate, is often difficult for animals. Presumably it rarely becomes necessary in nature, so that one is testing the learning machinery in a task that it can hardly have been evolved to deal with. The reversed task is a new task and one immediate result must be the search for an appropriate analyser. The

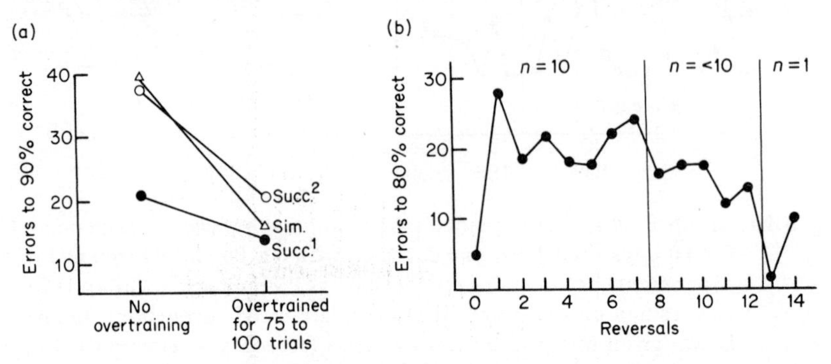

Fig. 8.11 (a) The effects of overtraining on subsequent reversal of a learned discrimination between black and white rectangles. The rectangles were presented simultaneously (Sim.) or successively (Succ.[1]) or successively with irrelevant changes in orientation (Succ.[2]) $n = 4$ for each point plotted. Each animal was trained until it made 18 out of 20 correct responses in a day of 20 trials (data from Mackintosh and Mackintosh, 1963). (b) Scores made in successive reversals of learned responses to black and white rectangles shown simultaneously with orientation as an irrevelant cue. Each animal was trained until it made 8 out of 10 correct responses in a day of 10 trials (after Mackintosh and Mackintosh 1964a).

'correct' strategy is to persist with the analyser already in use and the probability of the animal doing so will presumably depend on the amount of training that it has already had. Overtaining – training after attainment of some specified criterion of correct response – should and does make reversal easier (Fig. 8.11(a)). Repeated reversals should become progressively easier. The evidence here is less certain. The first experiments made (Young 1962c; Mackintosh, 1962) indicated that octopuses show no progressive improvement when tested in this way, indeed that their performance becomes progresssively worse. With a change to simultaneous rather than successive presentation of the figures to be distinguished, the situation seems to alter, and some progressive improvement is found (Fig. 8.11b, Mackintosh, 1965; Mackintosh and Mackintosh, 1963, 1964a). It is arguable that progressive improvement is most likely to be shown if the situation is complex enough to admit the possibility of the animals' selecting inappropriate analysers in the first instance, while simultaneous rather than successive presentation minimizes the tendency for the animals to give up responding as a result of repeated shocks.

8.1.16 *Colour vision*

Octopus is a colourful animal, capable of rapid colour change. It succeeds in matching its background and it employs colour changes in a range of displays directed at its own kind and at other animals. *Prima facie* one would expect octopuses to be able to distinguish between colours.

The evidence available nearly all suggests that they cannot. Experiments have included discrimination learning, electroretinogram studies, and an examination of the optomotor responses of animals in a striped drum; a list of these studies is given in Table 8.9. Most, including all those reporting positive results, were inadequately controlled for light intensity so that the results are explicable without having to invoke wavelength discrimination.

Table 8.9 History of research into colour vision in cephalopods (after Messenger, Wilson and Hedge, 1973).

Date	*Author*	*Species*	*Method*	*Conclusion*
1910	Hess	*Sepia officinalis*	Pupillary response	Colour-blind
1914	Fröhlich	*Eledone moschata* *Octopus vulgaris* *O. macropus*	ERG isolated eyes	Colour vision
1914	Piéron	*O. vulgaris*	Learning experiment	Colour-blind
1917a, c	Goldsmith	*O. vulgaris*	Learning experiment	Colour vision
1921	Mikhailoff	*Eledone moschata*	Learning experiment	Colour vision
1926	Bierens de Haan	*Octopus vulgaris*	Learning experiment	Colour-blind
1950	Kühn	*O. vulgaris* *Sepia officinalis*	Learning experiment	Colour vision
1961 1962	Orlov & Byzov "	*Octopus dofleini* *Ommastrephes sloanei-pacifcus*	ERG isolated retina	Colour-blind
1968a, b	Hamasaki	*Octopus vulgaris* *O. briareus*	ERG intact animal	Colour-blind

Octopus vulgaris carries a rhodopsin with an absorption spectrum surprisingly like that of man (Fig. 8.12). Messenger, Wilson and Hedge (1973) took advantage of this similarity to prepare rectangles painted red, green or blue, matched for subjective brightness when

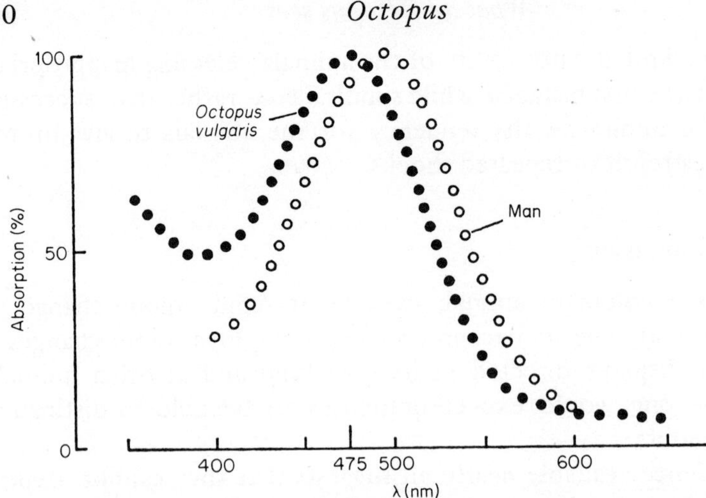

Fig. 8.12 Absorption spectra for *Octopus vulgaris* and human rhodopsins. The λ_{max} lies further over into the blue in octopus, as is usual in marine animals, but the shape of the curve is very similar to man's. The apparent increase in sensitivity at very short wavelengths is almost certainly not biologically significant (from Messenger, Wilson and Hedge, 1973).

seen through red, green or blue filters. They then tried to train octopuses to discriminate between the coloured shapes and contrasted their failures in colour discrimination with their successes in discriminating between shapes of the same colour but differing brightness (Fig. 8.13). In a further series of experiments animals were shown rectangles differing in both brightness and colour, but failed to learn to distinguish between them any more rapidly than when offered brightness cues alone – a state of affairs that can be contrasted with the results of Messenger and Sanders (1972) experiments, in which octopuses learned to distinguish between rectangles differing in brightness and orientation more rapidly than between rectangles differing only in one or another of these respects.

An alternative approach to the colour vision problem is to test for optomotor responses. Octopuses in a striped drum make clear compensatory movements of their eyes and body when the drum rotates. If coloured stripes are alternated with greys or other colours matched for brightness, optomotor responses should cease if the animals are incapable of wavelength discrimination. Messenger *et al.* (1973, *loc cit*) found that they could turn nystagmus on and off at will by changing the stripes in a continuously rotating drum; with brightness-matched coloured stripes, eye movements ceased, to return as soon as underlying black and white stripes were revealed.

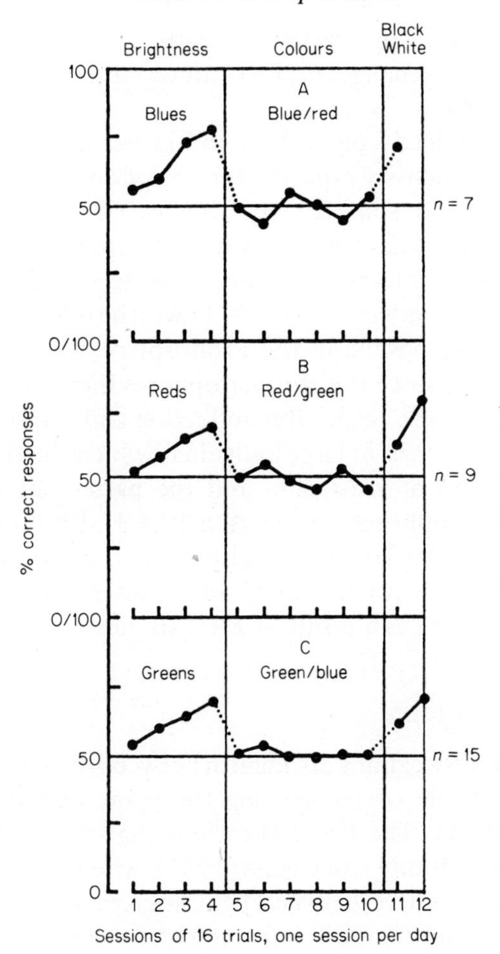

Fig. 8.13 Attempts to train octopuses to distinguish between coloured vertical rectangles, matched for brightness, failed, while the same animals previously and subsequently learned to discriminate between similar rectangles of different brightness (from Messenger, Wilson and Hedge, 1973).

Studies of the *Octopus* retinogram (Hamasaki, 1968a, b; Orlov and Byzov, 1961, 1962 – see Messenger *et al.* 1973) confirm these behavioural findings. Hamasaki found no evidence of the discontinuity in the dark adaptation curves for *Octopus briareus* or *O. vulgaris,* that one would expect to find in an animal having more than one type of retinal receptor. Orlov and Byzov (1961, 1962) found no change in the ERG of *O. dofleini* when they altered the wavelength of the light used, provided that the successive lights were matched for brightness. In fact, the animal also appeared to be rather insensitive to changes in brightness, with no change in ERG detectable with light intensity changes of up to 8%.

If *Octopus* is indeed colourblind and rather insensitive to changes in light intensity, then its capacity for background matching is all the more remarkable. To some extent the match obtained is achieved passively. Small animals are somewhat transparent and even larger ones owe some of their hues to iridophores and leucophores (Messenger, 1974; Packard and Sanders, 1969, 1971) which reflect all wavelengths. The greenish colours visible in the Frontispiece, for example, must be attributable to this, since the chromatophores themselves are all in the yellow-red-brown-black range. But reflection and transparency cannot wholly explain the crypsis of large individuals which clearly depends very largely on the state of expansion and the pattern of the nervously-controlled chromatophores (see Section 10.3.3). We do not know how the animals manage to do it; control of tone is clearly important, but appears inconsistent with the comparative insensitivity to changes in light intensity claimed as a result of ERG studies.

8.1.17 *Polarization plane*

Octopuses apparently cannot distinguish between different wavelengths. But they are capable of recognizing the plane of vibration of light falling on the retina. The first indications that this might be so were given in a paper by Moody and Parriss (1960), who succeeded in training octopuses to discriminate between light sources seen through discs of polaroid arranged to transmit light at right angles. A more detailed analysis of their results was given in Moody and Parriss (1961); Fig. 8.14 shows the scores of the animals concerned in this study. The octopuses were able to distinguish between screens oriented at 0° and 90°, and, less effectively (Fig. 8.14), between the two oblique orientations 45° and 135°; 0° *vs* 90° proved to be quite an easy discrimination.

Other animals manage to distinguish differences based on polarization plane in two ways; resolution is either a property of their receptors, or a property of the environment. Polarized light will produce reflections from particles in the air or water that will be most obvious at right angles to the electric vector of the light source. There should be corresponding differences in the amount of reflection from the walls and water surface in the tanks in which octopuses are kept. Alternatively, and in this instance more probably, discrimination of

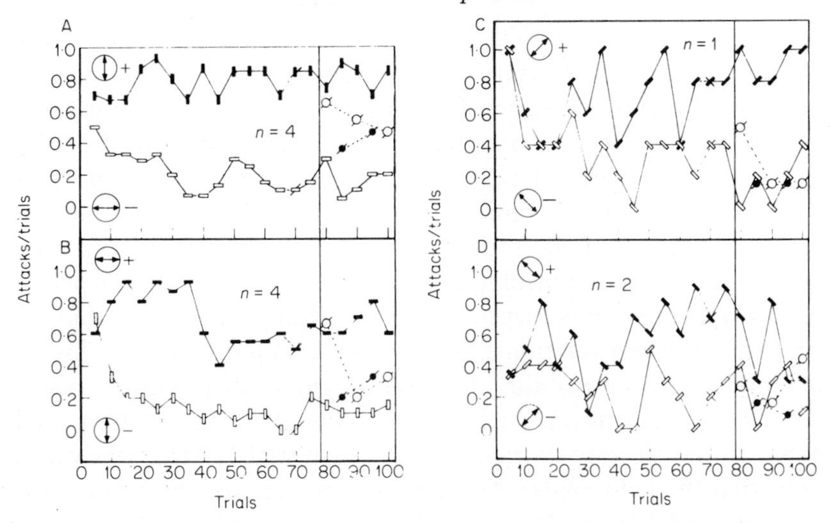

Fig. 8.14 Shows the mean number of attacks per trial for each of four groups of animals, A, B, C and D, discriminating between different positions of the electric vector of plane polarised light. Open shapes indicate proportions of attacks at the vector for which attacks were discouraged by shocking the animals; filled shapes indicate proportions of attacks at the vector for which attacks were rewarded. The angles of the vectors being discriminated are represented by the orientation of the shapes; 0°, 90°, 45° and 135° respectively. The oblique line through the shapes at 70 trials indicated that these attacks were neither rewarded nor shocked. ⌀ and ● indicate mean attacks for unrewarded trials with unpolarised light and no light respectively (from Moody and Parriss, 1961).

polarization plane may be truly intraocular, a property of the retinal elements themselves.

Moody and Parriss (1961) argued that reflection patterns were unlikely to be the cue used by octopuses on two grounds. First, that such differences were scarcely visible to themselves, looking through a glass panel in an aquarium similar to those used in training, even in a darkened room; the octopuses, of course, were trained in lighted tanks, which would make detection of any patterns more difficult. Secondly, it was argued that the capacity to distinguish between light plane-polarized at 45° and 135° could not be based on reflection patterns, because these would be oblique mirror images and octopuses were known from Sutherland's (1957b) work to be unable to distinguish figures differing in this way.

An alternative approach to the reflection pattern problem was adopted by Rowell and Wells (1961). When the statocysts of *Octopus* are removed, retinal orientation is lost. The animals are subsequently unable to distinguish between figures that differ in orientation, unless these are matched to the orientation of the retina at each trial (Wells, 1960a). A learned discrimination between sources of plane-polarized light can be maintained similarly by matching the plane of vibration to the orientation of the retina (Fig. 8.15). Since reflection patterns from the tank surfaces will then vary from trial to trial, depending upon how the octopus happens to be sitting, they are unlikely to bear any consistent relationship to reward and punishment. This does not eliminate the possibility that each light source causes a distinct halo of reflection from particles in the water, since this pattern will be rotated to match retinal orientation at each trial. All that one can say here is that the halo explanation appears unlikely for the reasons given by Moody and Parriss (1961, above). In the Rowell and Wells experiments, very small (10–30 g) octopuses were used, with the light source presented close enough for the animals to reach out and touch it without altering the position of the head and eyes. The tanks had to be opened quite widely in order to observe the animals. The rather short light path between octopus and polaroid, and the presence of relatively intense background lighting would combine to make an explanation based on haloes even more improbable than in Moody and Parriss' experiments with 200 g octopuses in metre-long aquaria.

Moody and Parriss' study included an examination of the fine structure of the retinal elements and their arrangement in the retina. This work has been reviewed in Chapter 7. It seems probable that polarization-plane discrimination depends upon the orientation of the rhodopsin molecules perpendicular to the long axes of the tubules of the rhabdomeres. Absorption would then be about twice as great when the e-vector lay parallel to the tubules than when at right angles to this. Since the rhabdomeres are orientated mainly horizontally and vertically, polarization in these planes should be readily distinguished, provided that the distinction between the orientations is preserved in the CNS. The system must in fact be more complex than this, since oblique orientations are distinguished, though less readily; Moody and Parriss' discuss some theoretical possibilities.

The main puzzle about polarized light discrimination is that *Octopus* is capable of making this sort of distinction at all. What is it for? Differences – indeed the fact that light is polarized – will only be detectable to the animals if the visual system is wired to preserve a

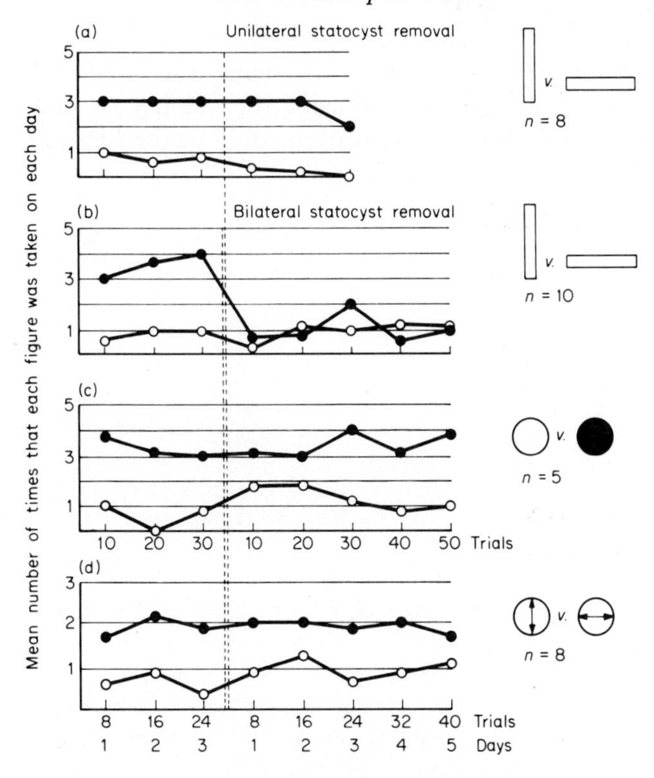

Fig. 8.15 Statocyst removal and visual discrimination. ● shows the number of attacks on the +ve situation and ○ the number of attacks in the −ve in each day of 8 or 10 trials (4+, 4− or 5+, 5−). In each instance only the results of the last 3 days of preoperational training are shown. Unilateral statocyst removal does not upset visual discrimination of orientation but bilateral removeal does, while not disrupting a black-white discrimination. In (d) discrimination of the plane of polarization has been maintained after bilateral statocyst removal by matching the plane of the polaroid to the orientation of the retina at each trial (a–c) after Wells, 1960; (d) after Rowell and Wells, 1961).

separation between the input from classes of rhabdomere that differ in orientation. It seems too much to believe that the separation has been retained as a matter of chance; it must confer some selective advantage. Light in the sea is polarized to an extent that varies considerably with turbidity, cloud cover and the nature and proximity of the bottom. Close to the bottom, where *Octopus* lives in 10–40 m, about 12% of the light falling into the animals' eyes at 500 nm will be linearly polarized (Ivanoff and Waterman, 1958). The capacity to distinguish between surfaces that reflect this oriented component to a greater or lesser extent might go some way to substitute for the inability to distinguish between wavelengths. Alternatively, or additionally, the animals might be using their ability to find their way about, since the degree of polarization could be used to determine the direction of the sun or moon.

CHAPTER NINE

Touch and the role of proprioception in learning

9.1 Tactile discrimination

9.1.1 *Feeling and learning*

The layman's octopus is an animal that consists of arms. In amongst the arms, somewhere in the middle, is a head with almost human eyes; children and cartoonists usually forget about the body. In a way the laymen, children and cartoonists are right; the arms make the octopus. They comprise the greater part of the weight of the body and they contain most of the nervous system. Young (1963a) estimates that the nerve cords of the eight arms together carry some 3.5×10^8 neurones, while the brain itself is composed of less than half of this number of nerve cells. Each arm is equipped with prehensile suckers, 200 or more in an animal of 500 g, arranged in two alternating rows.

The structure of the sense organs in the suckers, and the little that is yet known about the electrophysiology of the arm nerve cords, have already been discussed in Chapter 7.

Octopus uses its arms to move about, collect food, defend itself, and examine its surroundings. The animal is clearly able to discriminate between objects that it touches and, as we shall see below, it can learn to distinguish between objects by touch as readily as it can learn to distinguish between figures by sight. This has made it possible to investigate the tactile world of the octopus in much the same way as the animal's visual world, discussed already in Chapter 8. The tactile world in which *Octopus* lives is particularly interesting because while many, perhaps most, invertebrate animals depend overwhelmingly on chemotactile senses, very few learn readily enough under laboratory conditions for a systematic analysis of their tactile capabilities to be made. The octopus touch world is strangely different from our own; the animal can taste by touch, but it cannot distinguish shapes, weights and sizes by touch. Texture is recognized, but along a one-dimensional scale of irregularity in which pattern plays no part. It is arguable that what is true of the octopus must also be true of other soft-bodied animals, so that a study of *Octopus* gives insight into the perceptual world of a much wider range of organisms.

9.1.2 *Training to discriminate*

An intact octopus will reach out to grab any small object dropped near it as it sits in its tank. The object is grasped with the suckers and generally passed under the interbrachial web towards the mouth. Some minutes later it will be rejected if inedible, passed out from under the web by the suckers, which hand it on from one to another. Finally the arm may extend to thrust the object away to a distance.

The decision to take or reject an object touched can be made without passing it under the web. The suckers grasp and examine the object by moving over its surface and the octopus takes or rejects it, extending or contracting the arm and passing the object along the rows of suckers. A familiar object may be dropped after a cursory examination by one or two suckers only, the arm that has extended to take it simply withdrawing without moving the thing it has examined. If a new object is now introduced it will be taken, but soon rejected if it resembles the

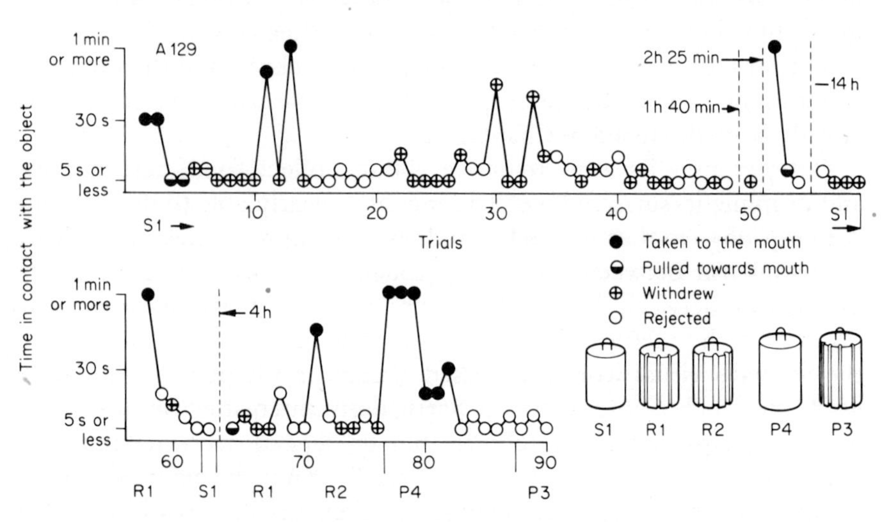

Fig. 9.1 Responses of an unblinded octopus to objects repeatedly presented by dropping them close to the animal, which would reach out to examine each cylinder with its arms. Trials were at 2 min intervals except where otherwise specified, the second begun with a change of object 2 min after the end of the first (from Wells and Wells, 1956).

first at all closely; the animals both discriminate and generalize (Fig. 9.1).

An octopus will continue to take an inedible object if it is rewarded for doing so by giving it a crab or a small piece of fish when the object is grasped and passed out of sight beneath the web. Regular rejection of a second object, to be distinguished from the first, can be accelerated by giving the octopus a small electric shock if it passes this 'negative' object beneath the web. The result of a typical experiment is shown in Fig. 9.2; under these conditions normal animals generally begin by

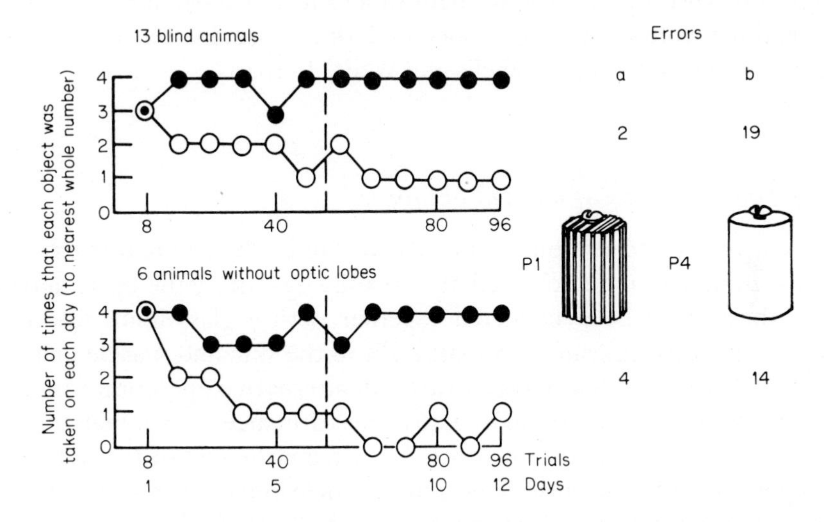

Fig. 9.2 Results of training to discriminate between the objects P1 and P4 (either as P1+/P4− or as P4+/P1−). The average performance of thirteen blind octopuses compared with that of six animals lacking the optic lobes. Eight trials per day (4+, 4−). ●, positive object; ○, negative object. A broken line indicates a break in training of 36 h. Errors: *a* = mean number of errors due to rejecting the positive object, and *b* = mean number of errors due to accepting the negative object (from Wells and Wells, 1957b).

taking most or all of the test objects presented to them, so that most errors in discrimination arise by taking the 'negative'.

Since it is desirable to recover the objects as soon as possible after delivery of the reward or punishment, and at all events before the beginning of the next trial, they are usually presented suspended on a nylon line. This can be let slack as soon as the octopus has grasped the object, and later used to jerk the object away after the food or shock had been given. Withdrawal is easy provided the octopus has no forewarning that the jerk is coming; if the experimenter hesitates, or tightens the line first, the animal will grip the object too tightly for it to be recovered without a struggle. As training proceeds, one gets the impression that the octopus actually learns to release the inedible object as soon as the food has been received; but this would be difficult to quantify and no specific studies have been made of the matter. The animals certainly *do* learn to reject the 'negative' object, snatching their arms away after examination with the suckers, or thrusting it away to arms length; only occasional individuals tend to remain grasping the object, neither taking nor rejecting it, and these can readily be taught not to do so by giving them shocks if they have not succeeded in breaking contact within a specified time – usually ten or twenty seconds from first contact (Wells and Wells, 1956).

9.1.3 *The behaviour of blinded octopuses*

It is convenient to use blind animals for tactile discrimination experiments. Blinding can be achieved most readily by cutting the optic nerves after these have been collected together with a glass hook inserted into the postorbital sinus through a slit in the skin and muscle behind the eye. The nerves do not regenerate. An alternative operation (carried out on some of the animals whose performance is summarized in Fig. 9.2) is to disconnect the optic lobes from the rest of the brain by cuts peripheral to the optic gland and peduncle lobe; rather surprisingly, since they form such a large part of the bulk of the brain, the optic lobes seem to play no part in touch learning (Wells and Wells, 1957b). Both optic nerve section and optic lobe removal are quite serious operations and these animals take longer to resume feeding than after most more central brain lesions; generally, it is necessary to keep the animals for several days after blinding before the beginning of training. During this time they at first sit huddled up with the arms curled tightly around them. If disturbed they may roll over to present the suckered underside, as an octopus does if poked in its hole in the sea. Within a day or so the blinded animals begin to feed regularly, accepting small pieces of fish or crabs touched against the arms. Within a week they will normally be found sitting on the walls or floor of their tanks, with the arms outstretched (Plate 2, opposite p. 141). They now move towards any source of disturbance and will once again grasp and pass small objects towards the mouth.

The outstretched posture is convenient. It means that the experimenter can usually select whichever part of whichever arm he chooses in presenting the test objects, and this is useful in relation to transfer and brain lesion experiments. The arms are normally held sucker side down, against the tank surface and this again is useful. The test objects can be lowered on their lines and touched gently against the upper surface of the arm selected, which will twist to grip with the suckers. Under good conditions this makes for a very standardardized presentation, in which the octopus and not the experimenter can determine how contacts between the suckers and the test surface are achieved. In less happy circumstances, where posture has been affected by brain lesions, or the animal has become withdrawn as a result of making large numbers of errors and receiving many shocks in the course of training, it is not so easy. Considerable practice may then be required to recognize which arm is which in the tangle, and some delicacy may be needed in presenting the test objects since the suckers are very sensitive and the octopus may shy away from contacts. In these circumstances it is clearly very desirable that the people actually doing the training are as far as possible unaware of the 'expected' result or of the nature of the lesions in the animals, since it is plainly possible to influence the probability of a 'take' by being gentle or otherwise in the presentation of the test objects. As with visual discrimination training, these problems tend to disappear when one is dealing with confident animals, engaged in easy or well-learned discriminations. Such octopuses will continue to discriminate reliably more or less regardless of the ham-handedness of the experimenter, just as experienced animals will attack or avoid visual figures, more or less regardless of the manner in which these are moved about (see Section 8.1.3). In the interpretation of experiments with animals that are being trained to make difficult tactile discriminations, it is important to watch the overall level of takes. If (as is usual) the octopus make their mistakes mainly by taking the 'negative' objects rather than by rejecting the 'positives', one can be reasonably sure that there is little likelihood of the result being influenced by the experimenter. A 50% take level is plainly optimal, if the two objects to be distinguished are presented equally often. But anything much below this is suspect. It probably means that the animals were sitting with their arms withdrawn, scared or sick, and it is precisely under these conditions that any errors or irregularities in presentation could influence the results.

9.1.4 *Taste by touch*

In Section 7.3 it was pointed out that the discs and rims of the suckers are riddled with presumed chemoreceptors. The presence of abundant contact chemoreceptors on the arms would be expected from the behaviour of the animals; pieces of fish and other sources of diffusing material placed close to the under surface of the arm cause individual suckers to expand and reach out in the general direction of the stimulus source (Giersberg, 1926). If food is placed near the back of the arm, this will twist to present its suckered surface, apparently as soon as the diffusing materials reach the rim of the suckers.

Systematic studies of the chemical sense can be made using the usual reward and punishment technique: the only difficulty is regulating the stimuli. Test objects covered in a spongy material can be soaked in solutions to be distinguished and presented by suspending them close to the arms of blinded octopuses, which grasp, take and reject them like other small moveable objects. Discrimination between tastes that we would classify as sweet (sucrose), sour (hydrochloric acid), and bitter (quinine) can be established within forty or fifty trials, while the minimal concentrations detectable (in seawater) seem to be 10 to 1000 times more dilute than those detectable (in distilled water) by ourselves. The animals are, in addition, able to distinguish between normal seawater, and seawater that is more concentrated or dilute than usual (Wells, 1963b).

In a further series of experiments, an attempt was made to discover the minimum differences that octopuses are able to distinguish. The animals readily learn to discriminate between seawater and seawater with M/10 potassium chloride added. Having learned this the additional KCl can be reduced progressively; $M/10^5$ KCl was still reliably detectable and there were indications that some individuals could do yet better. When KCl was added to both of the solutions to be distinguished, discrimination broke down when the ratio of the two concentrations was about 1:1.3 (Fig. 9.3).

Experiments with acids showed that these were recognized on the basis of their pH with acetic acid apparently tasting noticeably more acidic than hydrochloric or sulphuric acids at the same pH. As with the 'just noticeable difference' experiments, these results closely parallel those that have been obtained in experiments on chemoreception with mammals and insects (Wells, Freeman and Ashburner, 1965).

All of these results depend on the assumption that the concentration of the substance to be distinguished is the same when the octopus grasps

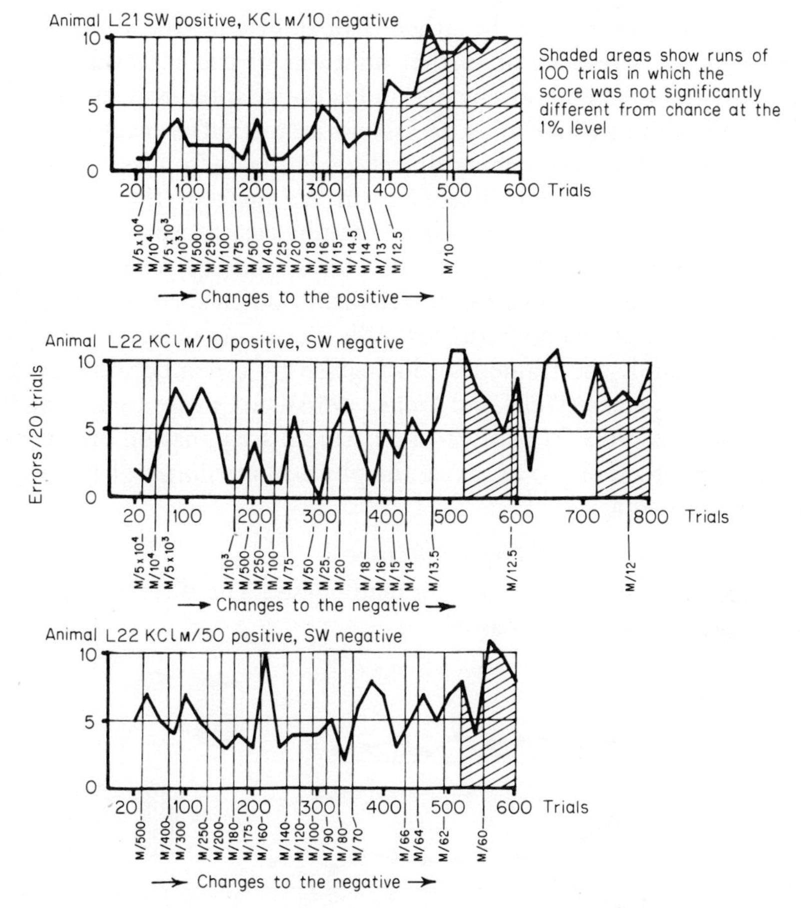

Fig. 9.3 Chemotactile discrimination. The animal can readily be trained to distinguish between spongy objects soaked in seawater and seawater with M/10 or M/50 potassium chloride added. If KCl is then added to the object in plain seawater, discrimination breaks down as the ratio of the two concentrations approaches 1:1.3 (from Wells, Freeman and Ashburner, 1965).

the spongy solution-laden object as when the object is lifted from the stock solution in which it has remained between trials. In practice, of course, the concentration in contact with the sucker chemoreceptors will only be close to that of the stock if the octopus grasps and squeezes the test object; under all other conditions it must be less. Add to this the complication that material must diffuse out from the objects continuously, so that suckers approaching contact will move up a concentration gradient as they approach; we know nothing about the rapidity with which the chemoreceptors adapt, but their capacity to register an event cannot be other than dulled by the obligation to operate in the presence of further dilutions of the chemical they are supposed to be detecting. The method is not adapted to yield accurate information about thresholds, or just noticeable differences in concentration. What the experiments do tell us is that octopuses are able to detect exceedingly small differences in the taste of the objects that they handle. In experiments where the physical attributes of objects are to be distinguished one must be correspondingly careful to avoid contamination of the positive object with food given as the reward for taking it. In experiments with naturally occurring objects – octopuses can, for example, be trained to distinguish by touch between bivalves of different species (Wells and Wells, 1956) – it is wise to assume that the octopus is quite as likely to be basing discrimination on chemical stimuli, undetectable to us, as on the physical attributes of the objects which we present.

9.1.5 *Discrimination of surface texture*

Surface chemistry is difficult to manipulate. The physical properties of objects are easier to measure.

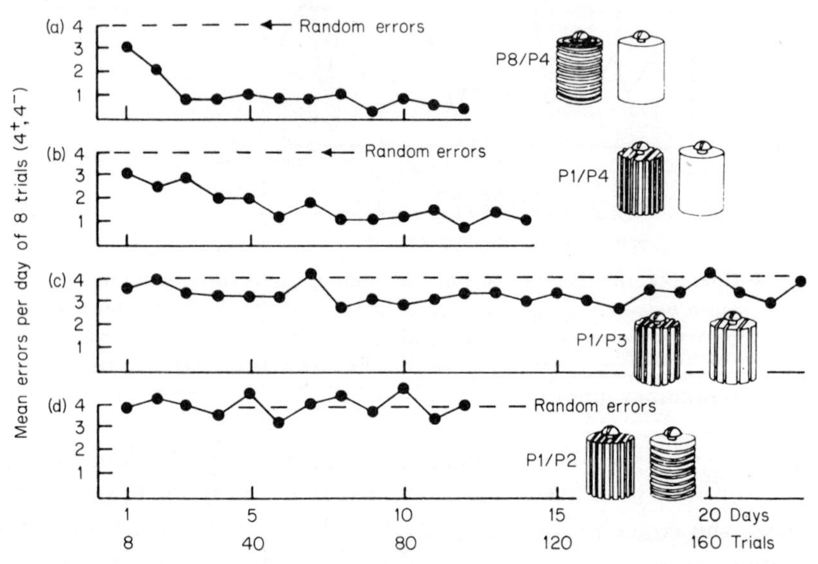

Fig. 9.4 A summary of the errors made in the course of four series of training experiments in each of which six (or twelve in the case of P8/P4) octopuses were trained to distinguish between two Perspex cylinders. The four series are arranged in order of difficulty for *Octopus*, P8/P4 being the easiest discrimination for them and P1/P2 the most difficult (from Wells and Wells, 1957a).

In the earliest attempts to survey the sorts of physical difference that *Octopus* was able to detect the 'Perspex' cylinders illustrated in Fig. 9.5 were used. Octopuses were trained to distinguish between pairs of these, and the relative difficulty of the discriminations assessed from the scores made in a 40-trial period after training for 8 days at 8 trials per day; some typical results are shown in Fig. 9.4. It was immediately obvious that *Octopus* does not classify these objects in the same way as we do. P1 and P2, with the grooves running along the length and around the circumference of the cylinders, are very different to us, apparently identical so far as an octopus is concerned. P6 had a square pattern of grooves that the octopuses found about as difficult to distinguish from P1 as P3, with longitudinal grooves alone; P1/P3 is by far the more difficult discrimination so far as we are concerned.

The proportion of errors made in discrimination correlates with the difference in proportion of grooved surface between the cylinders to be distinguished (Fig. 9.5). The dimension along which the octopus is arranging the objects is evidently related to roughness of texture, with the pattern and/or orientation of the grooves on the cylinders either undetectable or treated as irrelevant by the animal. The simplest hypothesis is that what matters to the octopus is the degree of distortion imposed on the suckers in contact with the object. At a point halfway along the arm, the expanded suckers of an octopus of 300 g will each span four or five of the grooves on cylinder P1. The rims of the suckers can be seen to bend into the grooves, so that the rim will be distorted in several places. The suckers are known to have mechanoreceptors (Section 7.3) and it is presumably these that would generate signals the magnitude of which might be the measure that *Octopus* remembers as characterizing the object (Wells and Wells, 1957a). Tests for stimulus generalization, run after training to distinguish between spheres of different texture, show that objects intermediate in roughness between those used in training are taken or rejected with frequencies predictable from their degree of irregularity (Fig. 9.6a). When the objects used were very similar in roughness, transfer tests showed maximum response rates to objects more extreme than those used in training (Fig. 9.6b). These results have been interpreted as suggesting that the system responsible for recognition of specific degrees of roughness consists of a series of mutually inhibitory units connected to form a linear array (Wells and Young, 1970b). The location of this array is discussed in Section 11.3.5.

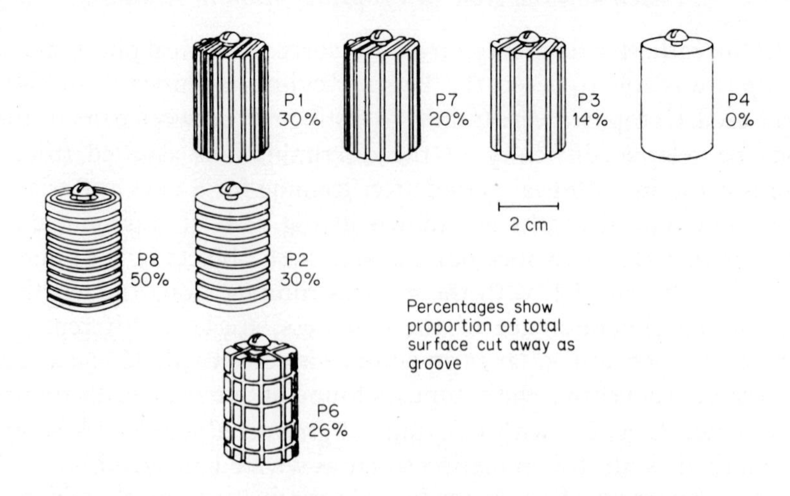
P1
30%
P7
20%
P3
14%
P4
0%
2 cm
P8
50%
P2
30%
Percentages show proportion of total surface cut away as groove
P6
26%

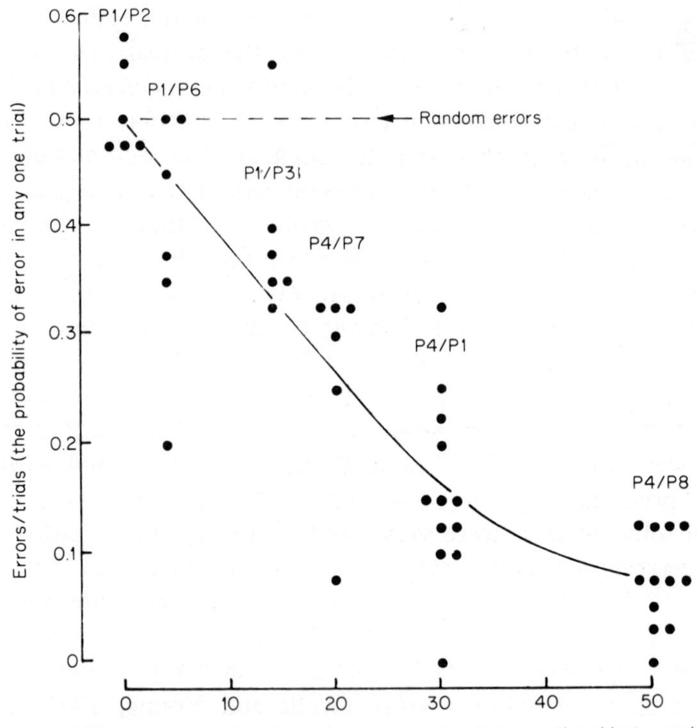
P1/P2
P1/P6
Random errors
P1/P3
P4/P7
P4/P1
P4/P8
Errors/trials (the probability of error in any one trial)
0
0.1
0.2
0.3
0.4
0.5
0.6
0
10
20
30
40
50
Difference in the percentage of grooved surface between the objects used in the experiment

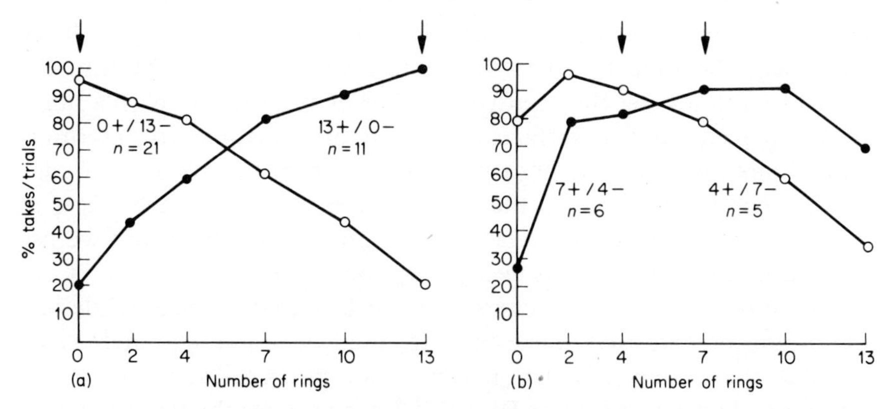

Fig. 9.6 (a) The results of transfer tests carried out with octopuses trained to discriminate between a smooth sphere (0 rings) and a similar sphere roughened by annular grooves (13 rings). The animals were first trained to a criterion of less than three errors in each of 2 successive sessions of 16 trials. Training was then continued at 8 trials per session, the remaining 8 trials being replaced by unrewarded tests with objects having numbers of rings intermediate between 0 and 13. Each animal had 32 tests with each object. Arrows indicate the spheres used in training. (b) Transfer tests carried out after training to discriminate between spheres having 4 and 7 rings (as in (a)). Note that maximum response rates are displaced away from the spheres used in positive training in the direction away from the spheres that the animals were shocked for taking (from Wells and Young, 1970b).

9.1.6. *Sizes and shapes*

The observation that octopuses apparently ignore differences in surface patterns which are readily detectable to ourselves, led to a re-examination of the manner in which the octopus handles objects that it touches. If a man is handed one of the cylinders used for the octopus experiments, and asked to examine it with his eyes shut, he will scan the object, moving his fingers repeatedly over the surface. Evidently we integrate surface contact information with proprio-

---

Fig. 9.5 Perspex cylinders used in tactile training experiments and the relation between errors made and the similarity of objects to be discriminated. Each point shows the probability of error by one animal trained to make a stated discrimination (indicated P1/P2 etc). In each case this figure is derived from the forty trials during days 8—12 of training at 8 trials per day. The progress of some typical experiments is shown in Fig. 9.4 (from Wells and Wells, 1957a).

ceptive inputs recording the successive positions of our fingers. If we just grasp the object it is still possible to identify which way the grooves go, knowing the relative position of the fingers. With a single finger, placed on the surface, it is still just possible to determine groove direction (and perhaps pattern); we seem to have a knowledge of the relative position of the distortion receptors in each finger as well as knowledge of the relative position of the fingers and we use both to interpret the signals we receive from the contact.

An octopus moves its suckers more or less continuously while it is examining objects that it touches. It could be scanning, much as we do. But the evidence is that it is not; octopuses failed to detect the difference beween cylinders P1 and P2 in the series shown in Fig. 9.5. One cylinder had longitudinal, the other circumferential grooves, a difference so gross that the animal could hardly have failed to notice if it were capable of scanning at all, or possessed any sort of detailed internal representation of the distribution of the mechanoreceptors in its suckers.

Octopus makes scanning movements, but fails to scan; it fails to integrate proprioceptive with surface contact information. We know that the animals *have* proprioceptors because the arms respond to passive stretch, and abundant receptors of an appropriate structure can be found in the musculature of the arms (Section 7.4). The simplest explanation of the state of affairs revealed by touch discrimination experiments is that the information these proprioceptors provide somehow fails to penetrate to the touch learning mechanisms in the central nervous system (Wells and Wells, 1957a).

If this explanation is correct, octopuses should be incapable of making distinctions between objects of similar texture but different size or shape. A problem is that differing sizes or shapes nearly always do result in textural differences, if these are defined in terms of the distortion likely to be imposed on suckers in contact. A corner bends suckers, a flat surface does not, a small radius of curvature causes more distortion than a large, and so on. It is possible to design shapes that could not be distinguished on a basis of sucker distortion, provided one assumes that the octopus will sample the whole surface. Thus, rectangular slabs can be created with the same edge to flat surface ratio as a cube of any given size. But *Octopus* does not generally sample the whole surface before a decision to take or reject an object touched and a slab will (*ex hypothesis*) appear either 'rougher' or 'less rough' than a cube depending upon the ratio of edge to flat surface in the particular area sampled. Size discrimination is only independent of these 'textural' differences in the special case of spheres, which should all appear alike to octopuses, since any sphere will appear to be a flat surface so far as the circular rims of the suckers are concerned.

In practice it is therefore easier to test the distortion hypothesis by using objects of different sizes and geometric shapes that will cause different degrees of distortion. The ease with which they are distinguished should correlate with the difference in degree of bend imposed upon the suckers in contact.

Some of the objects that have been used for this sort of experiment are shown in Fig. 9.7. The series includes simple cylinders of differing

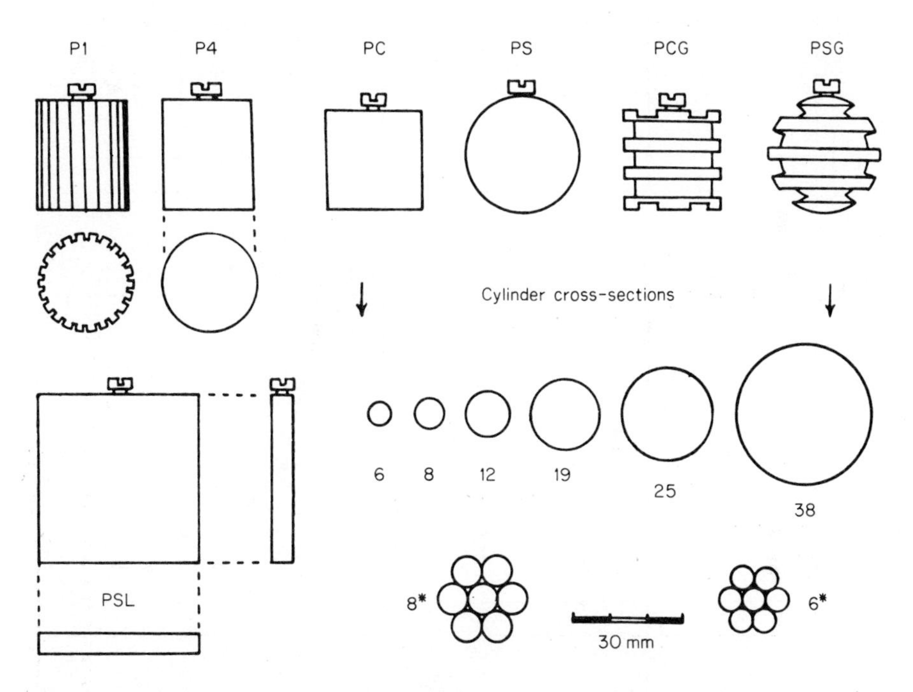

Fig. 9.7 Test objects used in the experiments. All are drawn to the same scale, and all were made of Perspex, with a flat-headed bolt threaded into the top for attaching a nylon line by which the object could be lowered to the octopus. Cylinders here shown as cross-sections only, are described in Figs. 9.8 and 9.9 by their diameter in millimetres, or by their unit diameter followed by an asterisk in the case of the compound cylinders 6* and 8*. All were of the same length, 50 mm, excluding suspension bolt. Octopuses of the size used in the experiments have suckers ranging downwards from about 10 mm diameter, measured when applied to a flat glass surface (from Wells, 1964a).

diameter and compound cylinders composed of bundles of rods. Training experiments (Figs. 9.8 and 9.9) show that the ease of discrimination is not predictable from the ratio of the diameters of the cylinders, but clearly depends upon the differences in curvature of the cylinder surfaces. Compound cylinders are treated as being of the

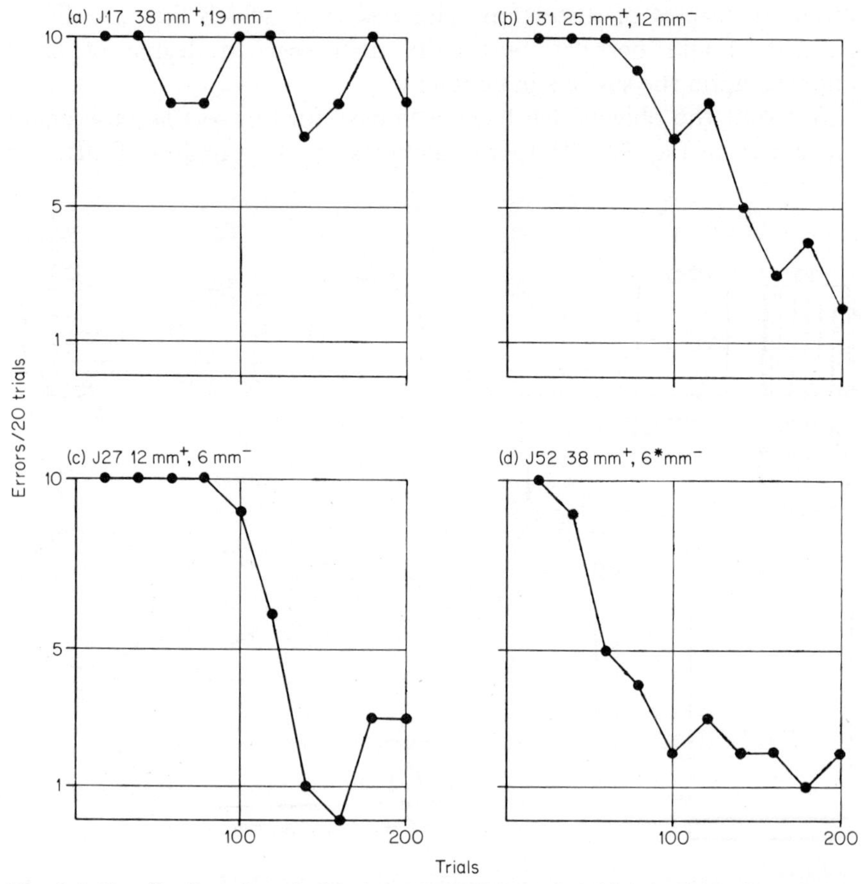

Fig. 9.8 Details of some typical training experiments in which individual octopuses learned to discriminate between cylinders of different diameter. There were 40 trials per day in two groups of 20 (10+, 10−). In experiments (a), (b) and (c) the ratio of the diameters of the cylinders to be distinguished was similar. In (d) the smaller cylinder had an overall diameter of 18 mm, being made up from 6 mm units (see Fig. 9.7). Ease of discrimination is clearly not related to the ratio of the diameters. The compound cylinder 6* mm in (d) is much more readily distinguished from the 38 mm cylinder than is the simple 19 mm cylinder in (a), despite its having very nearly the same overall diameter (from Wells, 1964a).

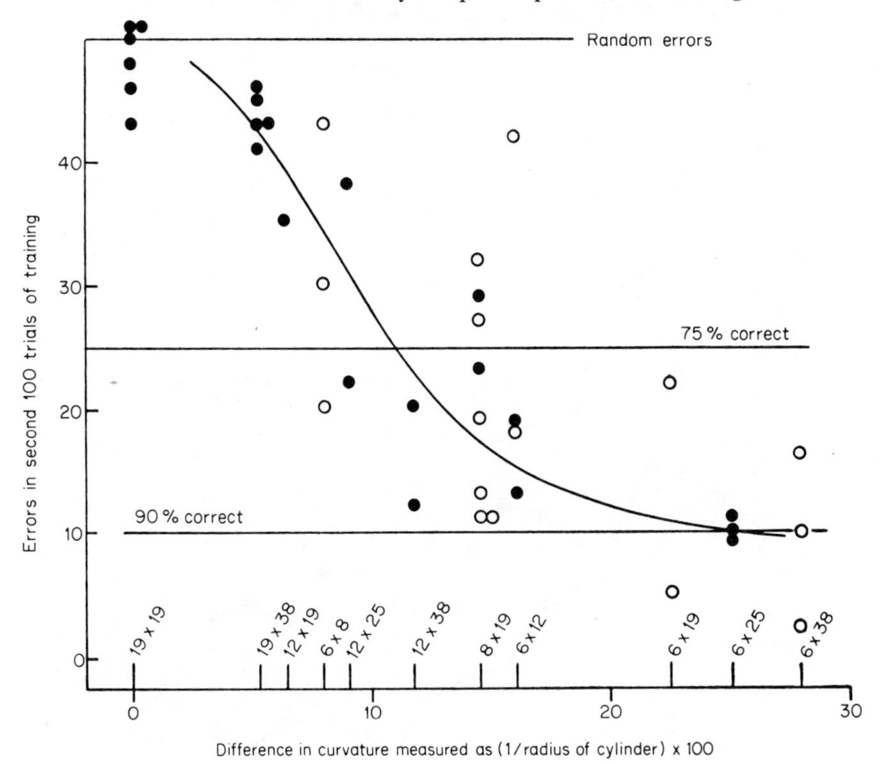

Fig. 9.9 Showing that the proportion of errors made varies with the difference in surface curvature of the objects to be distinguished. Each point plotted shows the result of an experiment in which an octopus was trained to discriminate between two cylinders of different diameter. The diameter of the cylinders used in each group of experiments are given along the abscissa; clearly the ratio of the diameters bears little relation to the number of errors made. The abscissa also shows the difference between the cylinders measured in terms of surface curvature; the proportion of errors made declines as the difference in curvature increases. Plots ● and ○ respectively show the results of experiments with simple and compound cylinders (see Fig. 9.7), the stated diameter of the latter being the unit and not the overall diameter, which would appear to be irrelevant (from Wells, 1964a).

same overall diameter as their component rods. This can only mean that arm bend and the relative positions of the suckers in space are both irrelevant; the animals can have no idea of the relative sizes of the cylinders to be distinguished and one is driven, once again, to the conclusion that the only parameter that matters is the degree of distortion of the grasping surfaces of the suckers (Wells, 1964a).

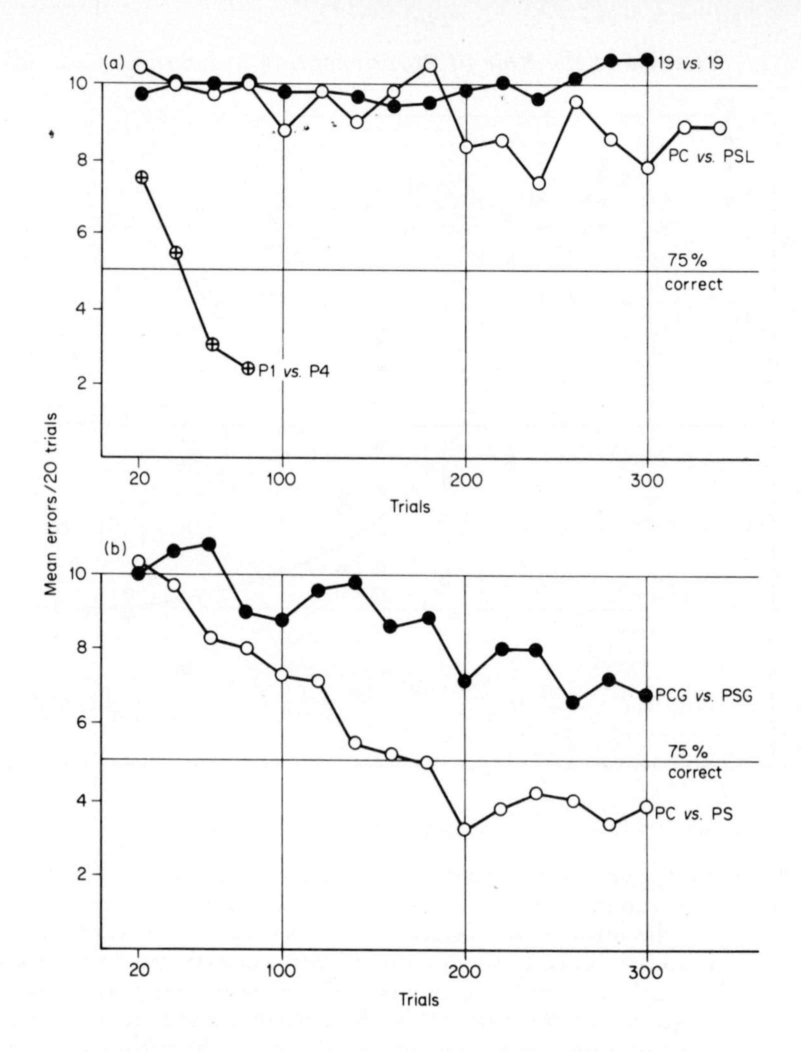

Fig. 9.10 Plots showing the course of five series of training experiments. Results (a) show: ●, the errors made when an attempt was made to train six octopuses to distinguish between two identical 19 mm cyclinders. This was done to control the possibility that the testing sequence was itself responsible for 'discriminatory' scores; ○, the errors made when octopuses were trained to distinguish between a cube and a flat slab having the same area to corner ratio (4 animals), and ⊕, the errors made in a simple textural discrimination between rough and smooth cylinders (26 animals). Results (b) similarly show; ●, errors made in learning to distinguish between a cube and a sphere roughened by grooves (5 octopuses), and ○, between a smooth cube and a smooth sphere (6 animals). Pictures of the objects used in these experiments are included in Fig. 9.7 (from Wells, 1964a).

Further evidence supporting this conclusion comes from results obtained with the other objects shown in Fig. 9.7. The cube and sphere (PC and PS) are distinguishable, as one would expect since the cube causes distortion of any sucker in contact with a corner, while the sphere is 'flat' (see p. 228). The distinction is learned less readily than the straightforward textural discrimination between P1 and P4 and more readily than the discrimination between the coarsely grooved cube and sphere PCG and PSG (Fig. 9.10). P1 and P4 will yield unequivocal signals wherever a sample is taken, PCG and PSG might be expected to yield very similar signals most of the time, with PCG, which has additional corners lacking in PSG, only sometimes yielding a 'rougher' signal than the grooved sphere. The slab PSL has the same edge to area ratio as PC and is scarcely distinguishable from it, as one would expect since signals from PSL will only from time to time fall outside the range arising from PC (Fig. 9.10; Wells, 1964a).

In a second series of experiments with the smooth sphere and cube, octopuses were trained and subsequently offered 'cubes' modified either by rounding the corners, or by eliminating the flat surfaces – the nearest practicable approach to a cube without flat surfaces being a narrow rod. Having attained a standard of 75% correct responses on the cube PC/sphere PS discrimination after considerable training, they were offered the modified cube PC2 (with rounded corners) in place of PC. The proportion of errors made rose, predictably. Five of the animals were then presented with PR, the narrow rod, in place of PS; two were offered PR in place of PC2, the cube with rounded corners. After 20 trials the situations were reversed, the PR/PC2 animals being trained with PS/PR and vice versa. The results are summarized in Fig. 9.11.

PR, the narrow rod, is quite clearly a cube, so far as *Octopus* is concerned. When substituted for PC2, the rounded corners of which had the same radius as PR, errors declined. When substituted for PS, errors rose to the point where the negative object was actually being taken more than the positive. PR is not only a cube, it is a 'better' cube than PC2. A reasonable conclusion from these experiments is that the distinguishing feature of a cube is its corners, as one would predict if octopus shape recognition is based on sucker distortion (Wells, 1964c).

9.1.7 *Weight discrimination*

When an octopus picks up a heavy object it must exert more muscle tension than when handling a light one. One can see the effects of

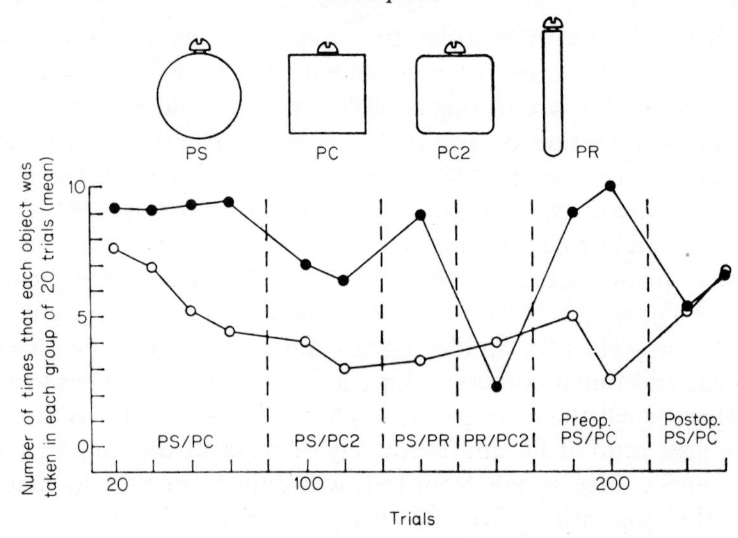

Fig. 9.11 Summarizes the last 80 trials of training to distinguish between a cube and a sphere, and the effect of substituting further objects for one or other of these. ● shows the number of times the positive object, and ○ the number of times the negative object was taken. There were 40 trials per day in two groups of 20 (each 10+, 10–). Finally the animals were retrained and lesions made to the inferior frontal system, after which discrimination broke down (see Chapter 11) (from Wells, 1964c).

this quite clearly if a blinded animal, sitting as usual on the side of its tank, is allowed to grasp a heavy cylinder; as the experimenter releases the weight the animal's arm extends, then stiffens to support the weight and finally handles the cylinder much as before but with the arm visibly more tense than usual.

The animal itself seems unable to learn to recognize this difference. Attempts to teach octopuses to recognize differences between objects that differ only in weight seem always to fail; the octopuses continue to take the light and heavy objects in similar proportions for as long as training is continued (Fig. 9.12).

This failure again shows that proprioceptive inputs are not used in touch learning (Wells, 1961a).

9.2 On the absence of proprioception in learning

Octopuses have muscular stretch receptors in their arms (Graziadei, 1965b) but do not use the information from these in touch learning.

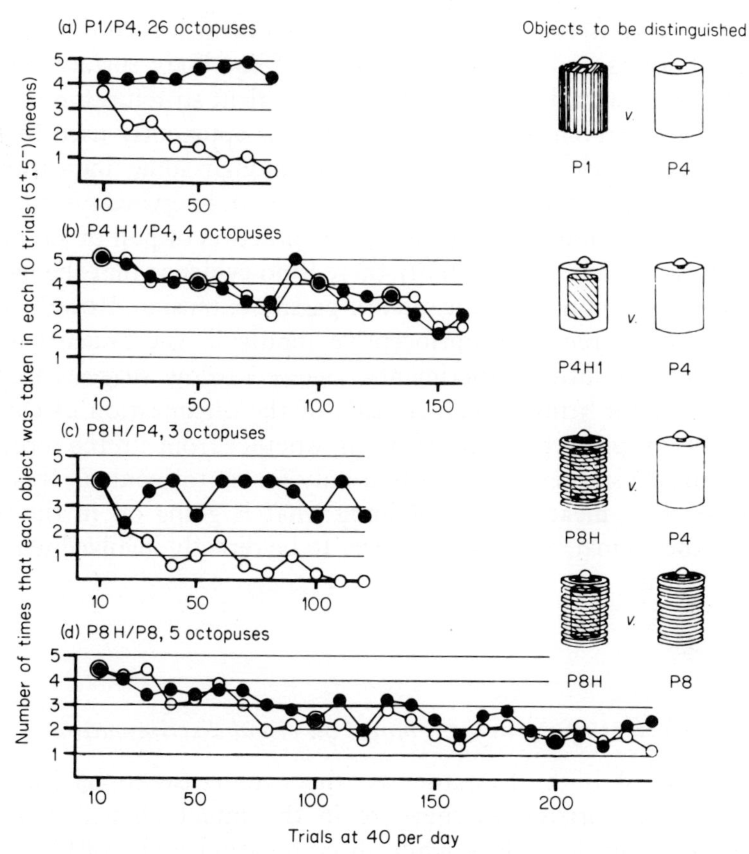

Fig. 9.12 Weight discrimination. In (a) twenty-six octopuses were trained to discriminate between the smooth P4 and the grooved P1, both objects being of the same weight (5 g). In (b) an attempt was made to train four octopuses to distinguish between P4 and P4H1, a second smooth object nine times as heavy (in seawater) as the first. In (c) three animals were trained to discriminate between P8H and P4, which differed in both weight and texture. In (d) an attempt was made to train five animals, two of them already trained successfully under (c) to discriminate between P8H (weighing 25 g) and P8 (5 g). Experimental results plotted as in Fig. 9.11 (from Wells, 1961a).

This is not very surprising. Stretch receptors record tension not position achieved, and are thus incapable of providing unequivocal data on the movement or the relative positions of parts of the animal's sensory surface – variations in load will always confound the record. In vertebrates and arthropods, unequivocal information about bodily position seems to be derived from a second set of proprioceptors, hair plates or

Paccinian corpuscles, located in positions that will record the movements at joints and pressures on parts of the body surface (Pringle, 1963). There would appear to be no equivalent structures in the arms of *Octopus.* Stretch receptor information is apparently used for local control of muscle tension only, and although some indications of stretch must penetrate up the arm to the central nervous system, since the arms as a whole will align along the direction of pull of any one of them (Boycott and Young, 1950), there is no evidence that the message is more than a minimal indication of passive extension. Rowell (1966) was unable to record proprioceptive inputs in the brachial nerves.

A variety of other experiments reviewed below suggests that the situation in the arms is characteristic of the organization of *Octopus* generally. Proprioceptive information, whether from stretch receptors, or the statocysts, seems to be used only for motor control. It is never available as a means of determining what is going on in the world around the animal, and it seems never to be directly involved when the animal learns.

9.2.1 *The effect of statocyst removal on visual discrimination*

After removal of both statocysts, the eyes cease to be locked to an 'artificial horizon' as they are in the intact animal (Fig. 7.7). Visual learning of a black/white discrimination was unaffected, but the ability to make a learned discrimination between vertical and horizontal rectangles was lost until the orientation of the figures was matched to the orientation of the retina at each trial (Wells, 1960a). After statocyst removal, retinal orientation depends upon how the animal is sitting, notably on its attachment to the walls or floor of its tank. A horizontal rectangle, for instance, will appear vertical on the retina, when the animal is sitting on the sides rather than on the floor of its aquarium. But the animal continues to act as if the retina were right way up. It seems to be quite incapable of correlating the results that it achieves with the posture that it happens to have adopted at the time. The matter is discussed in relation to the organization of the visual analysing system in Section 8.1.8 and in relation to the intraocular discrimination of light polarization plane in Section 8.1.17. As with the tactile experiments, one conclusion is that *Octopus* cannot use proprioceptive information in learning.

9.2.2 *Maze experiments*

These indicate an almost total inability to remember movements carried out to achieve an end. Bierens de Haan (1949), Boycott (1954) and Buytendijk (1933) all tried to teach octopuses to detour around a partition to reach crabs; the animals sometimes succeeded but showed no improvement in performance in successive trials. In a somewhat more elaborate apparatus (Fig. 9.13) Schiller (1949a, b) and subse-

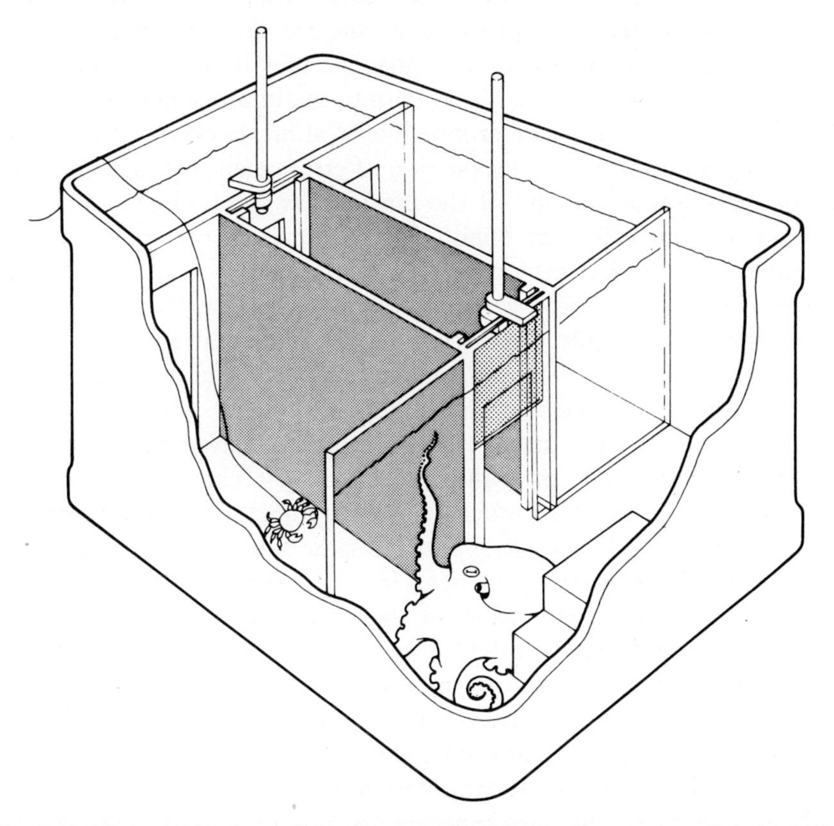

Fig. 9.13 The apparatus used for detour experiments. The octopus lives in a pile of bricks in a 'home' compartment and was usually confined to this compartment between trials by lowering a shutter. There was a second shutter at the far end of the central corridor. On leaving the corridor at this end the octopuses entered a choice compartment, like the home compartment but without bricks and with doorways leading into the two feeding compartments, on either side of the passage. If the octopuses went into the correct feeding compartment it was rewarded. The crab was removed as soon as the octopus entered the corridor (from Wells, 1964b).

quently Wells (1964b) found that octopuses would move through a maze to get crabs visible but not directly accessable to them. But the maintenance of the animals orientation towards the goal depended upon their remaining in tactile contact with the intervening wall (Schiller) or, when this contact was lost, upon visual tracking along the wall (Wells). With repeated trials there was some improvement in performance; the animals spent less time attacking through the glass before going into the corridor, and made fewer abortive entries into the corridor. Three sorts of evidence suggest that the animals learn little or nothing about the movements that they have made in running through the apparatus. One is that despite the improvement in running speed and reduction in the proportion of abortive entries, there is no sign of a reduction in the proportion of errors made by turning in the wrong direction at the end of the corridor. A second line of evidence arises from the effect of blinding the animals in one eye. If visual tracking is upset in this manner (which makes it more difficult to retain visual contact for a detour in one direction than the other) persistent errors result. There is no indication that the unilaterally blinded animals are able to remember that successful detours towards the 'difficult' side begin with a 180° turn on entry into the corridor. Animals without their statocysts, in contrast, detour successfully, if somewhat unsteadily. In a third check on the capacity to learn to repeat a specific series of movements, two intact animals were trained for 30 trials each to run always to the same side. These animals made only one mistake between them in a subsequent 10 trials when tested by showing them crabs on the previously untried side of the maze; there would seem to be no question of their having learned to carry out a specific series of movements to reach the goal (Wells, 1964b). Further experiments with this apparatus are described in Sections 10.2.2 (split brains and the control of movement) and 12.1.12 (short-term learning, and delayed detours).

In a further series of maze experiments, Walker, Longo and Bitterman (1970) taught *O. maya* to crawl to one side of the T maze shown in Fig. 9.14. The conditions were peculiar in that the incentive to run was created by dumping the animal in its 'home' box into the maze, so that the water ran out, leaving the octopus high and damp. To get back into a box full of water the octopus had to run to one side, the other being closed off by a grid. Rather surprisingly, the animals learned to do this, achieving 80% correct responses after about 80 trials; reversal took a little longer. Although there is no record of special

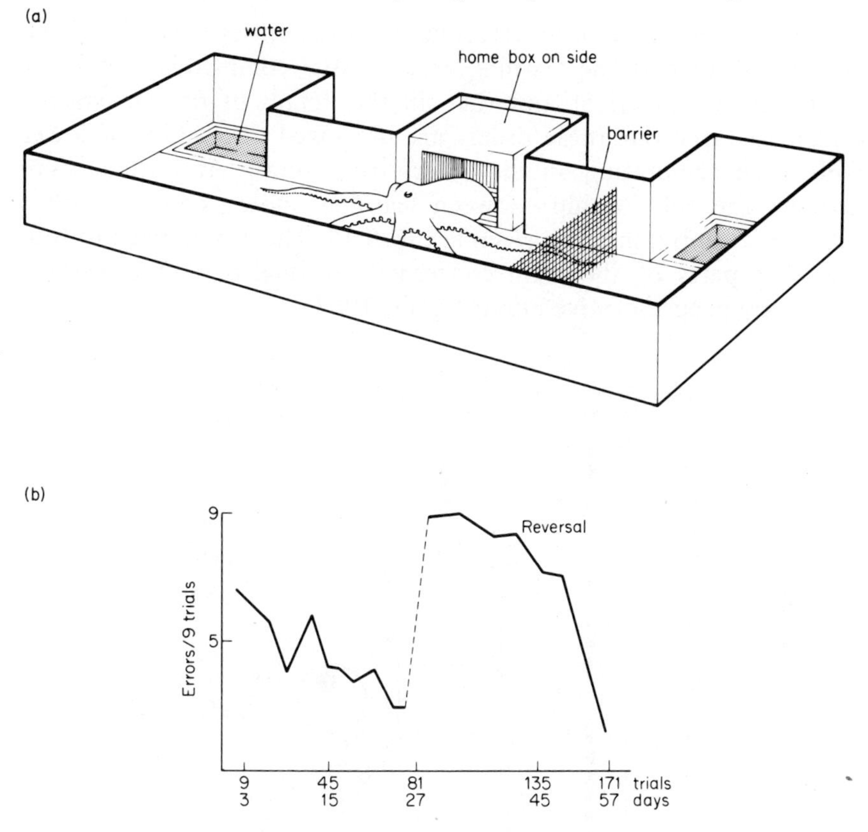

Fig. 9.14 (a) Apparatus. (b) The result of an experiment in which *O. maya* was trained to run a T-maze (after Walker, Longo and Bitterman, 1970).

precautions being taken to eliminate external cues, this performance could indicate that *Octopus* can learn to carry out a specific series of movements. It does not of course prove that it is done proprioceptively and in view of what we know about the relation between eyes and statocysts, it seems entirely possible that the animal was learning from the shift of stimulation left or right across the visual field as it came out of its box and made the choice. To prove the matter one would have to run trials in the dark.

9.2.3 *Brain lesion experiments and proprioception*

The effects of brain lesions, described in Chapter 10, show that removal of those parts of the brain concerned with sensory analysis and learning has little or no effect upon movement or posture. All the indications are that the learning centres issue commands to take or reject, attack or retreat and that all the details of the movements needed to carry out these orders are organized at a series of lower levels, each elaborating on the orders from above. An outline, summarizing the relationship between learning, motor control centres and proprioceptive inputs is given in Fig. 9.15. There is no need to suppose that parts of the brain concerned in visual or tactile learning receive any proprioceptive inputs (Wells, 1963a).

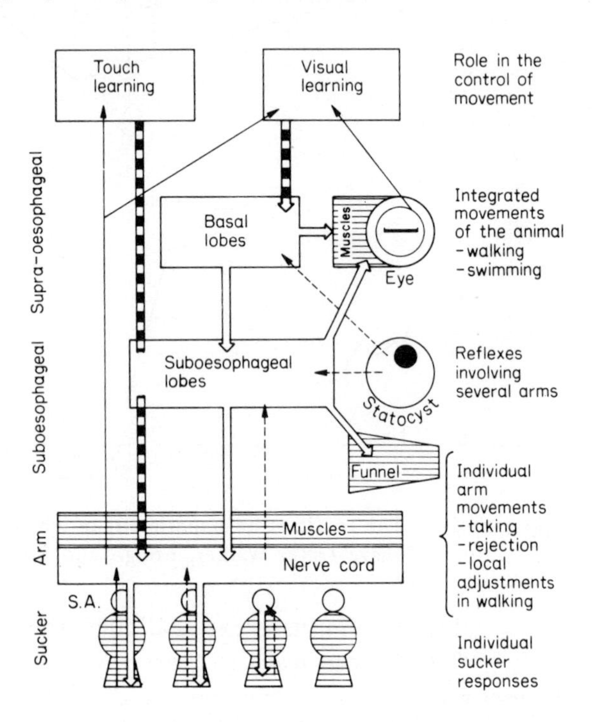

Fig. 9.15 A summary of the relations between the various levels at which movement is controlled and learning in the octopus. Dashed lines indicate sensory inputs carrying proprioceptive information, solid lines visual or contact information. Motor output channels are indicated by double solid lines and command fibres from the learning centres by similar pairs of lines with cross hatching. S.A., subacetabular ganglion at the base of each sucker. The diagram shows the minimum pattern of connexions indicated by a variety of brain lesion and learning experiments (see Chapters 10–12; recent work has shown an additional region of interaction between the visual and statocyst inputs in the peduncle lobes, which are not shown here–see Section 10.2.3) (after Wells, 1963a).

9.2.4 *Manipulation of objects in the environment*

Octopuses will pull stones together and thus restrict the entrances to their homes. Sometimes this looks as if the octopus has methodically constructed a wall of stones, but this is probably an illusion - if the animal grasps stones with the suckers and draws back to its normal defense position with the suckered side of the arms raised to protect the soft body, it will inevitably form such a barrier. The structure collapses when the octopus leaves its lair.

In the laboratory, octopuses pull all the movable objects, including any breakable apparatus, into a heap around them, but they never show any tendency to balance bricks and other objects upon one another to construct homes, here or in the sea.

A variety of experiments on manipulation has been made. Crabs can be wrested from containers (Piéron, 1911; Schiller, 1948; Cousteau and Diolé, 1973) but the results appear to be achieved by chance and there is little indication that the octopus can learn to deal with the situation more efficiently with practice. The animal approaches and struggles with the apparatus until something happens; if it learns anything as a result of its experience it is only to be more persistent and vigorous. Octopuses will approach and grasp any new object placed in their tanks. A lever is an obvious target, and the response - approach and grasp the lever - can be maintained if the animal is rewarded. Coates, Hussey and Nixon (1965) devised an automatic crab dispenser that Nixon (1969a) subsequently used to study the rate at which octopuses would feed themselves. The animals were pretrained to attack a white vertical rectangle that was then attached to the lever activating the apparatus.

Dews (1959) trained octopuses to pull a lever, which switched on a light under which the animals were fed; three animals were trained, the response being shaped by feeding under the light and then under the lever. Extinction occurred when the animals were no longer fed, and with one animal when it was fed for a period only at every third trial. Dews had difficulty persuading the animals to let go of the lever once they had grasped it and found the method unsuitable for a study of response rates. Individual behaviour varied, with changes that were not always obviously related to the reward; one of Dew's octopuses ('Charlie'), became more and more preoccupied with attempts to pull the light into the water, took to squirting the experimenter whenever he appeared, and finally broke the lever gear altogether, terminating the experiments.

In the course of a further unsuccessful attempt to train *Octopus* to pull levers Crancher, King, Bennet and Montgomery (1972) noticed that *O. cyanea* would sometimes insert an arm into a tube used to deliver food into its covered tank. The response was readily shaped; their six octopuses all learned to insert an arm into the tube and out through the water surface. The response was maintained if the animals were fed for doing this, and extinguished when feeding was stopped. Three of the animals learned when fed at alternate trials, while one learned well with variable ratio reinforcement. High rates of response were recorded because the end of the arm, extended above the water,

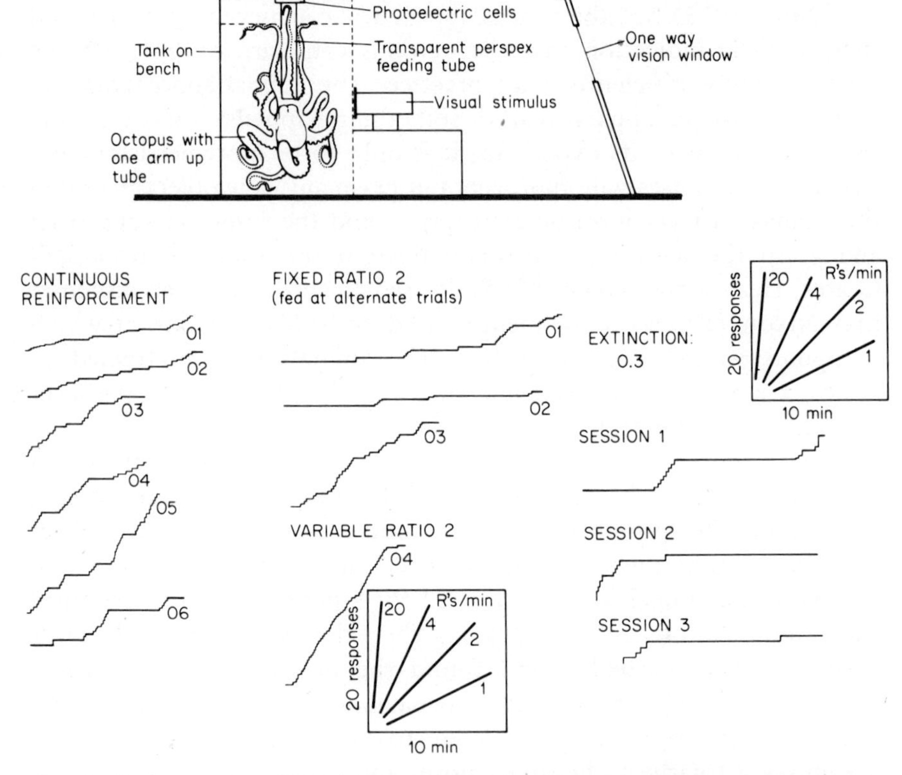

Fig. 9.16 Operant conditioning. Animals 01–06 were trained to insert an arm through a tube. Shaping the response required 30 min to 3 h, after which results such as those shown were obtained with continuous and intermittent reinforcement schedules (from Crancher *et al.*, 1972).

cannot apparently sustain this position for long and flops back (Fig. 9.16).

It should be noted that the lever and tube experiments reinforce actions that are within the animals normal repertoire. As with the tactile and visual training experiments, there is no question of teaching the animal an act that it would not normally carry out, once in a while, without training. It is not learning a new response.

It has sometimes been alleged that octopuses will place stones between the valves of shellfish, to prevent them from closing. While the ability to do this would, again, not necessarily imply learning to carry out a new action, it would at least indicate a fine control of manipulation by the suckers. The action would have to be visually directed, since contact would cause the bivalve to close. The story appears to have originated with Pliny (*Naturalis Historia* 9:48:2) and is still repeated, for example, in Naples where both Boycott and myself have heard it from different fishermen. It seems to be nonsense. Bierens de Haan (1926) and Boycott (1954) among others, have encouraged octopuses to repeat the performance under laboratory conditions, providing a variety of suitable pebbles and bivalves such as *Pinna.* Nothing relevant has ever occurred; octopuses either wrench bivalves apart or, as we have seen in Section 4.2.5, they bore holes to get inside.

9.2.5 *Proprioception and flexibility*

Taken all around, the performance of *Octopus* in manipulative tasks is disappointing. The animal has the motor equipment to carry out very complex operations with its prehensile arms. It is capable of rapid learning in a variety of situations, yet fails to solve manipulative problems that would present no great difficulty to animals like rats and cats with limbs that are, on the face of it, much less suited to such tasks.

There are at least three possible reasons for the octopus's failure. One is that the animal is probably unable to discover precisely how it has achieved any desired effect because the arm nerve cords (themselves incapable of learning, as we shall see in the chapters that follow) always elaborate on the orders received from higher levels in the central nervous system. The arm cords adapt the execution of these commands to local conditions without apparently at the same time signalling the details of what they have done back to the central command structures. In learning, the animal cannot possibly incorporate details of handling techniques that have been added in and fully controlled downstream of the source of the original orders. The hierarchial arrangement of the motor control system would thus set an upper limit to the complexity of the acts that an octopus can learn to perform, since it has no way of building up components by trial and error and then linking them together. To generate chains of manipulative behaviour (or, indeed, to learn to run a complex maze) the CNS would need to receive signals showing that each set of muscular responses was completed as ordered before the next was set in motion.

A second possible reason for the failure to manipulate is that the animal almost certainly lacks sense organs capable of providing information of the sort that it would need to do so. Octopuses have stretch receptors in their muscles (Section 7.4). But muscular stretch receptors record tension, not movement. In the absence of receptors capable of recording arm and sucker position independently of muscle tension, there is no way for the octopus to discern the position of its arms other than by observation.

It is in any case arguable that quite apart from the absence of suitable sense organs an octopus could not possibly process all the information that it would have to take into account to define the position of its arms (let alone the suckers) at all accurately. Since there are no joints to limit movement, the position of any section of an arm can only be defined in terms of the muscle lengths in all the more proximal sections. The number of sections for this purpose is potentially infinite, since the arm can bend, extend or contract anywhere along its length. Add to this the flexibility of the two hundred or so suckers on each arm, each independently free to expand and contract, extend and move about under the control of a dozen or more individual muscles (Guérin, 1908) and it is plain that something quite exceptional in the way of a computing mechanism would be needed to establish the relative positions of the ends of the arms, let alone of the individual suckers. There is no evidence that such a system exists in *Octopus*, or in any other soft-bodied animal.

The *Octopus* case thus highlights a state of affairs that must divide the animal kingdom rather cleanly into two groups. On the one hand are the arthropods and the vertebrates, jointed and therefore potentially possessed of a sense of bodily position that would permit them to learn to adopt postures and repeat movements, navigate their way through complex mazes, and develop manipulative skills. On the other hand are the soft-bodied invertebrates, forever debarred from such behaviour by the nature of their motor control, by their proprioceptive sensory equipment and by the near impossibility of computing with any degree of accuracy the whereabouts of their flexible ends. The octopus is a remarkable animal in many ways, not least in being a rapidly learning soft-bodied animal. Its most obvious sense organs and its behaviour are so 'vertebrate' that we tend to assume that it lives in a very similar perceptual world. It is important to realize that it does not. The parts of its brain concerned with sensory analysis and learning are related to its body in a manner quite different from our own. One can only approximate to the octopus condition, by using some complex piece of machinery (or another animal) as an extension of one's body. A man driving a car or riding a camel has only very limited feedback from the motor system he is controlling. In the main he must judge the success of his commands through his exteroceptors. The analogy doesn't quite hold, because the man will also use proprioceptive information from his own body, and in this once again has the advantage over his soft-bodied model. The world of the unjointed is a very large and varied world. Most sorts of animals live in it. If we are to understand their behaviour and the effects of their behaviour on their ecology, we have somehow to attempt to comprehend the surroundings in which they live, from their point of view. One of the many virtues of the octopus as an experimental animal is that it allows us a glimpse of these strange surroundings.

CHAPTER
TEN

Effectors and motor control

Octopuses move in a mysterious way. Being flexible the movements that they make are often difficult to specify and correspondingly difficult to investigate. The literature does not contain a description of octopod walking comparable with descriptions of the six-legged, tripod gait of insects, or the stereotyped locomotor patterns of snails or polychaetes. Descriptions of posture run into very similar difficulties and perhaps partly because of this, research into motor control in cephalopods has proved a less attractive proposition than research on sensory analysis and learning.

Certain peripheral elements of the control system have been examined and described. These, for the most part, are dealt with elsewhere in this book, in relation to the function that they serve. Thus the control of the hearts and blood vessels (and of respiratory movements) is discussed in Chapter 3, the control of ingestion and digestion in Chapter 4, while the question of sense organ innervation from the CNS is covered in each case when dealing with the organs concerned, mainly in Chapter 7. Here we are concerned with the hierarchy of centres controlling the movements of the arms and suckers, and with the chromatophores and skin musculature.

10.1 Mapping brain function

10.1.1 *An outline of nervous anatomy*

The nervous system of *Octopus* is a very diffuse system, by vertebrate standards. Most of the neurones are outside the brain, in a series of peripheral ganglia. Of these, the arm nerve cords alone contain more than twice as many nerve cells as the central supra- and suboesophageal ganglia; 3.5×10^8 compared with 1.7×10^8 in the brain (Young, 1963a). The cell bodies of the sensory neurones throughout the body are, of course, peripheral even to these outlying ganglia. There is an extensive plexus, including both sensory and motor elements around the gut and many of the blood vessels.

In overall control of the activities of these outlying parts are the thirty or so lobes of the central sub- and supraoesophageal brain. Some of these control the patterning, though often not the details, of movement. Others seem to be concerned only with sensory analysis or with learning, a separation of sensory integrative and motor executive operations that has enormous potential convenience for anybody trying to relate structure and function.

A great deal of information is available on the anatomy of the brain and about some of the peripheral ganglia. The brain of *Octopus* is certainly the best-mapped of all invertebrate brains, and is quite possibly as completely known as that of any vertebrate at the present time. Young's 'Anatomy of the nervous system of *Octopus vulgaris*' (1971) is a far larger book than this and elegantly demonstrates how much can be deduced from structure, granted an anatomist with first hand knowledge of the behaviour of the system he is studying. It is clearly quite out of the question to include anything but the most superficial summary of this work here.

Very broadly speaking, then, the brain consists of a trio of suboesophageal lobes, the brachial, pedal and palliovisceral, which house neurones that include motor units forming the penultimate link between brain and effectors. Together with these are regions of cells with motor neurones that run directly to the chromatophores and to many of the blood vessels. All these suboesophageal lobes are plainly motor control centres, and they include a proportion of very large neurones, as one might expect in any final common pathway (Table 10.1).

Connected to the suboesophageal lobes and thus only indirectly to the periphery are a further series of centres of complex structure, the magnocellular and basal lobes around and above the gut. The former has a small number of neurones with nuclei in the 15–20 μm diameter class but in general these regions lack the very big nerve cells characteristic of the suboesophageal brain. The same is true of the superior buccal lobe, which controls the mouthparts and salivary glands via the inferior buccal ganglion.

In addition to these regions, which it would be reasonable to classify as motor control centres on the grounds of their anatomy and connexions alone, there are several supraoesophageal lobes of an entirely different structure, characterized by relatively enormous numbers of

Table 10.1 The numbers and sizes of nerve cells in certain lobes of the *Octopus* brain (after Young, 1963a). Figures show numbers in thousands and nuclear diameters in a brain from an animal of about 450 g. Nerve cell numbers increase considerably as the animal grows (Packard and Albergoni, 1970).

	Total nuclei	*Diameter of nerve cell nuclei*					*Probable function*
		% <5 μm	% 5–10 μm	% 10–15 μm	% 15–20 μm	% 30–25μm	
Suboesophageal lobes, lower and intermediate motor centres							
Brachial, pedal and palliovisceral	692	26	44	24	5	1	Final or penultimate pathways to arms, guts and body musculature
Vasomotor	1307	18	74	8%	—	—	Control of the circulation
Chromatophore	526	19	50	27	3	—	Colour and pattern changes
Circumoesophageal and supraoesophageal lobes							
Superior buccal	150	23	53	21	1	—	Mouthparts and eating
Magnocellular	581	44	54	2	1	—	Relay to lower centres, defence and escape reactions
Higher motor centres							
Anterior basal	380	28	71	2	—	—	Head and arm movements
Medial basal	245	37	56	8	—	—	Swimming, respiration; Dorsal part – visceral activities
Lateral basal	127	43	54	2	—	—	Chromatophores; relay from optic lobes
Sensory and learning centres							
Inferior frontal	1085	91	7	2	—	—	Distribution and sorting of tactile input
Subfrontal	5308	100	—	—	—	—	Tactile memory
Superior frontal	1854	96	2	2	—	—	Distributes input from the optic lobes and touch centres
Vertical	25066	100	—	—	—	—	Visual and tactile learning
Optic	128940	95	5	—	—	—	Visual analysis and learning

very small neurones. Some of these, the inferior frontal and optic lobes, receive sensory inputs directly from the arms and eyes. Others, like the superior frontal, are relay centres, passing inputs onwards from the regions that receive sensory inputs, to the small-cell areas of the subfrontal and vertical lobes. These last are only quite remotely connected to the periphery on either the sensory or the motor output side; most of their tiny neurones are amacrines, cells without an obvious axon, connected to their neighbours but lacking processes to lead outside their parent lobe. Brain lesion work, to be reviewed in Chapters 11 and 12, shows that they are concerned in learning while having no detectable effect on motor performance.

Figures 2.7 and 6.1 show, respectively, the appearance of the brain seen from the side and from above. Fig. 10.1 is a drawing of a longitudinal section through the supraoesophageal lobes, while Figs. 10.2 and 10.3 summarize some of the internal connexions and the sources or destinations of peripheral nerves. Plate 3, opposite p.252, shows a longitudinal section through the whole brain. Some further details

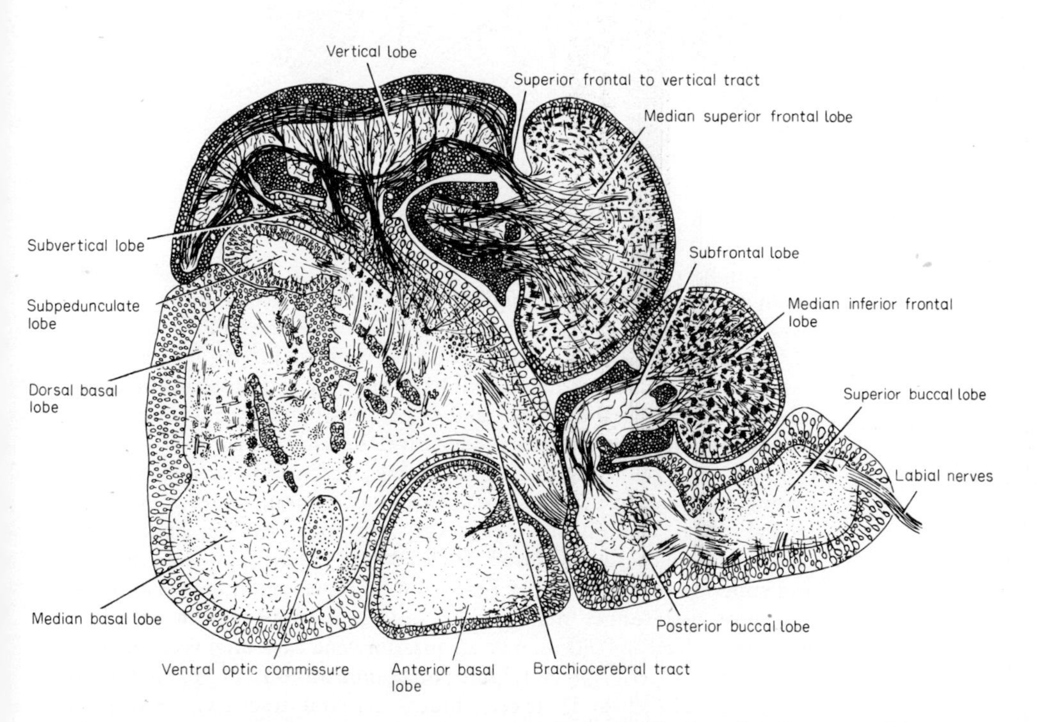

Fig. 10.1 Sagittal section through the supraoesophageal lobes, slightly to one side of the mid-line (from Young, 1971).

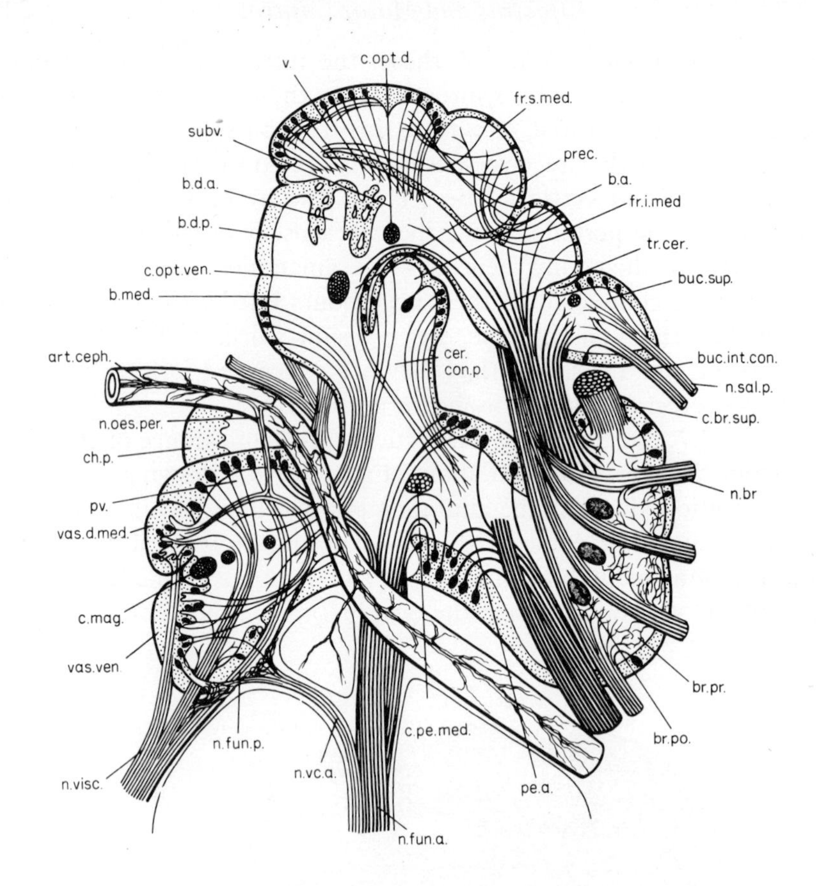

Fig. 10.2 Connexions in the brain of *Octopus.* The diagram represents a section taken to one side of the mid-line. The cephalic artery branches and passes through the suboesophageal brain between the palliovisceral and pedal lobes. 1., lobe; n., nerve; art.ceph., cephalic artery; b.a., anterior basal lobe; b.d.a. and b.d.p., dorsal basal anterior and posterior basal ls.; b.med., median basal 1.; br.po. and br.pr., postbrachial and prebrachial ls.; buc.int.con., interbuccal connective; buc.sup. superior buccal 1.; c.br.sup., suprabuccal commissure; c.mag., magnocellular commissure; c.opt.d. and c.opt.ven., dorsa and ventral optic commissures; c.pe.med., medial pedal commissure; cer.con.p., posterior cerebral connective; ch.p., posterior chromatophre l.; fr.i.med. and fr.s.med., median inferior and superior frontal ls.; n.br., brachial n.; n.fun.a. and n.fun.p., anterior and posterior funnel nerves; n.oes.per., perioesophageal n.; n.sal.p., posterior salivary gland n.; n.vc.a., anterior vena cava n.; n.visc., visceral nerve; pe.a., anterior pedal 1.; prec., precommissural 1.; pv., palliovisceral l.; subv., subvertical l.; tr.cer., brachiocerebral tract.; v., vertical l.; vas.d.med and vas.ven., median dorsal and ventral vasomotor ls. (from Young, 1971).

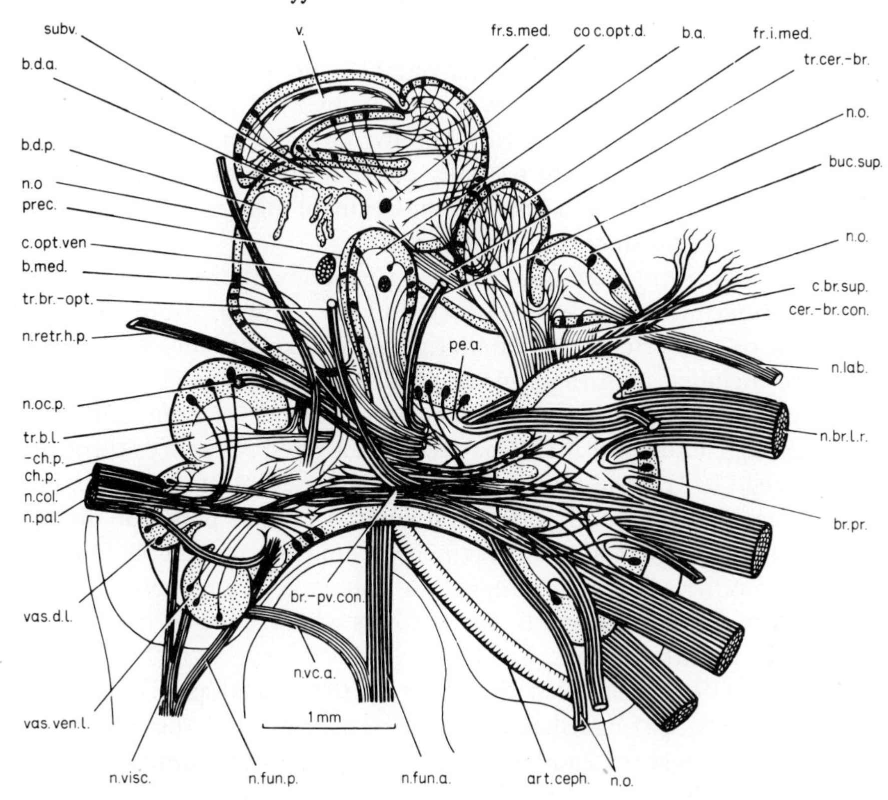

Fig. 10.3 Further connexions in the brain of *Octopus.* This diagram represents a section taken further laterally than 10.2. Abbreviations as in 10.2, with the following additions: br-pv con., brachio-palliovisceral connective; n.o., orbital, ophthalmic and oculomotor ns.; n.col., collar n.; n.retr.h.p., posterior head retractor n.; n.pal., pallial n.; tr.br.opt., brachial-optic 1. tract; tr.b.1.-ch.p., lateral basal 1. to post. chromatophore 1. tract (from Young, 1971).

of connexions within the supraoesophageal lobes are given in Figs. 11.8 and 12.1.

10.1.2 *Electrical stimulation of the brain*

If the brain of a cephalopod is exposed and stimulated through a wire electrode, movements of the arms and body, and chromatophore changes, can be elicited. This method was being used to map the parts of the brain concerned in motor control as early as 1867 when Bert attempted faradic stimulation of the brain of *Sepia*, found the supraoesophageal lobes electrically inexcitable – he called them 'silent' areas – and concluded that they were not involved in motor control. Von Uexküll (1895) used similar methods on the brain of *Eledone*, and showed that Bert was only partially correct. Some of the supraoesophageal lobes are truly 'silent', stimulation of others, including the optic lobes, can produce motor responses.

Since then, a number of stimulation experiments have been made with *Octopus*, mainly by Boycott and by Young. More detailed accounts of the effects of stimulating particular systems within the CNS are included in Altman (1968) and Messenger (1965). These two are Ph.D. theses; Boycott and Young's work has never been published and although much of it is summarized in scattered references throughout Young (1971), the most readily accessible account of brain stimulation in a cephalopod remains Boycott's (1961) paper on *Sepia*. Fortunately, the generalities that emerge from this study apply, almost without modification, to *Octopus*, so that it is not unreasonable to extrapolate from cuttlefish to the octopus when filling in the gaps in the published information.

Boycott (1961) classified the lobes of the cephalopod brain into five categories, distinguishable by their anatomy and the effects of stimulation. These categories begin with the *Lower motor centres*. These house the neurones that supply the muscles directly, without further intermediate synapses. Most such centres are peripheral. They include the arm nerve cords, the inferior buccal and gastric ganglia and the ganglia controlling the hearts. Within the brain the chromatophore and vasomotor lobes appear to include at least some neurones with axons that run directly to the muscles they control so that these two are also, by definition, lower motor centres. Electrical stimulation (1–6 V, 3 ms pulses at 60 Hz) of these final motor pathways will cause contraction of isolated groups of muscles, expanding the chromatophores in a limited area, for example, but never evoking integrated patterns recognizable as components of the normal behaviour of the animals.

The lower motor centres receive most of their input from a superimposed level of command, the *Intermediate motor centres*. These do not, in general, include final motor pathways and their stimulation excites whole groups of muscles; stimulation of the brachial lobe, for example, produces contraction in all of the arms at once; individual arms can be moved independently only by placing the electrodes close to the point of exit of one of the brachial nerves. The pedal lobe seems

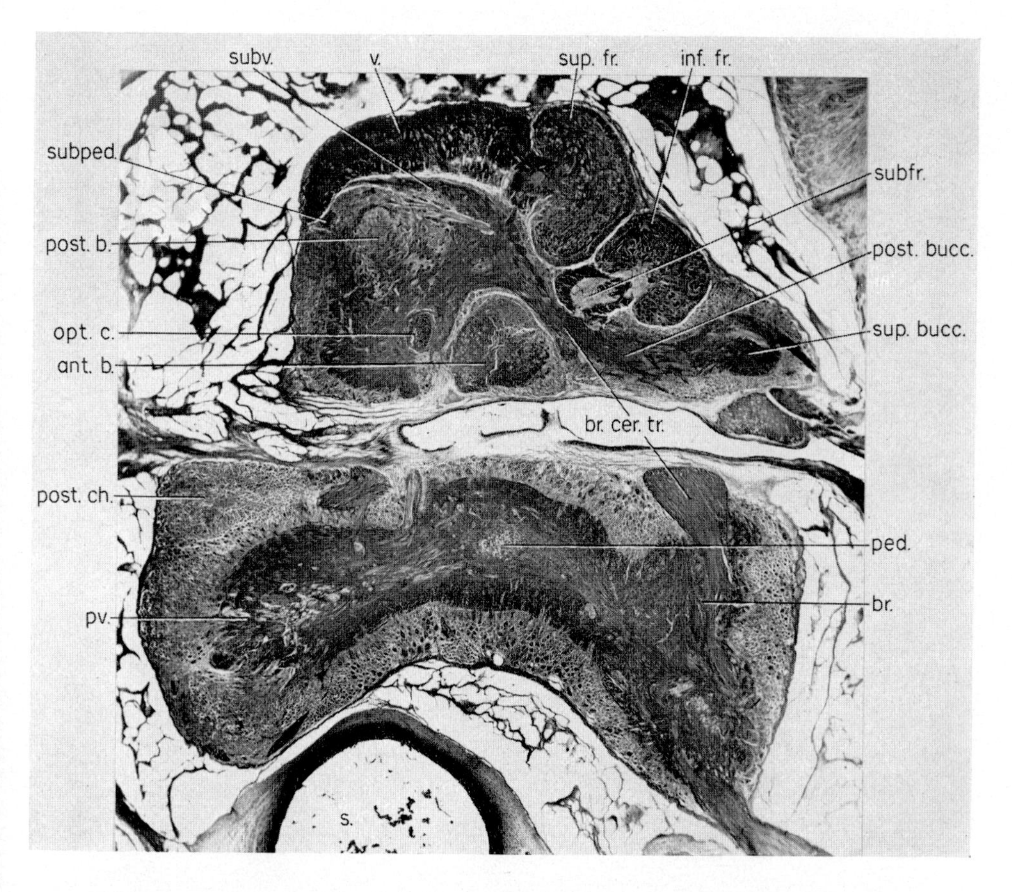

Plate 3 A longitudinal vertical section taken slightly to one side of the midline through the brain of *Octopus*. The supra- and sub-oesophageal parts of the brain are separated by the oesophagus. Lobes of the brain: ant. b. – anterior basal; br. – brachial; inf. fr. – inferior frontal; ped. – pedal; post. b. – posterior basal; post. bucc. – posterior buccal; post. ch. – posterior chromatophore; pv. – pallio-visceral; subfr. – subfrontal; subped. – subpedunculate; subv. – subvertical; sup. bucc. – superior buccal; sup. fr. – superior frontal; v. – vertical. br. cer. tr. – branchiocerebral tract; opt. c. – optic commissure; s. – statocyst. (Photograph, J. Z. Young.)

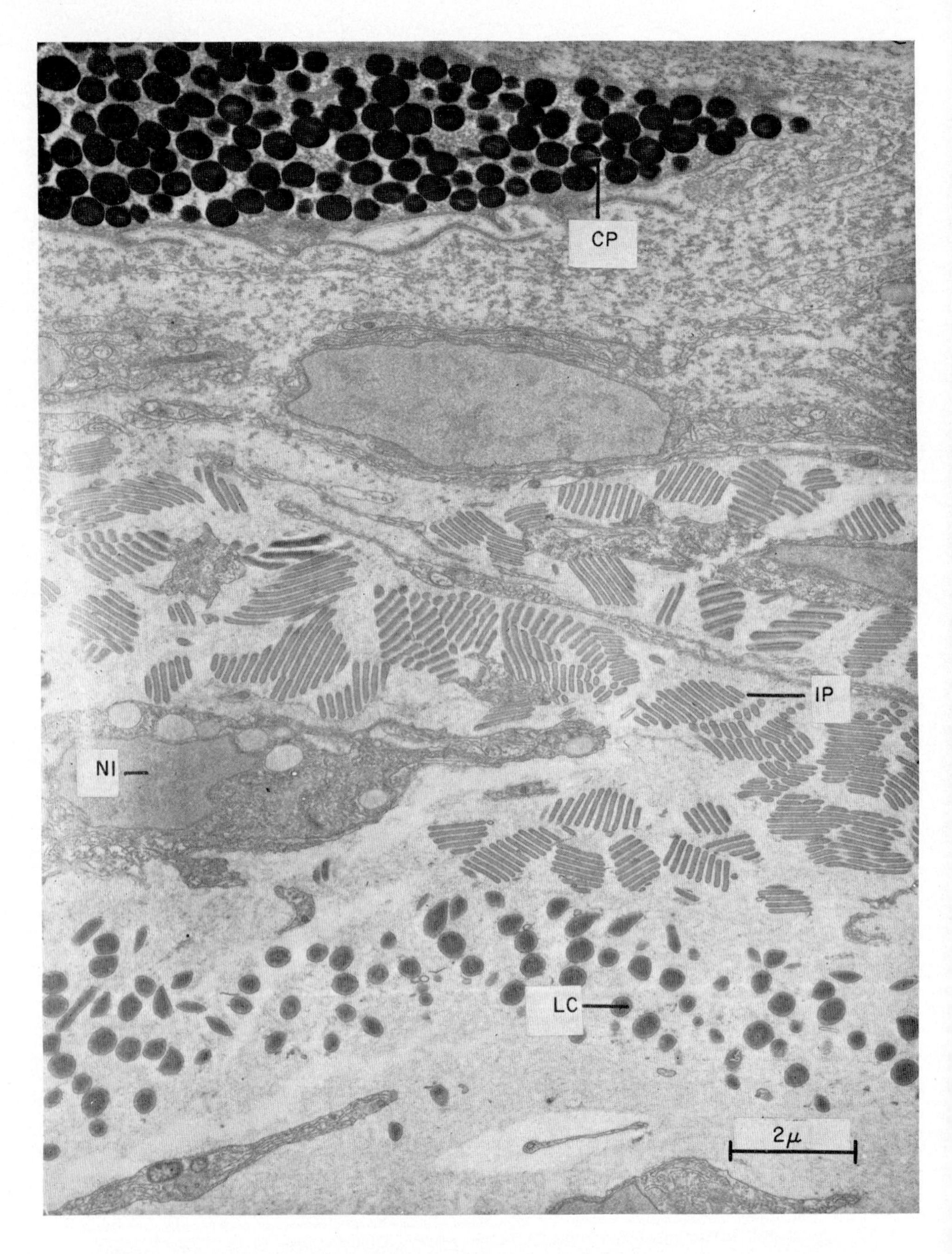

Plate 4 A section through the dorsal skin of *Octopus*. CP – part of the pigment-containing sac of a chromatophore (cf. Fig. 10.11, p. 278). Below this are the iridiophores, containing stacks of reflecting lamellae (IP): NI is the nucleus of an iridiophore. A second, lower, reflecting layer is formed by the leucophores, which contain granules of a white opaque material – LC. (Photograph, D. Froesch.)

also to contribute to arm control but is probably divided into several distinct functional regions since stimulation here can also evoke head, funnel and eye movements; it forms the link between statocyst inputs and compensatory eye movements (Section 7.2.3). The palliovisceral region, perhaps better thought of as the 'posterior suboesophageal mass' (Young, 1971) because it plainly consists of a number of structurally and functionally distinct lobes, is a mixture of lower and intermediate motor centres. Some nerves run directly to muscles in, for example, the funnel and collar, while those controlling mantle contraction have synapses in the stellate ganglia. Stimulation of the central part of the posterior suboesophageal mass may evoke contraction of the mantle and/or funnel and collar movements, but never the sort of co-ordinated action of all three necessary for the normal respiratory rhythm of the animal. At the intermediate motor centre level, one is dealing with regions that can organize recognizable elements of the movements that intact animals make, but these are always very simple components of behaviour.

The *Higher Motor Centres* form the third tier in the hierarchy of motor control. These are all supraoesophageal, the basal lobes of the brain. Stimulation of these regions can produce well-co-ordinated responses, such as a regular respiratory rhythm, with associated movements of funnel and collar. The median basal appears to control respiration and the more violent contractions of the mantle that occur when the animal swims. The anterior basal controls head, arm and eye movements – an *Octopus*, stimulated in this region, will walk using all the arms – while stimulation of the lateral basal evokes changes to the chromatophores and skin papillae extending all over the ipsilateral side of the body. Subdivision of function within the major lobes was not detectable.

The *Silent areas*, where stimulation has no visible motor consequences, include the inferior and superior frontals, the vertical and the subfrontal lobes. Brain lesion work, reviewed in Chapters 11 and 12, shows quite clearly that these parts are concerned in sensory analysis and in learning. Also silent are the dorsal basal, subvertical and precommissural lobes lying between the vertical lobe and the higher motor centres.

On the input side are a series of lobes perhaps better described as *Receptor analysers* (following Young, 1971) than as *Primary sensory centres* (Boycott, 1961). The inferior frontal/posterior buccal region handles inputs from the arms, and there is abundant evidence that these have already been processed very extensively in the arm nerve cords (see Chapter 7). The optic lobes must carry out correspondingly elaborate operations upon the visual input (see Chapter 8). The receptor analysers are for the most part electrically inexcitable; only the inner regions of the optic lobes (and the output fibres from the inferior frontal system) will produce responses if stimulated. These responses may be quite complex; stimulation of the base of the optic lobe can produce patterned chromatophore activity and may evoke any of the locomotory responses seen as a result of stimulating the basals.

Electrical stimulation by wire electrodes is a crude way of exciting a system as elaborate as that of *Sepia* or *Octopus.* Effects are liable to be generated at a considerable distance from the point of stimulation if tracts rather than cell bodies are excited, so that a fairly detailed knowledge of brain anatomy is required before reliable interpretations can be made. Boycott (1961) points out several instances where the results reported by previous workers were almost certainly attributable to this sort of spread of excitation, and for this reason alone it is best to be wary of findings from stimulation experiments that are not backed by brain lesion experiments.

The results are nevertheless useful in showing several general features of the organization of the brain. The most obvious are:

(1) There is plainly some sort of hierarchy of motor control, with regions superimposed upon one another. The lowest are final motor pathways, the highest organize whole patterns of activity in space and time.

(2) Certain areas, on the sensory input side of the higher motor control regions, seem to play no part in the organization of locomotion.

(3) Within lobes, there is rarely any indication of a topographical subdivision of function. The neurones controlling, say, respiratory movements, seem to be distributed throughout the lobes that play any part in their organization.

10.1.3 *Brain lesions, posture and movement*

An alternative, and almost equally crude means of investigating the function of different parts of the CNS is to destroy regions and see what effect this has on behaviour. The usefulness of this approach depends upon a rather precise specification of what has been destroyed and this necessitates the examination of serial sections from damaged brains. It is inevitably a slower way of mapping brain function than electrical stimulation, but at least it is possible to define what was done.

Lack of detailed maps of lesions made renders much of the early literature less valuable than it could have been, so that experiments like those of Fredericq (1878), von Uexküll (1895) or Buytendijk (1933) are difficult to interpret. Buytendijk, for example, claimed that octopuses are unable to climb out of their tanks after removal of the superior frontal and vertical lobes, and this is plainly untrue; he must have damaged the basal lobes as well, but there is no way of checking this now. So far as brain lesion work is concerned, it is safer to rely on work done from 1950 onwards when it became accepted as routine to section the brain and check the lesions in all the experimental animals used.

Even so, there are problems of interpretation, since the effect of a lesion changes with time. Behaviour within a few hours of operation may be influenced by discharges from damaged tissues, while that of the same animal several weeks later could be complicated by regeneration.

Sereni and Young (1932) studied regeneration in the mantle nerves and concluded that, at summer temperatures of 20–25 °C, the breakdown of axons separated from their nerves normally takes between one and three days. Regeneration from the central stump proceeds at 7–18 μm h^{-1}, a rate comparable with that of vertebrates at a similar temperature; functional connexions can be established if the regenerating nerves proceed through the scar tissue and into the peripheral stump. Sereni and Young and later Sanders and Young (1974) found a return of central control of chromatophore changes within two to four months after section of the mantle connective on the side concerned. The situation following surgical removal of parts from the brain is less certain. Degeneration studies have been very extensively used in mapping the brain. Regeneration certainly occurs as well and has been observed repeatedly when examining serial sections from animals with lesions but no evidence for the re-establishment of functional connexions has ever been obtained.

In the account that follows, the effect of brain lesions on the control of movement and posture is considered. The results of a variety of experiments confirm and extend the impressions gained from preliminary anatomical and electrophysiological mapping; sensory and motor regions of the brain *are* separated and there *is* a clear hierarchy of levels at which motor responses are controlled and elaborated. The special case of chromatophore control is considered separately in Section 10.3.

10.2 The control of locomotion

10.2.1 *Monocular vision and interocular transfer*

The visual fields of the two eyes overlap (Heidermanns, 1928), but the octopus rarely, if ever, approaches objects that it sees using both eyes to fixate the target. Normally the head is held sideways. Objects

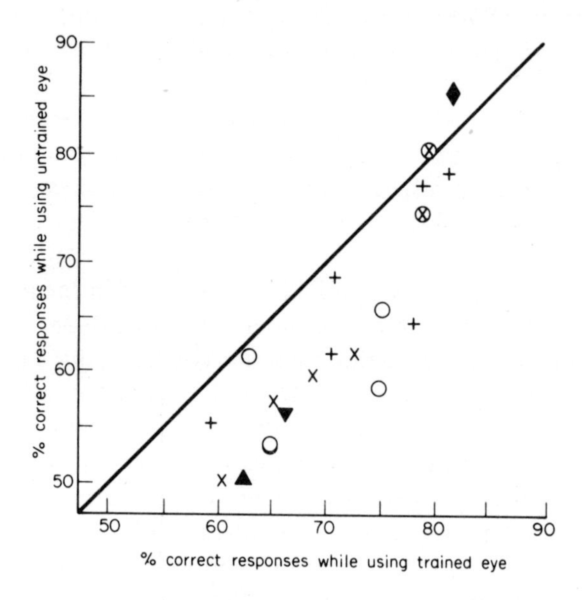

Fig. 10.4 Interocular transfer. Octopuses were trained using one eye and tested using the other; a summary of the results of 3 series of experiments. In series 1, 2 groups of animals were trained to discriminate between 10 × 2 rectangles, shown horizontally and vertically (results shown ⊗). A series of transfer tests, with figures that ranged from a 4.5 × 4.5 cm square, through rectangles to a 10 cm row of three 2 × 2 squares, was run alternately with further training trials (results shown ○). In a second series animals were trained on the rectangles (♦), T shapes (▲) and diamond vs. triangle (▲). In a third series, animals were trained on the rectangles and then subjected to a series of unrewarded tests, extinguishing their responses; plots + and × show the scores at different stages in learning and extinction. Except for the very simplest discriminations the performance using the untrained eye always lags behind that of the trained (from Muntz, 1961a).

must, nevertheless, pass from one visual field to another as the animal, or the object, moves about, and it would seem inevitable that some mechanism exists to ensure that the experience accumulated while the animal is using one eye can be applied when the same image is focussed on the other retina.

Muntz (1961a, c) has examined this matter, taking advantage of the monocular approach to train animals to discriminate using one eye (by presenting visual figures always to the same side) and then testing them using the other. Interocular transfer is rarely complete; in transfer tests, in training and in extinction (unrewarded responses) the performance using the untrained eye is always a little poorer. The effect is most marked for difficult visual discriminations (Fig. 10.4). Transfer was close to 100% only after prolonged training to discriminate between horizontal and vertical rectangles, a task known to be very easy for octopuses. Cutting the links between one side of the brain and the other after training on one side only, Muntz (1961b, c) was able to show that the memory trace is bilateral; once the animal is trained even the trained side optic lobe can be removed without abolishing subsequent correct responses by the untrained side.

In an alternative type of experiment Wells (1964b, 1967, 1970) showed that octopuses will detour through the apparatus shown in Fig. 9.13 in order to reach crabs seen through one or other of two alternative windows. In the course of these runs octopuses often changed from leading with one eye to leading with the other, without any consequent drop in the proportion of correct runs made (Table 10.2). This again indicates that interocular transfer can be complete, given favourable circumstances.

Table 10.2 Eye changes and errors made; an analysis of the results of completed runs through the apparatus shown in Fig. 9.13 (from Wells, 1967).

	Responses		*Proportion of errors (%)*
	Correct	*Incorrect*	
(1) Trials beginning with a change in the leading eye on entering corridor			
Controls	50	4	8
Animals with vertical lobe lesions	29	7	24
(2) Trials without a change in the leading eye			
Controls	317	24	8
Animals with vertical lobe lesions	197	42	21

10.2.2 *Split brains and the control of movement*

Muntz's (1961c) experiments showed that either side of the brain could, if necessary, control the whole animal. The same is found in detour experiments; octopuses with their supraoesophageal brains split by a longitudinal vertical cut move about quite normally most of the time, guided apparently by whichever eye happens to be leading at the time. There is no evidence that one side is normally dominant, in this or in any other experiment so far carried out; the octopus brain, so far as we can discover, is functionally symmetrical. Split brain animals, however, run into difficulties in cases where the input to the two eyes leads to conflicting demands by the two sides of the body. In detour experiments it is not unusual to observe the arms on one side of the animal reaching forward towards a crab, while those on the other are clutching at the brickwork of the 'home', apparently attempting to pull the animal in the opposite direction. One consequence is an unusually high proportion of abortive runs - the octopus sees the crab, moves into the corridor, and then returns home without completing a detour (Fig. 10.5). The decision as to which way to move is presumably

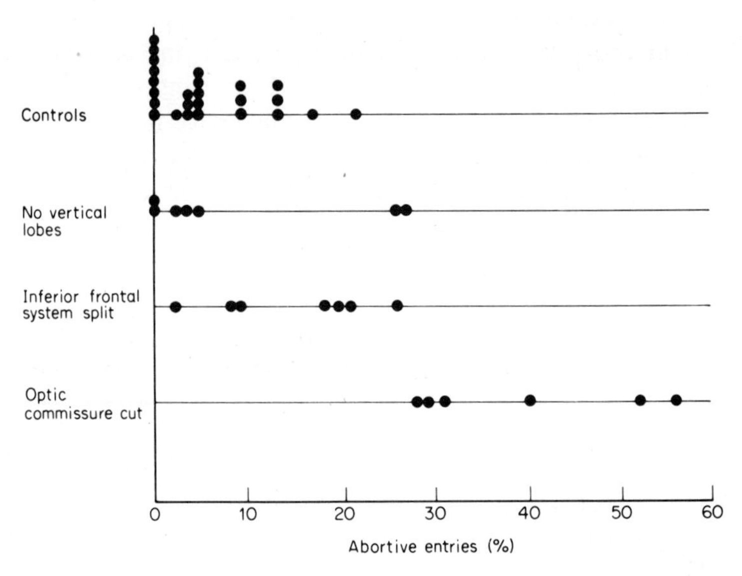

Fig. 10.5 Showing the proportion of trials at which animals with various lesions made only abortive entries into a detour apparatus. Each point shows the result obtained with a single individual having the stated type of lesion (after Wells, 1970).

made as a result of comparing the inputs in relation to some inbuilt scale of priorities, the result of the past genetic and personal history of the animal. A crab remembered is evidently less effective than a crab seen in generating commands to move onwards, so the signals arising from the attacking side of the body become stifled by the competing output from the contralateral command centres, operating on the principle that the animal should not leave the shelter of its home without good cause. The conflict can be resolved by blinding the split-brain octopuses in one eye, after which there is a marked reduction in the proportion of abortive runs (Table 10.3).

Table 10.3 Split brains and abortive runs (after Wells, 1970).

Animal	*Before blinding in one eye*			*After blinding in one eye*		
	Completed	*Abortive*	*%*	*Completed*	*Abortive*	*%*
N1	39	15	28	59	2	3
N3	7	14	67	14	15	52
N20	41	9	18	30	1	3

10.2.3 *The peduncle lobes*

In the intact animal, conflicts of command between the two sides of the brain must be resolved at supraoesophageal level, since there is evidently no mechanism for doing so in the suboesophageal brain. Octopuses can be taught to make visual discriminations after removal of their vertical and superior frontal lobes (Chapter 12). The sensory inputs that initiate attack or retreat must come in through the optic lobes on their way to the higher motor centres in the basal lobes and it would seem reasonable to search above the basal lobe level for the mechanisms that resolve conflicts between the two sides of the brain.

Anatomically, there are two distinct pathways from each optic lobe to the basal lobes and two distinct commissures connecting the optic summarized in Fig. 10.6. The fine structure of the peduncle lobe has been described by Woodhams (1977).

Surgical interference within this system produces results that are summarized in Table 10.4. Removal of up to the whole of one side of the visual system (lesions A to E) leaves animals that move about quite normally. The seeing eye leads, and when the animal is at rest it is often held uppermost and slightly forward. Unstable locomotion, with a tendency to oscillate from side to side or spin when the animal

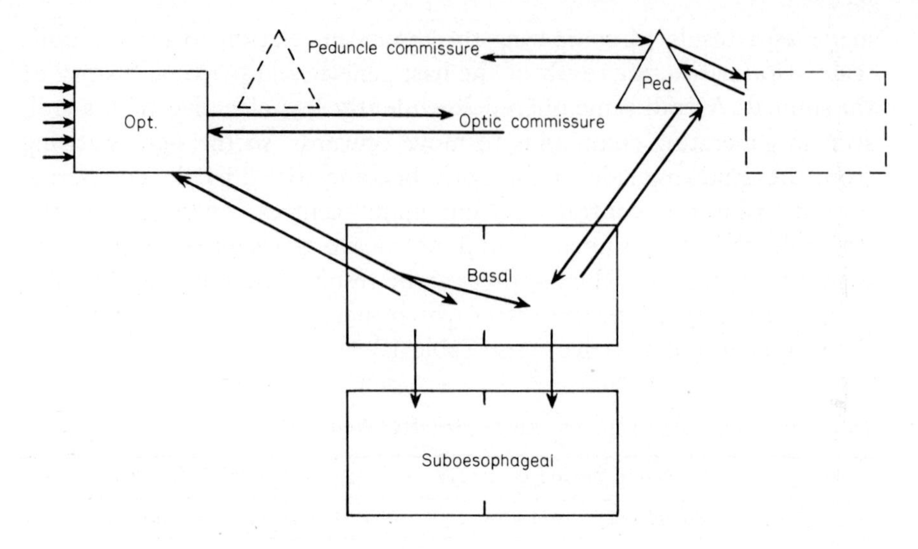

Fig. 10.6 Diagram showing the functional connexions between the optic, the peduncle and the basal lobes. For clarity the optic lobe only is shown on the left and the peduncle lobe only on the right (from Messenger, 1967b).

Table 10.4 Lesions of the visual locomotor control system (after Messenger, 1967b).

Lesion		*Posture*	*Unstable movement*	*Forced circling*
	A $n=3$	Normal, intact eye leads	–	–
	B $n=11$	Operated side forward, higher	–	–
	C $n=20$	Normal	–	In dark only, side without p.l. walks forward
	D $n=16$	As B	–	–
	E $n=4$	As B	–	–
	F $n=14$	Normal	–	–

Lesion	*Posture*	*Unstable movement*	*Forced circling*
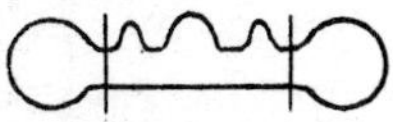 G $n=11$	Normal	–	–
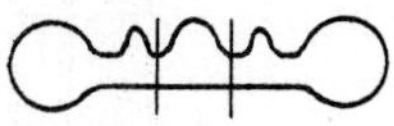 H $n=11$	None, immobile, limp	–	–
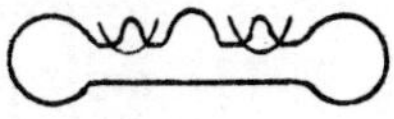 I $n=20$	Normal, but may sit with head up, down or sideways	Yes	–
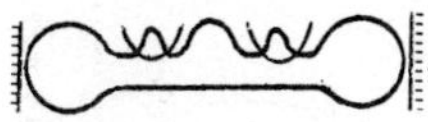 J $n=12$	Normal	–	–
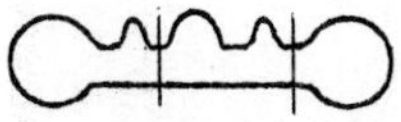 K $n=17$	Side with peduncle lobe down	Circles	Yes, side without p.l. walks forward
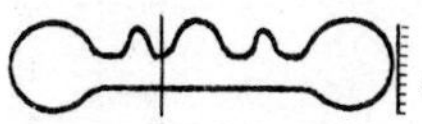 L $n=8$	As K	Circles	As K
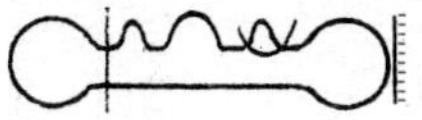 M $n=15$	As K	Circles	As K
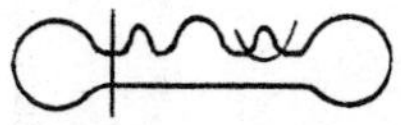 N $n=6$	Seeing side up, forward	Yes	–
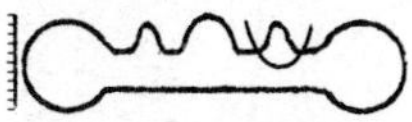 O $n=6$	As N, but less marked	Yes	–
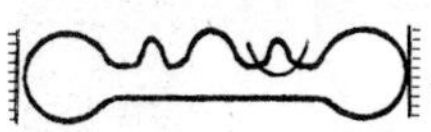 P $n=7$	Normal in walking, p.l. side leads	–	–
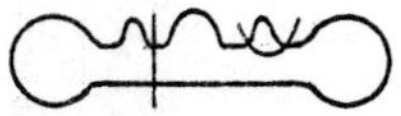 Q $n=6$	Normal, sometimes head down	Yes	–
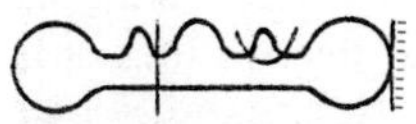 R $n=5$	Intact optic lobe up	–	–
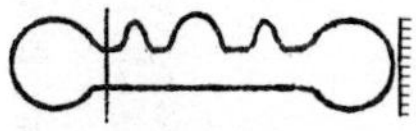 S $n=7$	Normal	–	–
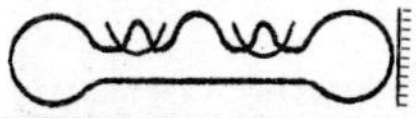 T $n=4$	Seeing eye up, forward	Yes	–

'Unstable movement' = Oscillations, particularly in the rolling plane, when walking or swimming.
'Normal' = head up if on side of tank, eyes up if on bottom, symmetrical, tone similar on the two sides.
p.l. = Peduncle lobe
Standard diagram and lesions:

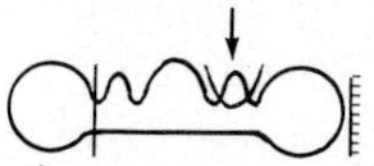

Peduncle lobe removed

optic tract cut optic nerves cut

attempts to walk or swim, results when the peduncle lobe is removed from the seeing side of animals blinded in one eye, whether by optic nerves section or by removal of the optic lobe on the contralateral side (lesions N, O, Q, T). It also occurs when both peduncle lobes are excised from animals with both sets of optic nerves intact (I).

Each peduncle lobe receives a direct input from the statocyst on the same side of the body (Fig. 10.7, Hobbs and Young, 1973) and the

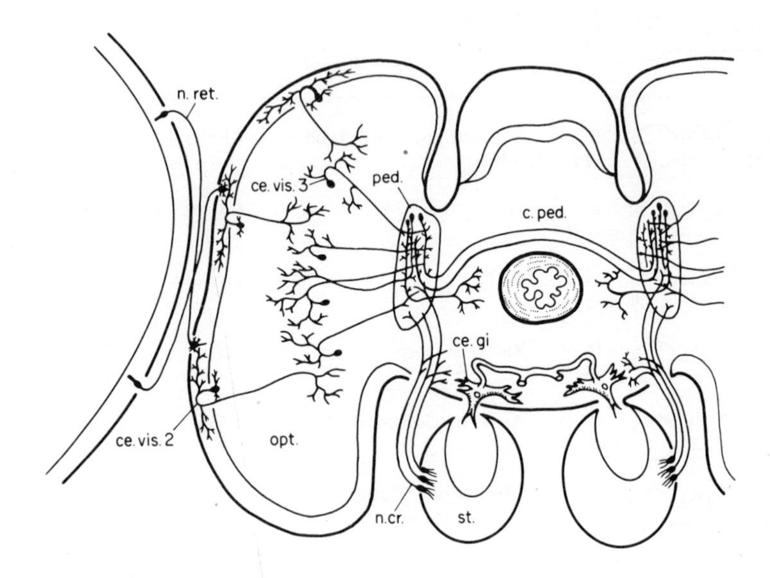

Fig. 10.7 Diagram of connexions of the peduncle lobe of *Loligo*, seen in transverse section; the condition in *Octopus* is believed to be similar. The axis lies in an oblique vertical plane. ce.gi., first order giant cell; c.ped., peduncle commissure; ce.vis.2, ce.vis.3., second and third order visual nerve cells; n.cr., crista nerve from the statocyst; n.ret., retinal nerve fibre; opt., optic lobe; ped., peduncle lobe; st., statocyst (after Hobbs and Young, 1973).

effects of peduncle lobe removal on walking and swimming certainly resemble those of removing the statocysts (Boycott, 1960). Removal of the statocyst from one side and the peduncle lobe from the other produces severe disturbances in locomotion, resembling the effects of bilateral peduncle lobe removal (Messenger, 1971).

The instability produced by interference with the peduncle lobes is not, however, attributable to cutting the statocyst input, since it is only found in animals that can see. Operations J, P and R (Table 10.4) produced blind, but stable animals, unlike their seeing counterparts I, O and Q.

The peduncle lobes seem to have two sorts of effect upon locomotion. One arises from their effect upon muscle tone; each lobe appears to have an excitatory effect on the musculature on the opposite side of the body, though the mechanism may in practice be inhibition of inhibition by the contralateral basal lobes (Messenger, 1971). The matter is complicated by the direct output from the optic lobes, which again seems to increase muscle tone, though here the effect is apparently bilateral, since removal of one optic lobe does not produce asymmetry. The full effect of peduncle lobe removal on muscle tone is only seen in bilaterally blinded animals that lack one or both of their optic lobes. These animals (operations K, L and M in the Table 10.4 series) circle around the side with an intact peduncle lobe. The same circling was observed in seeing animals with one optic/peduncle lobe complex removed (operation C) when visual cues were eliminated by switching out the lights.

Superimposed on these effects on tone are more effects that show up only when the seeing animals begin to move. The disturbance here is more subtle, related to the ability to make well co-ordinated responses. Animals lacking their peduncle lobes have no especial difficulties in recognizing prey, their problems relate to co-ordination of movement as soon as the visual field begins to shift on a side lacking a peduncle lobe. Here the statocyst input is clearly also relevant; in order to move rapidly in a smoothly co-ordinated manner, angular accelerations must be taken into account as well as changes in the visual input, and the peduncle lobe is evidently wired to do this. From the crossed-excitation effect, the commissures joining the two, and the effects of brain-splitting already described in Section 10.2.2 above, it is arguable that the peduncle lobes are concerned in balancing the motor commands from the two sides of the brain, so that disruptive conflicts do not occur between the two largely independent visual systems (Messenger, 1967b, 1971).

There remains, however, the problem that all of these effects are relatively short term. They are most obvious within hours of operation, but decline progressively so that within a week or ten days the postural and locomotor effects of the lesions figured in Table 10.4 have disappeared. This transience suggests that the optic and peduncle lobes may not themselves organize patterns of movements at all. Rather, they provide the sensory information to trigger patterns organized within the basal lobes, where, in contrast, lesions have permanent effects on locomotion. The fact that electrical stimulation of the base of the optic lobes can sometimes produce better-patterned locomotion than stimulation of the basal lobes themselves is a reflexion of the crudity of the means used to excite the basal lobes directly - we cannot produce an adequate pattern of stimuli in this way (Section 10.1.2 above). By exciting the tracts running from the optic lobes we are causing discharges in nerves that normally carry signals summarizing specific patterns of input (from the 'classifying' cells in the optic lobes, Sections 8.1.8 and 12.2.1) and these signals, in contrast to those generated by direct stimulation of the motor neurones in the basal lobes, may constitute an adequately patterned input. *Ex-hypothesis*, the transient disorganizing effect of damage to the visual pathways will be due both to the sudden withdrawal of normal sensory inputs and to the presence of irregular signals from wounds rather than to removal of a part of the motor system. Muscle tone, similarly, is lost with the normal source of exciting sensory input, but becomes replaced in time as the threshold of response to alternative, tactile, inputs drops. It is, on this view, wrong to describe the optic/peduncle lobe system as a 'visuo-motor' control system, as Messenger (1967a, 1971) has done. After it is removed, the motor control system remains, and produces as smooth movement as before, but based on an alternative set of inputs.

10.2.4 *Computing the visual attack*

Octopuses that are used to attacking crabs in their tanks will carry out a relatively stereotyped series of movements when they see prey at the far end of their aquaria. The octopus emerges from the home, walks forward, gathers itself together and makes a jet-propelled leap to cover the crab with its web. Detailed analysis of the time course of attacks can be used to establish which parts of this sequence depend upon visual feedback. Maldonado (1964) analysed cine film to establish the normal course of acceleration and deceleration during the leap and compared this with the pattern shown by animals in cases where the lights were extinguished for a period after the beginning of the jump. As Figs. 10.8a and b show, the leap is unaffected by interrupting the light, so it must be fully programmed before take-off. Tracing the attack back to its earlier phases Maldonado was able to show that the leap is computed during the walk forward from the home. Inter-

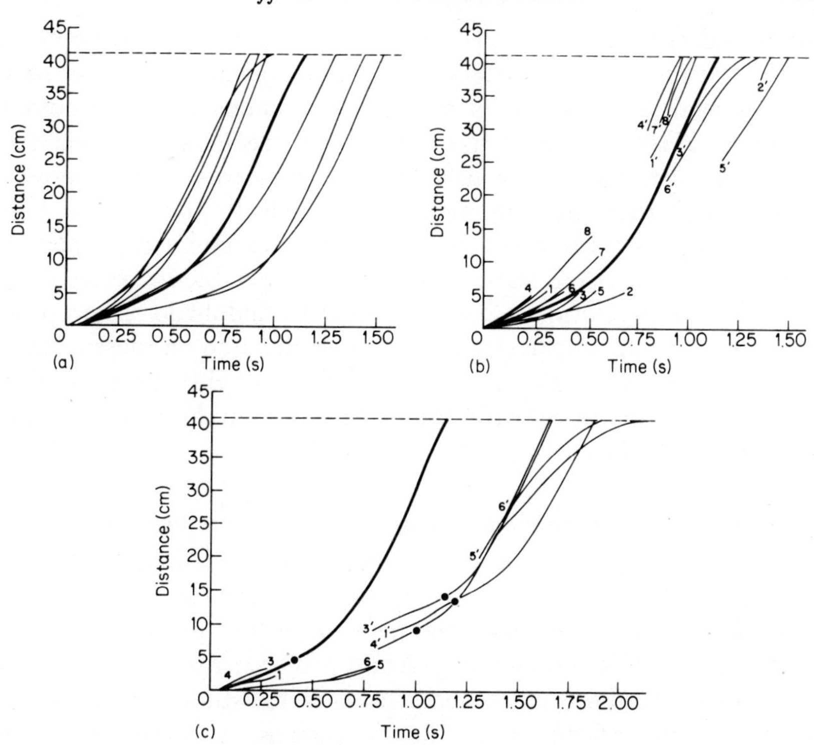

Fig. 10.8 Time and distance plots of the courses of attacks on crabs. (a) Course of seven attacks at trials with illumination throughout. Heavy line: average course. The curves are cut off at 41 cm from the home, the distance where tactile information may begin to play a part. (b) Course of 8 attacks at trials with periods during which the light was interrupted. Heavy line: average shown in (a). Numbers 1 to 8 stand for the instant at which the light was interrupted; numbers with primes stand for the instant at which the light was switched on again. (c) Course of 6 attacks in which the light was interrupted as the octopus was walking forwards from its home. Heavy line: average course from (a). A dot on curves 1, 3 and 4 indicates the beginning of final pattern of acceleration (from Maldonado, 1964).

ruption of the light before the final pattern of acceleration had begun delayed the final leap (Fig. 10.8c). Animals that were learning to attack characteristically approached closer to the target before take-off, took longer to walk forward and often made 'mistaken attacks' in which they overshot the target. Mistaken attacks were found to follow unusually short approach times (Maldonado, 1964). There is, it appears, a minimum time required to collect visual information and compute the leap. Trained animals reduce the approach time to this minimum, a little less than one second. The longer times (often several seconds) taken by untrained animals are attributable in part to a closer approach to the target, reducing the necessary accuracy of the leap, and in part to a tendency to rely upon present visual input rather than on memory. Lesions to the vertical lobe, which is known from other sorts of visual experiment to play an important part in visual learning, caused considerable increases in the time taken to attack (Maldonado, 1963b, c). Maldonado has proposed a theoretical model of the visual attack system based on these observations and other work on visual learning; some of the elements in the model can be identified with specific regions of the *Octopus* brain (Maldonado, 1963a).

10.2.5 *Tactile information and the organization of movement*

Any attempt to pursue the control of visually oriented movements into the basal lobes using behavioural and brain lesion techniques eventually founders because the necessary operations cut off the sensory input that is driving the system. Further analysis of the control of movement can, however, be made using tactile stimuli and observing the performance of the arms. Blinded animals, octopuses with their optic lobes removed and even octopuses with their optic tracts cut centrally to the peduncle lobes (Messenger, 1967a, b) will walk and swim stably if they are allowed a long enough period to recover from the short-term effects of the operations. Since the only sensory inputs now come from below – from the mouthparts via the superior buccal lobe, and via the suboesophageal lobes from the arms and mantle – the effect of lesions to all parts of the central supraoesophogeal brain can be investigated.

Destruction of the superior frontal and vertical lobes has no detectable effect on movement, whether the animals can see or not. Damage to the subvertical, dorsal basal and precommisural lobes similarly seem to have little effect (though this matter has never been considered in detail). Extension of the damage into the anterior or posterior basals, in contrast, has gross and apparently permanent effects on locomotion and posture.

10.2.6 *Basal lobes and the inferior frontal system*

Removal of all parts of the brain behind and above the median inferior frontal and subfrontal lobes (Fig. 10.1) produces an animal that lies on the floor of its aquarium in a tangled and rather flaccid heap. In time, muscle tone improves a little, and the animal may move slowly about if prodded, apparently as a result of small stepping movements by the suckers. These animals can feed (indeed they can be trained to make tactile discriminations; Wells, 1959b) and they will live and grow for many weeks. Apparently spontaneous activities increase with time – the animals are found to have moved between periods of observation – but there is never any return to normal walking or swimming. So far as locomotion is concerned an octopus with the basal lobes destroyed resembles one with the whole of the supraoesophageal brain removed. It exhibits certain reflexes. It will, for example, move arms to search in approximately the right position if it is scratched on the back of the abdomen. If the suckers of one arm are stimulated by contact, they will grip and hold on. If an attempt is made to withdraw the probe, neighbouring arms will come across to the area of contact, and if a pull continues, the arms on the opposite side of the body may align themselves along the direction of the strain. The whole preparation may then pull convulsively. Handling a preparation often initiates 'writhing' or 'cleaning' movements in which the arms are curled back into tight spirals and then twist back and forward from the base (Altman, 1968, 1971; Boycott and Young, 1950; von Uexküll, 1895). Placed upside down, the animal lacking basal lobes may right itself; but it is difficult to tell whether this is achieved 'deliberately' or simply because the rather feeble grip of the suckers holds the arms right way up once they have by chance achieved this position. Basal lobe animals are not infrequently found upside down, mouth and suckers upwards with the arms folded back and applied to the floor – a 'defence' position that is again, perhaps, achieved by chance.

If the front part of the supraoesophageal lobes is removed leaving the basal lobes intact, *Octopus* remains capable of walking and swimming. So far as locomotion is concerned, these animals are indistinguishable from 'normal' blinded octopuses with the whole of the supraoesophageal brain intact. Differences can only be detected when their behaviour is considered in detail. Thus lesions to the inferior frontal system are found to interfere with learning to discriminate by touch, and elimination of the whole of this region prevents touch learning altogether (Chapter 11). Destruction of the superior buccal interferes with motor control of the mouthparts so that the animals cannot chew food that they will still take with the arms (Chapter 4). There are also changes to the arm reflexes; extensive damage to the posterior buccal and subfrontal lobes leaves animals with 'sticky suckers', which seem to have difficulty in letting go of objects that they have grasped. The effect is most marked when any attempt is made to withdraw an object in contact with the suckers (Wells, 1961b; Altman, 1971). It does not, however, interfere with normal locomotion, so that one can sometimes observe the apparently paradoxical situation of an animal in training running away but evidently unable to shake off an object that it has grasped (Wells, 1961b). This implies a dual control system. The posterior buccal/subfrontal region can inhibit the grasping reflex. So can the locomotor orders from the basal lobes. The latter, by themselves, are ineffective only when very strong stimulation of the grasping reflex is experienced.

10.2.7 *Suboesophageal lobes and the interbrachial commissure*

The behaviour of animals with the whole of the supraoesophageal brain mass removed is predictable from the combined effect of removing the basal lobes and the inferior frontal system. These animals do not walk or swim, and they cannot learn to recognize objects by touch. In training experiments, to be reviewed later, they err mainly by taking nearly all small moveable objects that are presented to them. With the elimination of the inferior frontal system, suboesophageal animals have 'sticky suckers', with the result that responses dependent upon adhesion, such as the aligned pull reflex, are easier to demonstrate; the animals will hang on where preparations with basal lobe lesions let go and they can, for example, be lifted out of the water by the grip of their own suckers. Perhaps because of this their posture, sucker-side-down, tends to be a little more reliable than that of basal lobe animals, particularly during the first few days after operation.

There is one further level at which integration of the activities of the individual arms might be achieved. The interbrachial commissure links the arms just centrally to the last of the nerve cell bodies that form the ganglia of the arm nerve cords; central to this point there are only fibres running to or from the arms (Fig. 2.7). The nerves in the interbrachial commissure are all centripetal; they arise from longitudinal fibres running towards the brain, and from the neuropil of the first ganglion in each arm linking the arm with its immediate neighbours and probably directly or indirectly with all the other arms (Graziadei, 1971). Stimulation of the commissure between two arms produces sweeping and stepping movements of the arms on either side, as does stimulation of the arm nerve cord distal to the interbrachial commissure. With continued stimulation (3 V, 0.7 ms pulses at 100 Hz) the effect spreads to further arms on either side. The spread of excitation is prevented by cutting the commissure but not by sectioning the arm nerve cord on the brain side of the commissure (Altman, 1968).

10.2.8 *Movement control at arm nerve cord level*

The bulk of the muscle in the arms is enclosed in a tubular connective tissue sheath to form a solid mass surrounding an inner sheath around the axial nerve cord (Fig. 10.9). Within this core of intrinsic muscu-

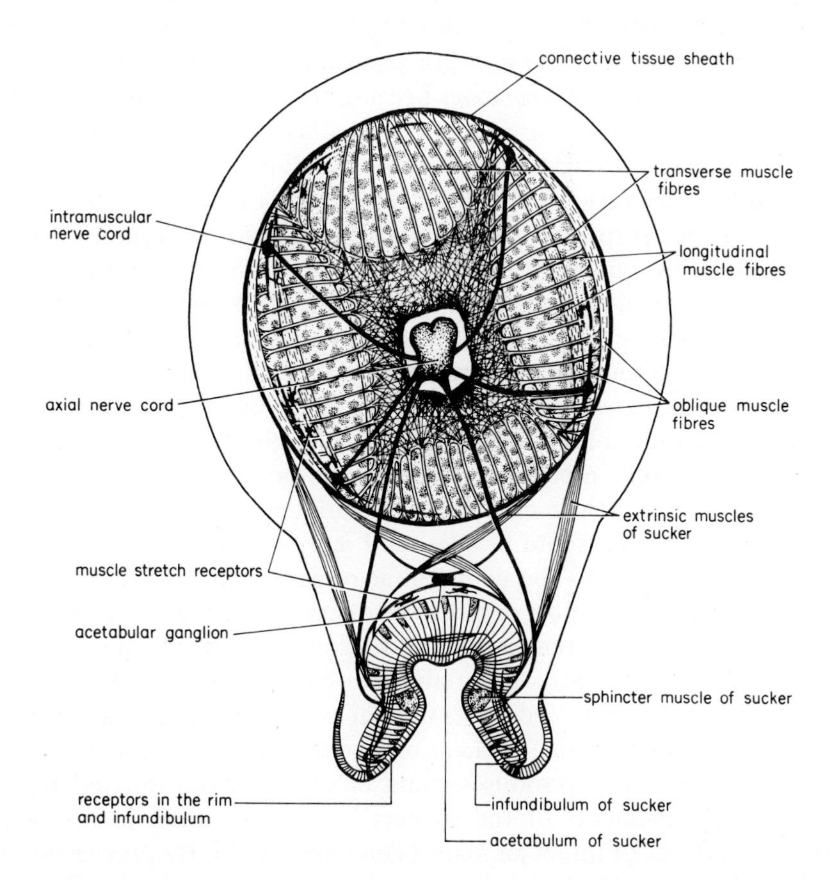

Fig. 10.9 Muscles and nerves in the arm, shown in transverse section. Muscles operating the skin have been omitted (after Graziadei, 1965b).

lature there are longitudinal and transverse fibres, and a complex system of oblique fibres arranged in three series that differ in orientation. Outside this core are the extrinsic muscles controlling the movement of the skin and suckers (Guérin, 1908).

The nerve cords in the eight arms are together estimated to contain between two and three times as many neurones as the brain. The bulk of these are to be found in the axial cords, but there are also four intramuscular cords in each arm, lying parallel to the axis (Fig. 10.9), and a subacetabular ganglion immediately above each sucker. Each of these elements contains motorneurones and receives sensory input from outlying nerve cells. The axial cord includes swellings, alternately to each side, related to the individual suckers. These are generally and conveniently referred to as ganglia, though the term is a little misleading since the nerve cell body wall of the axial cords is in fact continuous up to the level of the interbrachial commissure; the ganglionated appearance seen in sections arises because the cord flexes from side to side as it runs down the arm. Martoja and May (1956) and Graziadei (1971) have summarized the anatomical situation.

In view of the very considerable complexity of the arm nerve cord system it is not surprising to find that the behavioural responses even of isolated arms tend to be elaborate. Attempts have been made to list arm 'reflexes' (Ten Cate, 1928; von Uexküll, 1894) but the results are less than satisfying because the responses observed and reported are clearly not an exhaustive list, and because they are commonly overlain by movements that may be truly spontaneous, or longterm responses to amputation and manipulation (see Rowell, 1963).

Ten Cate (1928) used intact animals habituated to life in aquaria, and tested them in their own tanks, stimulating the skin of the arms and suckers with hairs and bristles, and more violently by pricking or pinching with forceps. The commonest response, to moderate (stroke with a bristle) stimulation of a sucker or its vicinity, was slight withdrawal followed by extension of the sucker disc towards the point of origin of the stimulus. Stronger stimulation caused withdrawal followed by flexing of the arm and extension of several suckers towards the stimulus. If the extending suckers touched anything, they grasped it.

The same series of responses could be elicited from isolated arms; 'The reflex movements of the suckers . . . do not differ in any way from the reflexes of intact animals' (Ten Cate, 1928). He proceeded to analyse the matter further and concluded that three levels of reflex arc could be detected in the arms. The first of these must operate through the subacetabular ganglion, since the rim of the sucker would still bend inwards and the disc still contract if touched after cutting all the peripheral nerves from the arm cord. Suckers cut off below the level of the subacetabular ganglion did not respond. A second level, with a much lower threshold, operated through the arm nerve cord ganglion immediately overlying the sucker concerned. Stimulation in the centre of the sucker disc led to flattening and expansion, stimulation of the more peripheral parts to extension and grasping. Cutting the cord on either side of the ganglion did not alter these responses. Still longer arcs could be traced in cases where stimulation led to recruitment of suckers, all turning towards the source of stimulation. Since stripping the intervening part of the arm nerve cord of its ganglion cells and peripheral connexions did not prevent the spread of such effects (von Uexküll, 1894) Ten Cate argues that these must be due to simple reflex arcs with long neurones connecting the more distant respondents to the stimulus source.

Rowell (1963) repeated some of Ten Cate's experiments, with essentially similar results. He failed, however, to observe the turning and grasping movements noted by Ten Cate (1928) and von Uexküll (1894). This, as he points out, is hardly surprising since he was working with direct electrical stimulation of isolated arms.

In tactile training experiments, the arms of blinded octopuses regularly twist and grasp objects touched against the suckers or upper surface of the arms. A ring of arms isolated from the rest of the body by a cut below the level of the buccal mass will do the same and, when fresh, can be relied upon to grasp, flex and pass pieces of fish in towards the base of the arms, where the mouth once was. Less commonly, such preparations will reject inedible objects (Altman, 1971; Wells, 1959b, see Table 10.5). Similar results (and most of the reflex responses described above) can be obtained with denervated arms, left attached to the octopus. These chronic preparations differ from excised arms only, it seems, in their muscle tone; they tend to be flaccid until excited and in general require more vigorous stimulation if they are to perform like freshly amputated limbs. The implication is that the brain has an excitatory effect in the intact animal that is mimicked by wound trauma in the acute preparations (Altman, 1968).

10.2.9 *Recordings from the arm nerve cords*

Recordings from the nerves entering the arm cords from the suckers

Table 10.5 Take and reject responses by preparations with increasing amounts of the tactile system intact (from Altman, 1971).

	Sardine			*Sardine soaked in quinine*		
	Accept	*Reject*	*Refuse*	*Accept*	*Reject*	*Refuse*
Isolated ring of arms $n = 9$	15	–	2	1	10	8
Denervated arm, chronic preparation $n = 7$	11	–	3	–	–	13*
Suboesophageal brain only ($n = 4$, 4 sets of tests)	32	–	–	23	1	8
Inferior frontal system intact ($n = 4$, 4 sets of tests)	30	1	2	7	22	1

2 Tests with each food were given by touching the pieces against an arm.
* but on 7 of these occasions the arm, having failed to grasp the bait, bent towards the mouth in a typical accept reflex.

are dominated by the discharges of rapidly adapting mechanoreceptors (see Section 7.3.2). Inside each arm ganglion, Rowell (1966) found units that responded to touch on the ganglion's own sucker, rim, cup or skin, and units that responded only to tactile stimulation of other suckers. Thus, for example, there were units that responded only to stimulation of either of the two suckers distal and contralateral to the recording point. A further series of neurones fired only when a sucker was moved, or chose to move itself. These presumably represented discharges from proprioceptors in the muscles (see Section 7.4.1).

Moving centrally, to record from the longitudinal tracts of the axial cords, it is evident that a very great deal of peripheral processing takes place before the input from the suckers is relayed to the brain. One would expect this on anatomical grounds, since the number of cells in each axial ganglion (about 1.3×10^5; Young, 1963a) is overwhelmingly greater than the number of axons in the peripheral nerves (about 5000). Rowell (1966) found interneurones that responded to areas as small as the rim of a single sucker, and as large as the whole tactile field distal or proximal to the recording point. Units were found that would respond to stimulation of the skin, or of the suckers only. There was a considerable range in threshold.

The great majority of interneurones in the axial cords were fast adapting, though a few would remain active for several seconds after the end of any obvious stimulation. Of the fast adapting nerves, many were apparently connected to detect differences in stimulation, adapting to a repeated stimulus pattern but immediately aroused by any change; it appears that these units habituate rather than fatigue, so that there is a measure of learning even at this level in the octopus CNS. In addition to the 'novelty' units there were interneurones that distinguished between contacts made by the experimenter and those arising as a result of the animal's own movements (Rowell, 1966).

10.2.10 *Stimulation of the arm nerve cords*

The complexity on the sensory side is matched by what is plainly a complex motor output system. Von Uexküll (1894) was the first to investigate this and more recent reports can be found in Altman (1968) and Rowell (1963); Fig. 10.10 is taken from Rowell's summary. Almost any movement of which a sucker is capable can be evoked by suitable stimulation of the ganglion controlling it. Widespread effects, with progressive recruitment of neighbouring suckers can be evoked by continued stimulation. Interestingly, Rowell (1963) was unable to excite nerves leading to the release of sucker grip, although from behavioural studies it is quite clear that the animal must have a detailed sucker-by-sucker control of this; the animals can hold small objects and move about, or pass objects from sucker to sucker in the course of take or rejection movements.

Stimulation of the middle region of the axial nerve cords can produce contraction or extension, twisting or sideways movements of the arm. Again repeated stimulation at threshold levels can lead to a slow build-up and progressive spread of excitation, indicating a range of polysynaptic pathways. The intramuscular nerve cords seem to control only the extrinsic dermal musculature (Rowell, 1963), which is perhaps surprising since they receive proprioceptive inputs from receptors buried in the axial musculature (Fig. 10.9).

10.2.11 *Hierarchic control, an attempt to summarize*

Behavioural studies show that most of the apparently complex things that an octopus does with its individual arms can be organized at arm nerve cord level. The function of the brain in relation to tactile discrimination, for example, would seem to be limited to triggering the sets of movements that result in taking or rejection. In locomotion all

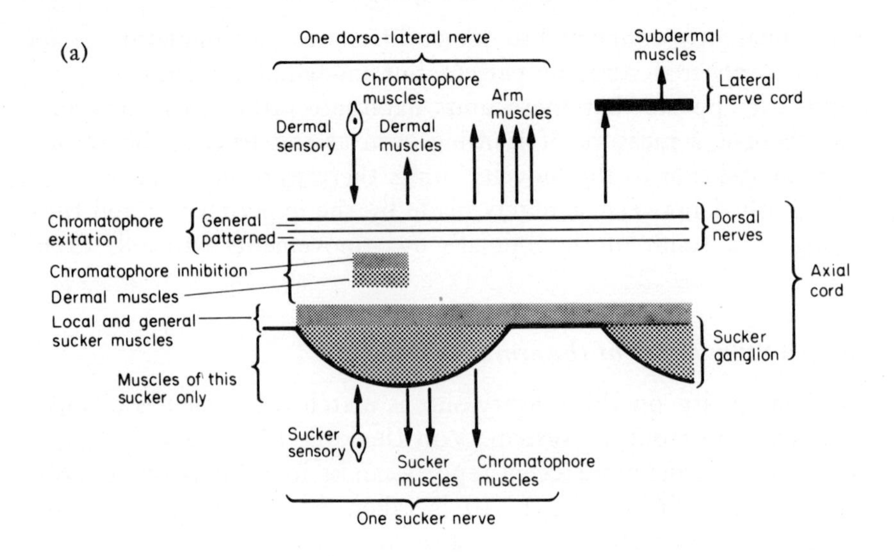

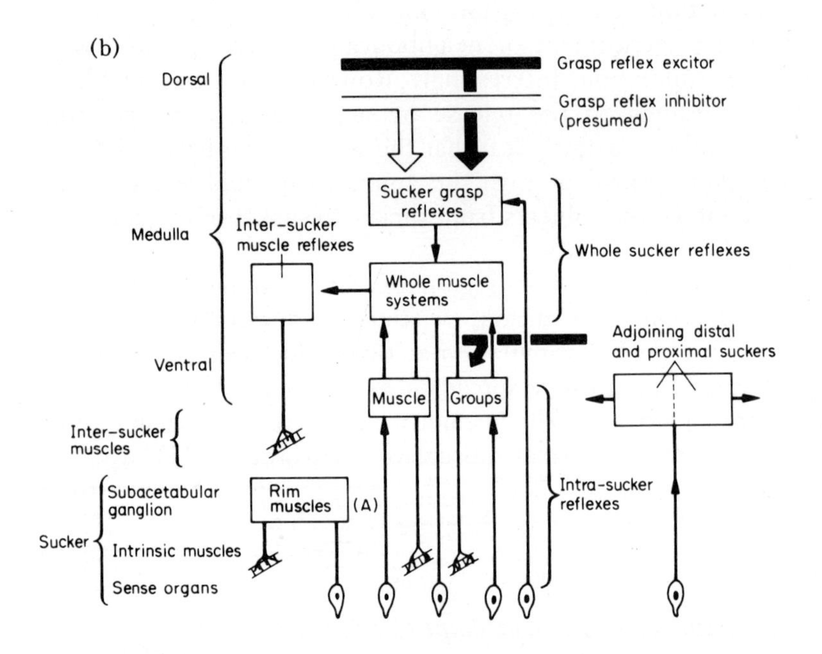

Fig 10.10 (a) A stimulator's guide to the brachial medulla of *Octopus*. A short length of medulla with one sucker 'ganglion' is shown and the effects obtained by electrical stimulation of different areas and of the nerves leaving it are indicated. Areas giving contraction of the dermal muscles and inhibition of the chromatophore muscles are small and poorly defined; the diagram shows only their dorso-ventral level on the

the detail of adapting the arms to the substrate over which the animal is walking could be dealt with at arm nerve cord level.

The fact that it might be does not prove that it normally is. The brain could override and control the details of what the arms are doing. Two lines of evidence argue against the possibility. One is the nature of the sensory information that is relayed from the arm nerve cords to the brain. Rowell (1966) found no proprioceptive component to the messages recordable from the brachial nerves, while the number of fibres in these nerves (about 30 000 compared with some 4×10^7 neurones in the axial cord of each arm; Young, 1965d) itself implies that only the most summary statements about conditions in the arms can ever be relayed to the brain. The second block of evidence that argues against central control of motor details is behavioural, and has been outlined in Chapter 9; very briefly, octopuses never seem to learn tasks that would require a detailed central control of their arm movements.

Integration of the movements of the individual arms seems to be determined at several levels. The interbrachial commissure plays a part, as does the suboesophageal part of the CNS (Section 10.2.7 above). But walking does not seem to take place unless the basal lobes are also present. Again, it is possible that the results obtained in the rather crude experiments so far carried out are misleading and that in fact, walking could be organized by the suboesophageal lobes, given an appropriate set of triggering stimuli. But suboesophageal preparations, and those with the inferior frontal system intact and the basal lobes missing will live for several weeks. This is more than enough time for the recovery of posture and locomotion by octopuses with their optic tracts cut (Section 10.2.3 above) where, too, the effects at first seem to be attributable to removal of a higher level of motor control. On balance, it would seem reasonable to assume that the organization of integrated stepping movements is a basal lobe function and that the

---

medulla, and not their lateral extent which is discontinuous. Unshaded areas control the general musculature of the arm. (b) Diagram of the nervous organization in the medulla affecting the muscles of one sucker of the arm. Anatomical levels are shown on the left-hand side. The two reflexes mechanisms labelled (A) and (B) are respectively those described by Ten Cate (1928) and von Uexküll (1894). The label 'whole muscle systems' indicates for example all the longitudinal muscles of the sucker, while 'muscle groups' would mean a small patch of the longitudinal muscles (from Rowell, 1963).

function of the suboesophageal brain, perhaps in association with statocyst input, is to balance the distribution of stresses by adjusting muscle tone and the position of the arms as the animal moves over uneven ground, or along the sides and bottom of its tank.

Above basal lobe level, the optic/peduncle lobe system is concerned only in the case of visually oriented locomotion. Blinded animals, even those with their optic tracts cut centrally to the peduncle lobes, move smoothly without it. The transience of effects of damage here suggest that we are dealing with a mechanism for feeding appropriate sensory information to the basal lobes rather than with a motor control override of the higher motor centres.

10.3 Chromatophores and their control

One of the startlingly beautiful things about an octopus is its ability to change colour. The same animal can, within a fraction of a second, convert itself from a dark brown glowering creature to a pale ghost; little specimens can become almost transparent. Accompanied, as it so often is, by a change in skin texture, so that the animal is at one time smooth and immediately afterwards prickly with extended papillae, the capacity for colour change can both camouflage the animal directly, and baffle would-be predators simply by virtue of its changeability. Search image is disrupted. If, as is also often the case, the retreating octopus discharges a puff of mucus-bound ink as it switches from dark to pale and jets away, a decoy is added that can regularly confuse even the most sophisticated human observer, who knows exactly what is going to happen.

Octopuses use their capacity for colour change in a variety of ways. For crypsis, as above; for displays between themselves (Section 5.1.5), for alarming potential predators (the 'dymantic' display, Frontispiece, Plates 2 and 5) and possibly for flushing prey that would otherwise remain motionless (Section 4.1.1 and Fig. 10.17c, below).

Until quite recently, observations of chromatophore structure and electrophysiology were limited by the resolution of the light microscope and the absence of microelectrodes capable of penetrating the very small muscle fibres that surround the bag of pigment. In the absence of unequivocal observations, there was much argument about the nature of the innervation of the muscles, about cell elasticity (radial muscles pulled it out, but how did the chromatophore contract again?) and about a whole range of problems arising from the behaviour of chromatophores in recently excised or denervated skin. Florey (1969) has reviewed this older work and the account that follows is based largely upon his and Kriebel's (1969) electrophysiological and pharmacological study and on Cloney and Florey's (1968) report on the ultrastructure of the chromatophore sac and its muscles. This work was done on *Loligo opalescens,* but there is no reason to believe that the story for *Octopus* differs from it to any substantial extent.

10.3.1 *Chromatophore structure*

Each chromatophore consists of a bag filled with granules of pigment, held in place on a framework of fibres. The bag is probably elastic, but it is also possible that the filaments around it are contractile; in *Octopus*, extracellular channels penetrate to the bag surface in a manner strongly reminiscent of the T-system in muscle cells (Froesch, 1973a). Attached to the bag is a ring of radial muscles, linked one to another by tight junctions, so that the possibility of electrical interaction between the muscles exists (the fact that the muscles sometimes pulsated in synchrony led some of the older workers to believe that the fibres formed a syncytium, which is now plainly not the case). The upper and lower surfaces of the pigment bag are enclosed in a single cell, the surface of which is thrown into a series of folds in the contracted state. Fig. 10.11a is a reconstruction of a pigment cell, based on electronmicrographs, while 10.11b shows the approximate relative dimensions of the cell when it is contracted and expanded.

The radial muscles are obliquely striated, with large and small (presumed myosin and actin) filaments surrounding a central core of mitochondria. There is no trace of large, paramyosin filaments which is perhaps surprising, since the muscles can apparently remain contracted for long periods and a molluscan 'catch' mechanism might have been expected. Along each muscle there is a nerve, covered in glia on its exposed surface. Each nerve contains between one and four axons, lying in a groove along the surface of the sarcolemma and forming a continuous synaptic junction along the whole of the length of the muscle (Mirow, 1972a). In the contracted state the nerve snakes along the muscle fibre in the manner shown in Fig. 10.11a in *Loligo*; in *Octopus,* the neuromuscular junctions are more localized (Fig. 10.11c).

(a)

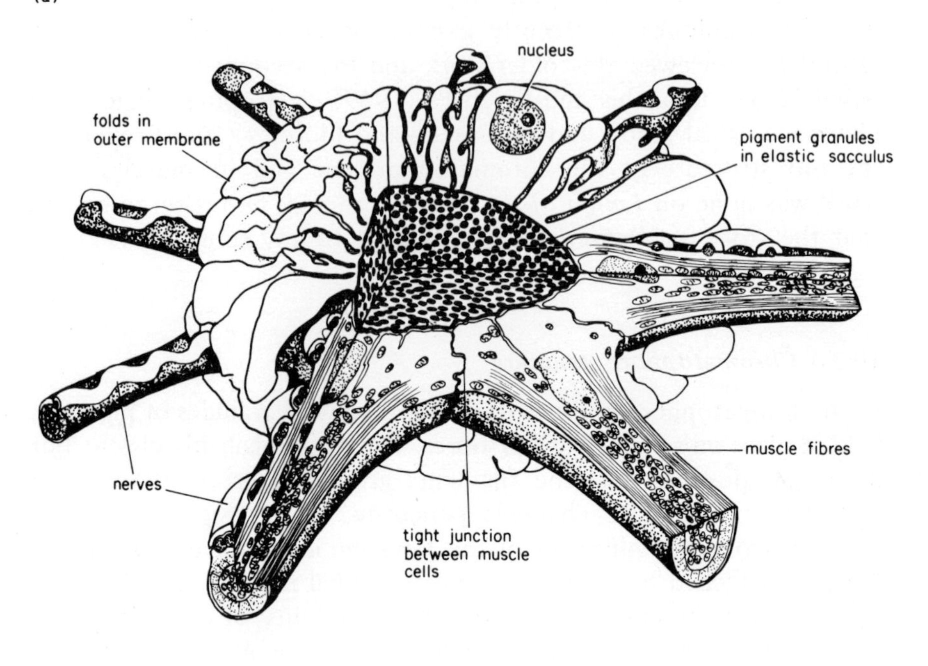
nucleus
folds in outer membrane
pigment granules in elastic sacculus
muscle fibres
nerves
tight junction between muscle cells

(b)

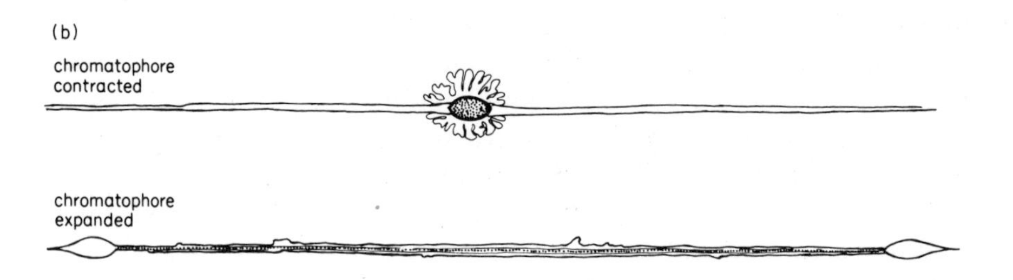
chromatophore contracted
chromatophore expanded

(c)

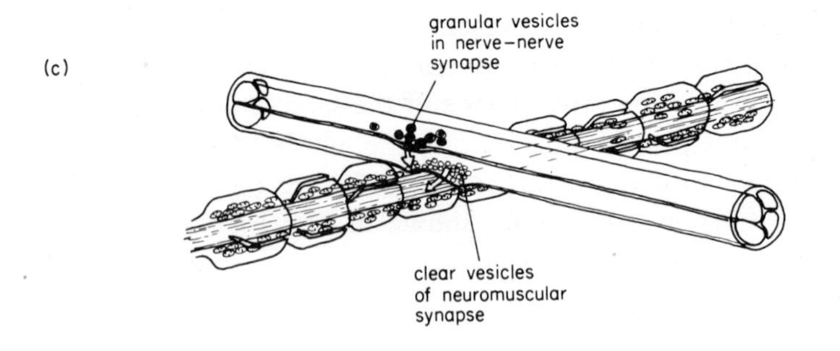
granular vesicles in nerve–nerve synapse
clear vesicles of neuromuscular synapse

10.3.2 *Innervation and contraction*

The expansion of a chromatophore reflects the combined activity of the radial muscles around it. Fig. 10.12 shows the effect of increasing

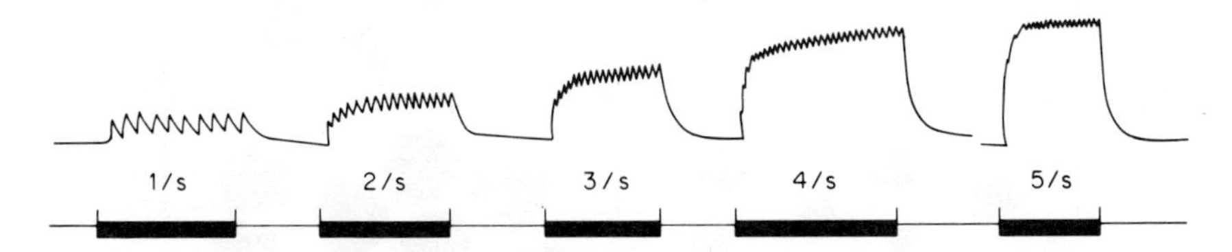

Fig. 10.12 Tracings of contractions of chromatophore muscle fibres resulting from stimulation of their motor axons. The upper channel records contraction, the lower channel the stimuli. The numbers indicate the frequency of stimulation in pulses per s (after Florey, 1969).

the frequency of stimulation through a nerve of several fibres; the muscles twitch, and the twitches combine to give a tetanus. Reducing the voltage produces a stepwise reduction in the strength of contraction; the chromatophores contract, but irregularly as one after another of the radial muscles drops out. Fig. 10.13 traces the outlines of a group of chromatophores during progressive expansion caused by progressive increases in the voltage applied at 20 pulses s^{-1}; full expansion in this instance follows recruitment of three separate motor units, each of which innervates some of the muscles from each of the chromatophores.

Intracellular microelectrodes inserted into individual muscles show that the effect of nerve stimulation is to produce local excitatory postsynaptic potentials. In *Loligo*, at least, there is no sign of spike generation, at whatever voltage. Instead, there is a stepwise increase in local postsynaptic potentials with increasing voltage that presumably

---

Fig. 10.11 Chromatophore structure. (a) Chromatophore of *Loligo opalescens* with the muscles relaxed. In this condition, the pigment sac would have a diameter of about 50 μm. (b) The same in section showing the relative dimensions with the chromatophore muscles relaxed and contracted. (c) Chromatophore muscle and nerves from *Octopus vulgaris.* Here the mitochondria lie around a central core of muscle fibres. The nerves do not lie along the fibres but synapse where they cross. As well as neuromuscular synapses there are presynaptic nerve-nerve junctions. (a) after Cloney and Florey, 1968; (b) after Florey, 1969; (c) after Froesch 1973a)).

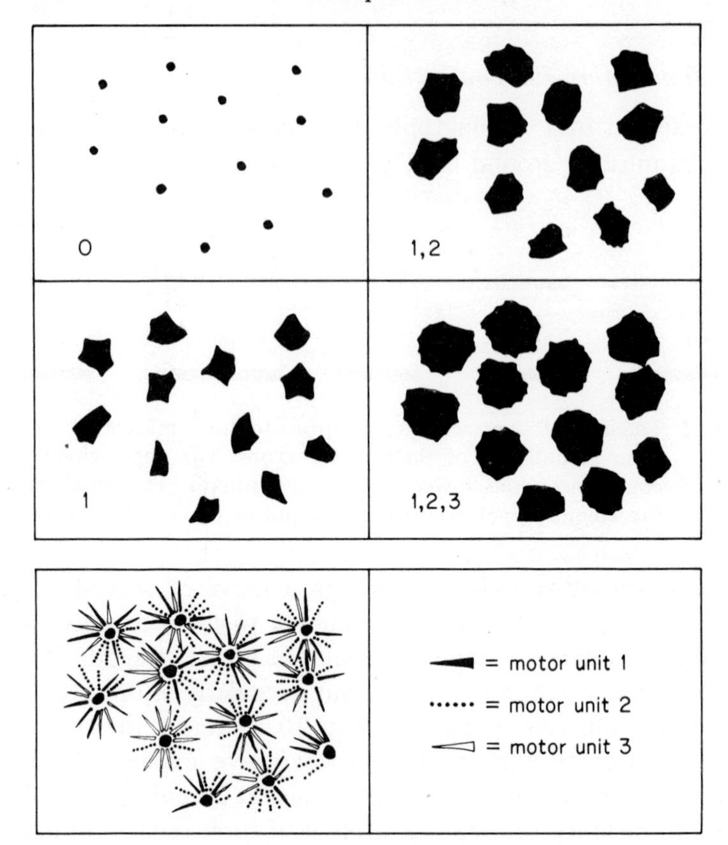

Fig. 10.13 Stepwise expansion of chromatophores due to recruitment of additional motor units. The upper set of four diagrams shows the appearance of the same group of brown chromatophores in the absence of stimulation (0), and when 1, 2 or 3 motor axons of the appropriate nerve bundle are stimulated at the same frequency of 20 pulses s^{-1}. The lower set of diagrams illustrates the distribution of the motor units. The diagrams are based on a series of photomicrographs (from Florey, 1969).

indicates polyneuronal innervation; the smaller potential changes show a slower rise time and a longer latency than the larger.

Isolated areas of skin, denervated on the otherwise intact animal, or removed and pinned out in seawater, are at first pale but later become coloured by waves of chromatophore contraction. The contractions appear spontaneously in one or a small group of chromatophores, spread outwards and return in a series of 'wandering clouds' which may continue for many hours *in vitro* or until nerve regeneration occurs in the intact animal (see Section 10.3.6 below). Muscles from neighbouring chromatophores appear to form bridges (Plate 2d) so that direct mechanical excitation is the most likely explanation of the spread.

A notable feature of the 'wandering clouds' contraction is that all the radial muscles in a given chromatophore may contract together. In chronic preparations, at least, it is obvious that this cannot be achieved nervously; the nerves have degenerated. Microelectrodes inserted into the spontaneously active muscles now reveal a change in physiology. The muscles generate spikes and can stimulate one another, presumably through the tight junctions already noted above (Florey and Kriebel, 1969).

The chromatophore muscles in isolated skin often show tonic contractions that may last for minutes on end. Florey and Kriebel (1969) examined muscles in this condition and found that tonic contraction was associated with prolonged showers of miniature potentials, apparently due to spontaneous release of transmitter by the chromatophore nerves. Transmitter release was not affected by tetrodotoxin, which would eliminate nerve impulses, so there is no possibility that the tonic effects are due to continuing activity in the nerves of the excised skin. Once set up, tonic contractions can persist despite the disappearance of the miniature potentials; the muscle stays contracted until something is done to release it, a molluscan 'catch' muscle without the paramyosin fibres often associated with the 'lock-on' capacity (Fig. 10.14).

The actions of acetylcholine (ACh) and 5-hydroxytryptamine (5-HT) strengthen the analogy with catch muscle. ACh induces tonic contraction and 5-HT supresses it (Twarog, 1954). Pharmacological experiments of this sort, the observation that excitation of the chromatophore nerves can cause relaxation in muscles tonically contracted (Fig. 10.14) and the existence of clear and dense-cored vesicles in separate nerves innervating the muscles (Mirow, 1972a; Froesch, 1973) all suggest a double innervation with excitatory and inhibitory components; the obvious implication is that ACh is the excitatory and 5-HT the inhibitory transmitter. Internal electrodes, however, never show hyperpolarization as a result of nerve stimulation or drug application. ACh enhances and 5-HT reduces the frequency of miniature potential changes, but the effect seems to be presynaptic, altering the rate of 'spontaneous' release of transmitter (see Fig. 10.11c). If either substance were a transmitter itself, one could expect it to have an

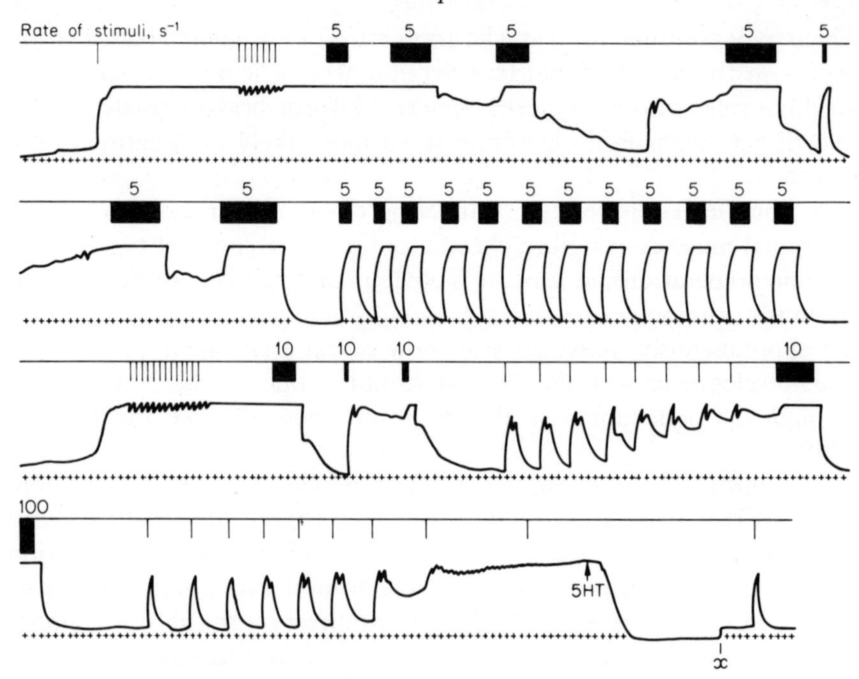

Fig. 10.14 Nerve-induced and tonic contractions and nerve-induced relaxations of chromatophore muscle fibres. Four sequences of 3-channel recordings are shown. In each case the upper channel marks the stimuli delivered to the motor nerve; the numbers indicate pulses s^{-1}. The middle channel records the contractions of about five muscle fibres as signalled by the darking of the optical field by the shifting boundary of a segment of the pigment mass of a single chromatophore. The lower channel provides time marks at 1 s intervals. Near the end of the fourth (lowest) sequence, 5-HT (10^{-6} g ml^{-1}) was applied to the preparation resulting in almost immediate loss of muscle tone. At x the trace was adjusted (after Florey and Kriebel, 1969).

effect on the permeability of the muscle membrane. In the absence of any such effects, Florey, and Kriebel (1969) propose that the continued release of transmitter from the nerves of ageing preparations is attributable to elevated free calcium levels, a failure to return to binding sites within the sarcoplasmic reticulum. A motor nerve impulse, or the application of 5-HT, reactivates the calcium transport system, reduces 'spontaneous' trasmitter release and allows the underlying muscle to relax. Additional support for this view comes from an additional action of 5-HT on the muscle itself, which contracts and

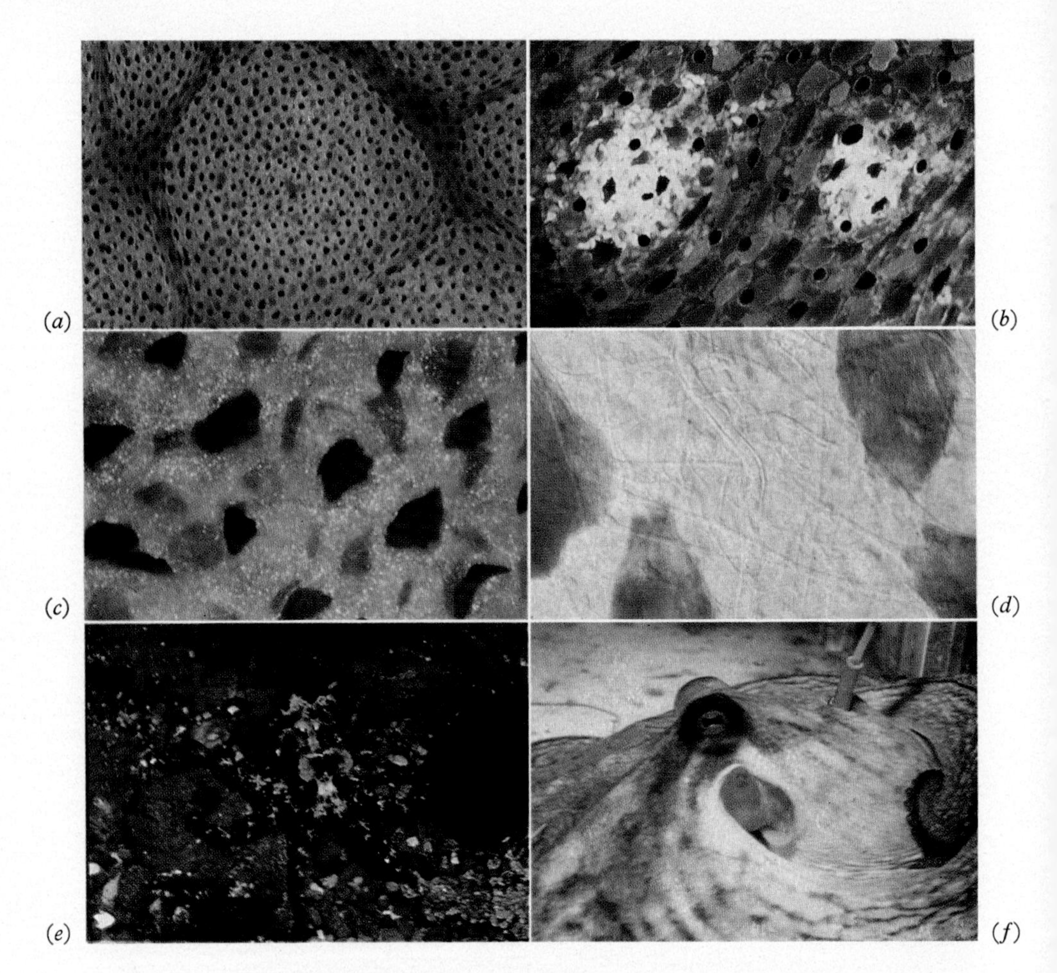

Plate 5 (*a*) Polygonal unit with the chromatophores largely contracted; the white leucophores can be seen against a greenish background provided by the reflecting iridiophores. Incident white flash. Frontal head skin. × 20.

(*b*) Chromatophores and leucophores from the web. Dark field. × 25.

(*c*) Chromatophores and iridiophores from the iris, where there are no leucophores. Incident white flash. × 50.

(*d*) Chromatophores and their muscles in the web. Interference contrast. × 120. (Photographs by D. Froesch.)

(*e*) Small octopus with the skin papillae raised. (Photograph from Packard and Sanders, 1969.)

(*f*) A larger animal with the skin smooth, dark hood and partial dymantic. Animal with tube to blood pressure recorder (see Chapter 3.) (Photograph by M.J.W.)

relaxes faster when 5-HT is present – again, *ex hypothesis,* a calcium-binding effect (Florey and Kriebel, 1969).

These experiments are interesting in relation to the physiology of muscle contraction in general. They are not immediately relevant to conditions in the intact cephalopod where, so far as we know, the chromatophore muscles never show tonic contractions. They appear always to be under direct nervous control, their state of contraction or expansion being attributable to variations in the frequency of a continuous shower of nerve impulses. Observed closely, the chromatophores of a living cephalopod are in perpetual shimmering movement, even when the animal, seen at a distance, remains a steady colour.

Hormonal effects, if they occur at all, are attributable to the action of substances at CNS rather than at chromatophore level (Section 6.2.2).

10.3.3 *Reflecting elements in the skin; colour and tone matching*

Octopus vulgaris hatches with about 70 red-brown chromatophores. As it grows the number increases steadily, so that an adult weighing 500–1000 g will carry between two and three million of these tiny organs, at a density of 100–200 mm^{-1} (Packard and Sanders, 1969).

By the time the little animals settle out of the plankton, the population of chromatophores has become subdivided into two groups, a very dark set, which can vary from black or red-brown, and a paler series which appears red when contracted and pale orange-yellow in extension. Plainly, these two alone cannot possibly account for the full range of colours that an octopus can show (Frontispiece and Plate 5). The explanation lies in the possession of two sorts of light-reflecting structure, the iridophores and the leucophores, lying beneath the chromatophores (Plates 4 and 5 and Packard and Hochberg, 1977).

Iridophores contain stacks of platelets of a chitinous material spaced out at about one quarter of the wavelength of visible light, acting as broad-band reflectors of all the colours in the environment around the cephalopod (Denton and Land, 1971). Leucophores are irregular in shape, lie below the iridophores and the chromatophores, and are filled with a creamy guanine-like substance. They are opaque, and reflect white light (Plate 5b). The ultrastructure of cephalopod iridophores has been described by Froesch (1973a – *Octopus*) and by Mirow (1972b – *Loligo*); nobody has yet reported on the fine structure of the leucophores.

All the available evidence suggests that *Octopus* is colour-blind (Section 8.1.16). It nevertheless manages to match the colours of the plants, animals and stones among which it lives to a quite amazing degree. Reflection is clearly a part of the explanation, and it is reflection rather than the chromatophores that is responsible for the cryptic coloration of an octopus at rest in the sea or matching blue and green colours in its aquarium. Reflection is passive. What the octopus can and does do actively is match the intensity of the light that it reflects to the intensity reflected from its background. The chromatophores can be expanded to screen the reflecting elements to any desired extent and the seeing animal clearly does this; a blinded octopus at rest does not (Messenger, 1974).

10.3.4 *Colour and tone patterns*

Chromatophores, iridophores and leucophores are not uniformly distributed. At a whole-animal level, there is a well-marked pattern of white spots, due to the leucophores (Fig. 10.15; Frontispiece and Plate 5). There are plainly more dark chromatophores on the dorsal than on the ventral surface. In detail, the whole upper surface of the animal is broken up into patchwork of polygons, each with a pale whitish centre (leucophores) surrounded by a reflecting region (iridophores) rimmed

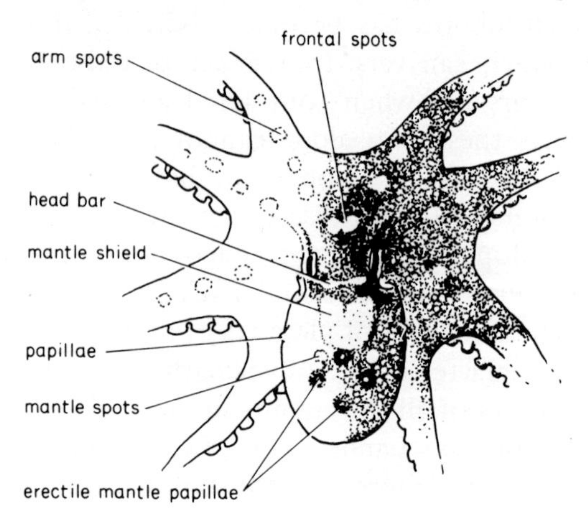

Fig. 10.15 Permanent white components of the patterns shown by *Octopus vulgaris.* These are always present and visible except when shielded by expansion of the chromatophores. The frontal spots are particularly conspicuous, and nearly always shown. The four long mantle papillae are included as landmarks (after Packard and Sanders, 1969, 1971).

by a dark groove (Frontispiece and Plate 5a); the groove colour is due to chromatophores, concentrated by the folds around the edges of the polygon; there are relatively few iridophores here. White spots and the polygonal network are permanent. As the animal grows each patch gets larger; the precise pattern may be as characteristic of the individual as a fingerprint, with no two octopuses quite alike. Permanent, too, are the locations of the erectile skin papillae on head and back and arms (Figs. 10.15, 10.16 and 10.17). These can affect colour

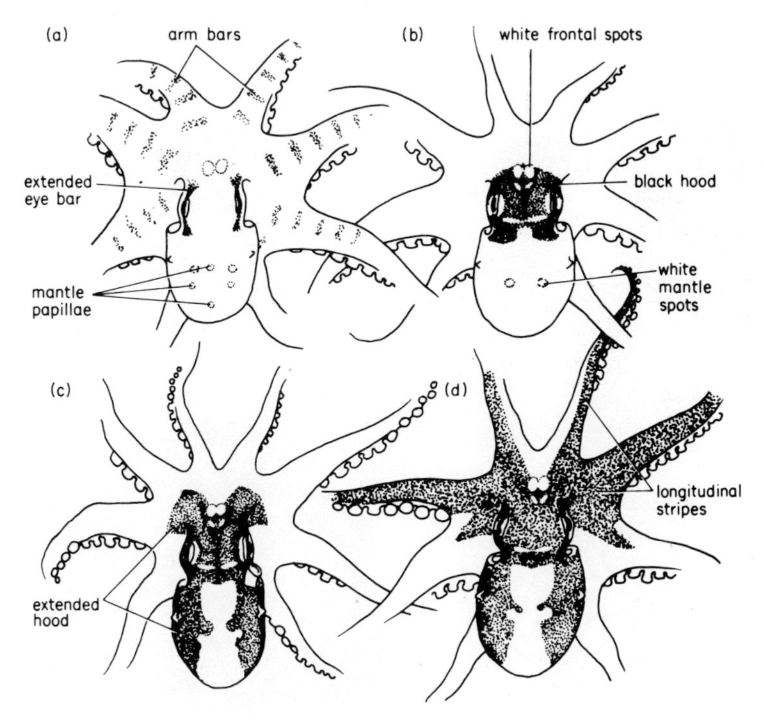

Fig. 10.16 Dark components of patterns shown by *Octopus vulgaris.* The white spots bear a constant relation to the rest of the patterns (after Packard and Sanders, 1971).

because contraction of the underlying muscles concentrates and reorients the iridophores. The animal, indeed, can hardly move without in some way affecting the colours that it shows to the world, so that the different postures and forms of locomotion that it adopts all play a part in determining the patterns that it displays (Packard and Sanders, 1969, 1971; Packard and Hochberg, 1977).

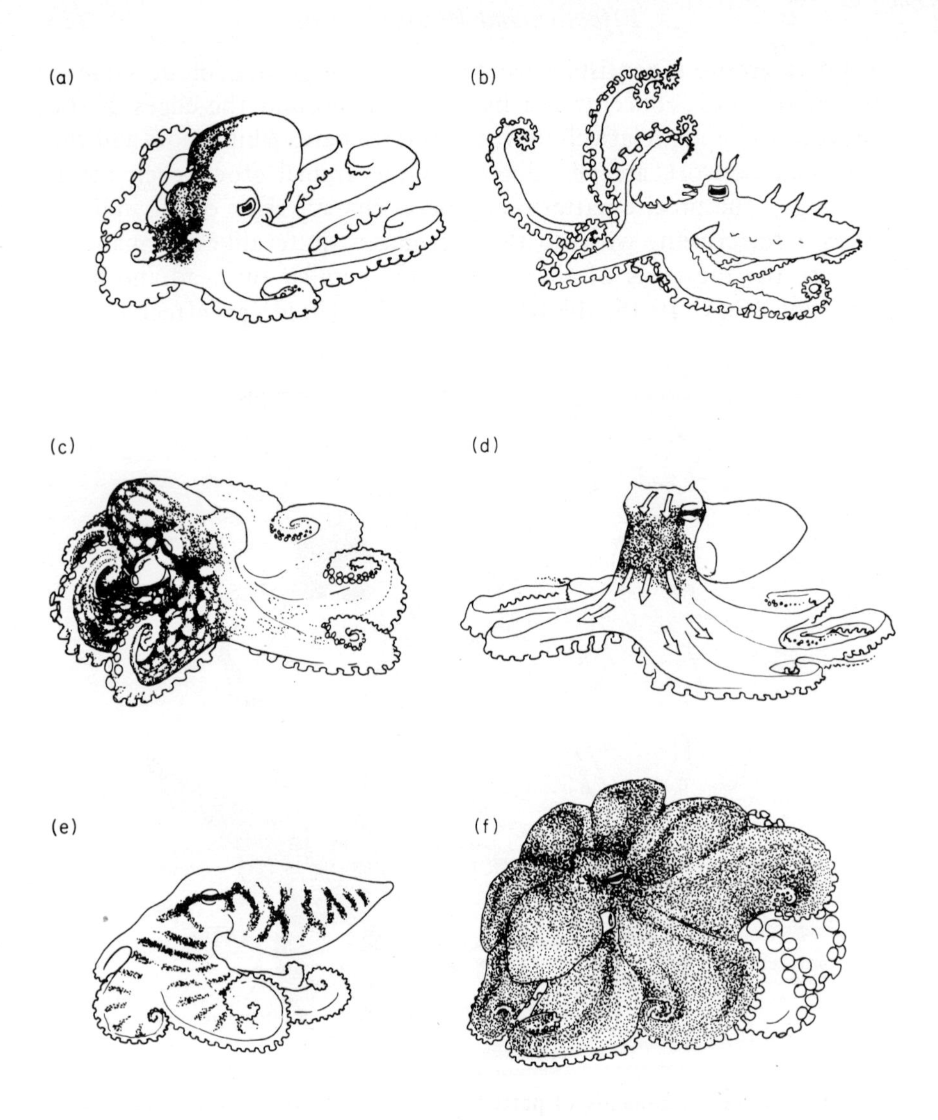

Fig. 10.17 Postures and patterns associated with particular situations. (a) Unilateral extended hood (see Fig. 10.16) displayed in response to an approach from the animal's right side. (b) The flamboyant posture adopted by small (< 5 g) animals. This is probably cryptic. The colour is uniformly dark in specimens of < 1 g, and later becomes replaced by black hood, broad mottle and lateral stripes. In animals larger than 5 g, the flamboyant response to disturbance is replaced by the dymantic (Plate 2 and (b) in Frontispiece). (c) Broad 'conflict' mottle here displayed unilaterally towards another octopus approaching from the animal's right side. (See also (a) in Frontispiece). (d) Moving flush, shown when a small object approaches to one side; this display may

Chromatophore expansion is directly controlled from the central nervous system and here too there is evidence of permanent patterning. What at first appears to be an endless repertoire proves on analysis to be built up from a strictly limited number of components, which recur regularly but in differing combinations and at differing intensities to give a wide variety, but by no means an infinite range of displays.

In an attempt to draw up an exhaustive list of the components of patterns of chromatophore expansion, Packard and Sanders (1969, 1971) have listed the following: Arm bars; eye bar and extended eye bar; eye ring, and dark edges to the suckers (perhaps always occuring together, as in the dymantic response, Plate 2); the dorsal network of polygons (see above), which can give a range of mottled effects; black hood and extended hood; longitudinal stripes; transverse stripes. These components are shown in Figs. 10.16 and 10.17 and can be recognized in some of the figures in the Frontispiece and Plate 5. They are of course superimposed upon the permanent white patches shown in Fig. 10.15.

The light areas and their dark overlays are used to build up patterns that may be sustained for hours or days, or switched on only for a matter of seconds. Sometimes specific components are associated with particular postures, as in the dymantic display. Others seem to occur only in specific situations; the centrifugal flush shown in Fig. 10.17d for example seems only to occur when a smallish object moves close to the octopus; the display is evidently a transient version of the uniform dark colouring that is switched on in fighting or when the octopus is about to pounce upon its prey (see Section 4.1.1). Mottled patterns, produced by varying degrees of expansion of the chromatophores in the dorsal network, are typically associated with conflict situations, when the octopus is uncertain whether to attack or retreat; they are often shown only on the side of the eye viewing the problem to be assessed (Fig. 10.17c).

Although a start has been made in this direction, it is quite clear that a great deal more work will be required before the full significance of many of the patterns shown by *Octopus* is understood. Different

---

serve to startle stationary prey into movement (see Section 4.1.1). (e) Zebra crouch by a young animal in the presence of another octopus. (f) Uniform dark red of an animal that is pushing back another in fighting or an attempt to copulate (after figures and photographs in Packard and Sanders, 1969, 1971).

species show different patterns, and may or may not use similar patterns for the same purposes. *Octopus cyanea*, for example, has a sexual display in which a dark longitudinal strip is shown down each of the arms – *O. vulgaris* only produces longitudinal stripes down arms 1 and 2 and never, so far as we know, displays these to advertise its sex (Wells and Wells, 1972a). A detailed study of the repertoire from a range of octopuses could well yield information about phylogenetic relationships, quite apart from its interest in relation to visual signalling.

10.3.5 *Central nervous control*

The chromatophores are controlled through ten of the thirty or so nerves that fan out from each side of the brain to the periphery. Section of any one of the ten de-enervates the chromatophores in a specific area of skin (Fig. 10.18; Froesch, 1973b).

There is no overlap in the fields controlled by different nerves, and it is obvious that most of the patterns shown in Figs. 10.16 and 10.17 must be determined by fibres in several nerves – there is no question of patterns arising from the activation of whole fields at a time since the maps of the projection areas (Fig. 10.18) cannot readily be correlated with the maps of chromatophore expansion (shown in Figs. 10.16 and 10.17).

Inside the brain, on each side, there are three regions that appear to be more or less exclusively concerned with chromatophore and skin texture control. The two lower centres, in the suboesophageal lobes, serve no other muscles. Mechanical stimulation of the walls of the posterior chromatophore lobe can produce expansion of the chromatophores in limited areas of the mantle. The cell bodies, in this lobe at least, seem to be laid out so that the motor neurones forming the cortex constitute a topographical projection of a part of the animal's surface (Boycott, 1961). The anterior chromatophore lobe controls the chromatophores in the arms. Again, the control appears to be direct, with no further synapses between the brain and the muscles (Young, 1971; but see Rowell, 1963).

The third, supraoesophageal chromatophore control centre lies in the lateral basal lobe, to the side of the hind part of the brain. The neurones here connect only indirectly to the periphery, via the anterior and posterior chromatophore lobes (Boycott, 1953).

Electrical stimulation of any of the three chromatophore lobes produces a dark flush on the side concerned, which may spread across to

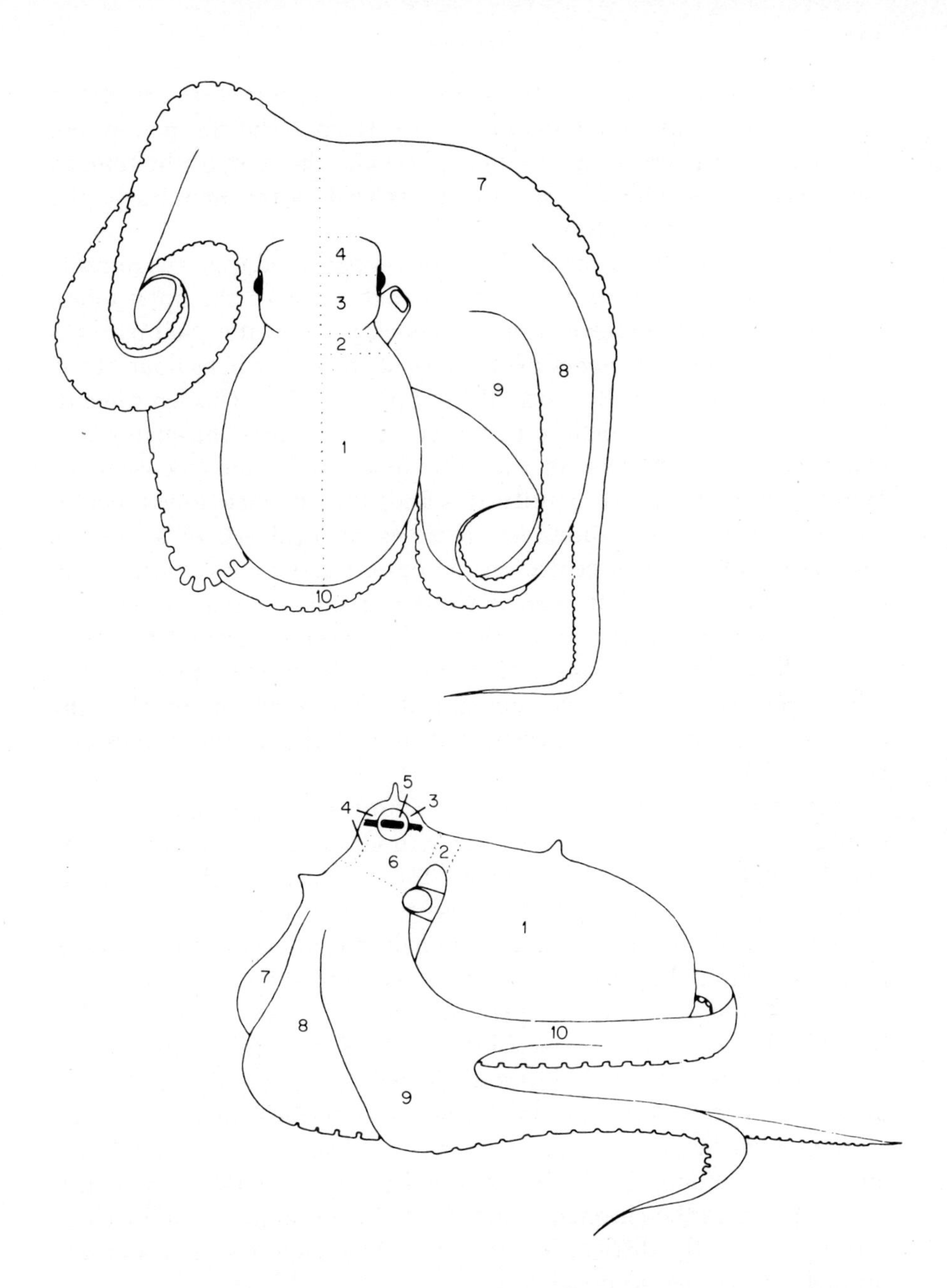

Fig. 10.18 Fields of chromatophores innervated by particular nerves. 1. pallial nerve; 2. collar nerve; 3. posterior superior ophthalmic (anterior root); 4. superior antorbital; 5. median superior ophthalmic; 6. anterior oculomotor; 7, 8, 9 and 10. arm nerves 1–4 (from Froesch, 1973b).

the other side of the animal; the suboesophageal centres are connected by commissures. Skin musculature is also affected; the chromatophore lobes control the erection of skin papillae and the state of the folds in the polygonal patchwork (see above) as well as the muscles of the chromatophores themselves.

Stimulation of the chromatophore lobes never produces recognizable patterns and it is only when a region at the base of the optic lobes, quite remote from the chromatophore lobes, is stimulated that patterns recognizable as components of the normal chromatic behaviour of the animal can be evoked (Boycott, 1961; Young, 1971). This led some of the earlier workers to believe that there was a further chromatophore control centre at the base of the optic lobes, and some (see Boycott, 1961) went so far as to identify this with the peduncle lobe (Section 10.2.3). Anatomically, however, there is no evidence of a discrete chromatophore control centre on the optic stalks or anywhere else in the optic lobes and what appears to be happening is that stimulation of the region excites the lower centres in a more natural manner than is possible by the direct application of electrodes to the lateral basal or suboesophageal lobes. A very similar state of affairs is found when the control of locomotion is investigated by stimulation techniques (see Section 10.1.2).

From a functional point of view the chromatophore control system is interesting in that it must operate without direct feedback from the muscles that it controls. There appear to be no sensory fibres from the chromatophores, and no stretch receptors have been reported from the muscles of the skin. Blinded animals do not show the complex patterns exhibited by their seeing controls, and the whole patterning and background matching operation must be controlled visually. Whether the octopus has to be able to see itself to do this is not known. The situation is particularly interesting because so many of the patterns shown are very stable and long-lasting. The chromatophore and skin muscles must be held accurately at specific degrees of contraction for long periods. Young (1971) has pointed out that the flickering that is so often seen in the chromatophores of the living animal could well be a reflection of the difficulties of controlling such a system entirely through an external feedback loop.

10.3.6 *Regeneration after pallial nerve section*

The existence of complex repeatable colour patterns, the lack of field overlap, and the bilateral nature of the mantle innervation makes the chromatophore control system peculiarly suitable for a study of nerve regeneration. If one of the pallial nerves is cut or crushed, the mantle becomes pale on that side. Within a few days spontaneous contractions of the chromatophore muscles begin, producing the 'wandering clouds' effect (Section 10.3.2 above), which continues until innervation is re-established 40 to 60 days later (at 23 °C). The animal then shows the same full repertoire of colour patterns on the control and the operated sides, a state of affairs that can only mean that the regenerating nerves have re-established functional connexions with the same groups of chromatophores that they innervated before section (Sanders and Young, 1974).

In this system it should also be possible to study regeneration after removal of areas of the skin, and just possibly after grafting in new regions. It will be most interesting to observe how regenerating nerves incorporate new chromatophores into the patterns that they re-establish; an octopus can more than double its weight in 40–60 days and will presumably have formed many more chromatophores in the interim. The skin of the mantle clearly has much to offer us as a preparation for studying the means by which nerves identify and home in on their muscular targets.

CHAPTER
ELEVEN

Learning and brain lesions: 1

MAINLY TACTILE LEARNING

11.1 Introduction: some terms and assumptions

Learning can be discussed in a variety of languages, and since the language that is used can reflect preconceptions about the likely mechanisms, it is as well to be clear about the sense in which any particular author is using words that mean different things to different people. I shall assume that any long-term (hours, days or months) change in behaviour is associated with structural changes in the nervous system. There are plenty of good precedents for this. In *Octopus*, as in other animals, learned changes in behaviour survive a variety of treatments, ranging from anaesthesic to electroconvulsive shock (provided that it is given several hours after training – see Table 11.1) that might be expected to disrupt any mechanism dependent upon on-going electrical activity. 'Structural change', in this context, could be at any level, from alterations in the external shapes or connexions of cells, to changes in their biochemistry. The structural changes that result in a long-term learned change in response to a specific stimulus constitute a 'memory store', 'trace' or 'representation' of that event. A memory store is at present a concept; nobody has ever observed one in any animal, only the consequences of its existence. These consequences are behavioural, 'memory' is demonstrable, the animal shows by its actions that it 'remembers'. A learned change in behaviour implies the continued existence of a structural alteration. In contrast a failure to remember is *not* evidence that the memory store has disappeared. It may have done, but all one can say for certain is that it no longer seems to be determining what the animal does.

In describing octopus behaviour it is convenient to use the terms 'short-term learning' and 'long-term learning' to distinguish between alterations in behaviour lasting seconds or minutes and changes that endure for days or months. The use of these terms is again behavioural; it does not necessarily imply either that there is a discontinuity between short- and long-term learning processes or that the two are

Table 11.1 Electroconvulsive shock and memory in the octopus.
The animals were confined in a box divided into two equal halves by an incomplete partition. One end was lit. The octopus was given repeated 10 V shocks if it remained in the dark compartment for more than 12 s. As soon as it had remained in the lit compartment for 25 s the light was switched to the other half of the shuttle-box. In this table *Anticipatory runs* are occasions when the animal moved into the lit half of the box before shocks began, the *Anticipatory run time* being the number of seconds between light off and the anticipatory run. An *Incorrect crossing* is made when the octopus moves from the lit to the dark compartment. *Error time* is the number of seconds spent in the wrong compartment after the onset of shocks. *Total error time* sums error time with the time spent getting shocks as a result of incorrect crossings. There were 20 trials in all, in two sessions, 10 on day 1 and 10 on day 4. Experimental animals were given electroconvulsive shocks (ECS) either 1 minute or 6 hours after session 1; controls had no shocks. (From Maldonado, 1969.)

Items	*Groups*	*Arithmetic means*		*D (Difference between day 1 and day 4)*	*P*	*Comparison of D of each item for ECS-1 min, and ECS-6 h groups with the corresponding values for the control group. P**
		Day 1	*Day 4*			
1 Number of anticipatory runs	Control	2.96	3.15	–0.19	NS	–
	ECS-1m	3.44	3.15	0.29	NS	NS
	ECS-6h	4.08	3.97	0.11	NS	NS
2 Anticipatory run time (in seconds)	Control	6.65	5.97	0.68	NS	–
	ECS-1m	5.45	5.90	–0.45	NS	NS
	ECS-6h	5.77	5.90	–0.13	NS	NS
3 Number of incorrect crossings	Control	3.17	1.48	1.69	<0.001	–
	ECS-1m	2.65	2.56	0.09	NS	<0.05
	ECS-6h	3.77	1.48	2.29	<0.005	NS
4 Error time (in seconds)	Control	13.74	11.81	1.93	<0.005	–
	ECS-1m	13.53	13.46	0.07	NS	<0.05
	ECS-6h	12.34	10.29	2.05	<0.025	NS
5 Total error time (in seconds)	Control	141.61	100.48	41.13	<0.005	–
	ECS-1m	120.68	128.34	–7.66	NS	<0.005
	ECS-6h	121.34	79.68	41.66	<0.005	NS

P = Probability from *t*-test, two tails, paired data. *P** = Probability from *t*-test, two tails, independent samples. NS = Difference is not signficant, i.e., probability greater than 0.05.

dependent upon different mechanisms. Conventional wisdom, based on experiments such as those summarized in Table 11.1, holds that they are and that in all probability short-term effects are due to ongoing electrical events, while long-term changes are structural – but this is hypothesis, not, as yet, observed fact.

Short- and long-term changes in learned behaviour may both be masked by changes in the overall level of responses that could be caused by yet another series of internal mechanisms. Feeding often lowers the threshold for positive responses to the point where an octopus appears to have forgotten a learned discrimination altogether – it attacks indiscriminately until starvation or the punishment that it receives raises the threshold and reveals that a memory store has existed all along. It should be noted, incidentally, that changes in 'motivation' (a term that will be avoided here) due to feeding can operate in what is, by mammalian standards, the unexpected direction; poikilotherms can afford to be more strictly opportunist and lie low when things are going badly.

11.1.1 *Early experiments, prawns and the cuttlefish*

Since the pioneer studies of Bert (1867) Fredericq (1878) and von Uexküll (1895), outlined in the last chapter, it has been known that large parts can be removed from the supraoesophageal brain of cephalopods without seriously upsetting locomotion. It also became obvious, as work continued, that these animals will learn rapidly in the laboratory (Chapters 8 and 9, and a number of earlier experiments by Ten Cate and Ten Cate 1938; Goldsmith, 1917a, b, c and Kühn, 1950 – see Boycott, 1954 for a summary). The first attempt to combine a study of learning with the effect of brain lesions came in 1940, when Sanders and Young were able to show that removal of the vertical lobe from *Sepia* had effects on the animal's hunting behaviour that could be attributed to memory failure. Cuttlefish will pursue prawns, and appear to continue to search for them when the prawns have disappeared from sight; this behaviour ceased after removal of the vertical lobe. Sanders and Young (1940) went on to demonstrate that *Sepia* would learn not to strike with the tentacles at prawns shown behind a glass sheet and that removal of the vertical lobe caused the cuttlefish to return to the attack; their two animals relearned but would strike at the prawns again within a matter of hours, where normal animals remembered their experience overnight.

11.1.2 *Crabs and squares*

Experimentation was stopped by the 1940 war and ten years elapsed before the appearance of the next of what was soon to become a long series of papers on learning and brain lesions in cephalopods. Boycott and Young (1950) used *Octopus*, rather than *Sepia*, because it lives better in aquaria and bleeds less readily when the cranium is opened to make brain operations; in the wild, octopuses seem to live a rougher, tougher existence than the comparatively delicate cuttlefish, and this is reflected in the animals' capacity for sealing off blood leaks, and wound repair. Boycott and Young found that in *Octopus* as in *Sepia* removal of the vertical lobe had no obvious effect upon movement or the capacity to recognize and attack prey; their study was not detailed enough to detect the small defects recognized by Maldonado (1964) in his subsequent analysis (see Section 10.2.4). Octopuses were shown to learn not to attack crabs if given a small (8 V) shock through an electrified plate whenever they attacked a crab/plate combination. Half a dozen trials, at intervals of 2 hours or less, was usually sufficient to change the animals' behaviour. They would then remember for two or three days, and could quite readily be trained to distinguish between crabs shown alone (which they were allowed to eat) and crabs shown together with the plate. Subsequent removal of the vertical lobe produced octopuses that forgot the discrimination, and were apparently unable to relearn it; with trials at intervals of 2 hours or more, attacks were made at every trial (Table 11.2). The animals were, however, not altogether incapable of learning; as Table 11.2 shows, they rarely made more than one mistake at each trial. Having got a shock, the octopus would retire into its home and sit watching

Table 11.2 Mean numbers of attacks made by eight separate *Octopus* at a series of presentations of a crab and electrified plate (from Young, 1951).

Day	*Trial*	*Before vertical lobe removal*	*After vertical lobe removal*
1	1	2.87	1.37
	2	0.50	1.12
	3	0.62	0.87
2	4	0.62	1.12
	5	0.50	1.62
	6	0.00	1.12
3	7	0.37	1.87
	8	0.25	1.00
	9	0.00	1.00

the crab and plate; provided the pair remained in sight, inhibition would last for 10 minutes or so. But as little as five minutes was sufficient for the animal to forget completely and attack again if the stimuli were removed and replaced.

These experiments are described in detail in Boycott and Young (1955a, b). They indicate that it is possible to abolish the capacity to set up long-lasting internal records ('memory stores') of visual events without at the same time destroying the ability to recognize prey or to remember for periods of a few minutes.

Subsequent experiments on learning and brain lesions in *Octopus* have evolved from this work in a number of directions. Because the tendency to attack crabs could be innate, rather than learned, it has generally become the practice to study visual learning using less emotive objects, commonly simple cut-out geometric shapes of the sorts discussed in Chapter 8. This has the additional advantage that it is possible to arrange to test the effect of brain lesions on a series of problems of graded difficulty. A parallel line of study, in which learning by touch has been considered, has proved to have advantages because the tactile input enters the brain from below. It is easier to avoid damage to this than to the corresponding visual inputs when operating upon the regions concerned in learning.

In this and the next chapter it has proved convenient to deal first with the results of tactile experiments and then with the rather larger number of experiments on visual learning. There is, as we shall see, some justification for dealing with the two series successively rather than simultaneously since different parts of the brain are concerned in tactile and visual learning (Fig. 11.1). The sorts of information available from brain lesion experiments also differ significantly. The principal conclusion from the tactile experiments is that we can track down the location of the structures that change when an octopus learns to discriminate to a small region with a specific type of cellular structure in the subfrontal-posterior buccal part of the brain. The corresponding mechanism within the visual system appears to lie mainly within the optic lobes. Because these also contain the cells and connexions responsible for the analysis of a complicated visual input (which is not presorted within the retina as it is in vertebrates), it has not so far been possible to narrow down the region responsible for the storage of visual memories in the way that has been done for the tactile. Instead, visual experiments have yielded a great deal of information about the function of the vertical lobe, a further region with a

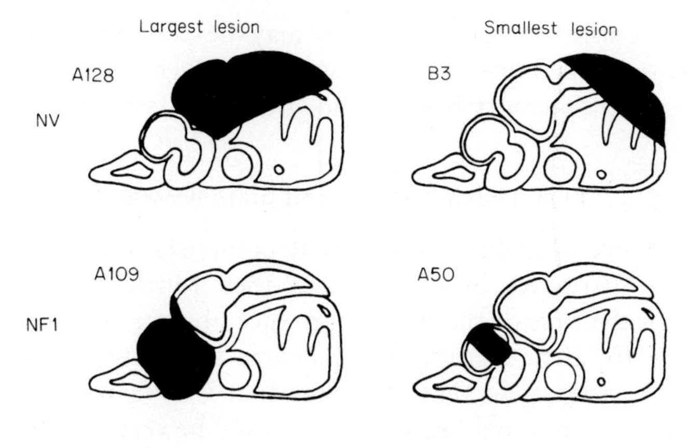

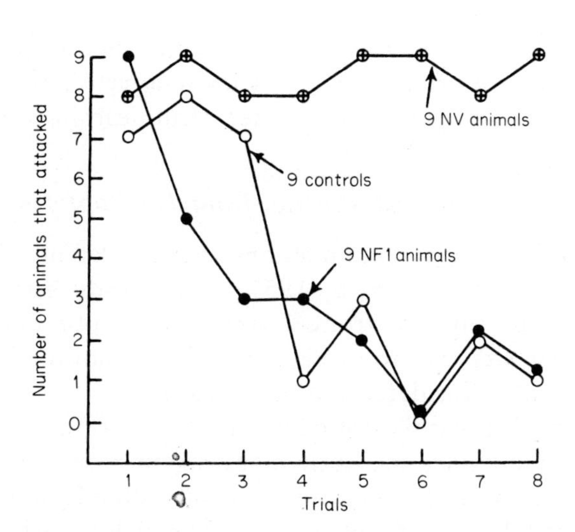

Fig. 11.1 A summary of the results of training octopuses not to attack crabs. Each trial lasted for 2 min and there were four trials per day. The nine 'NV' animals with lesions to the visual learning system were the same octopuses as the controls, training being continued after the operation. 'NFI' animals had lesions in the inferior frontal system. Lesion maps show the area removed from the brains of those individuals having the largest and smallest lesions in each group. They are plotted on a standard diagram representing a median longitudinal section through the supra-oesophageal part of the brain; large lesions, such as that shown for A109 in the diagrams above, stop tactile learning altogether (Section 11.2.2 below) (from Wells, 1961b).

highly characteristic structure, which may be concerned exclusively with short-term memory and the capacity to feed into and read out from a long-term memory bank in the optic lobes (see Sanders, 1975).

11.2 Touch learning and brain lesions

All the evidence available suggests that octopuses learn to recognize the physical attributes of objects that they touch in terms of the distortion of the rims of the suckers produced by the contact. There is no evidence for their learning to recognize the shapes of objects or their surface patterns, so that the signal that is being matched against the store in some internal memory bank could be very simple, perhaps no more than a frequency of discharge in the brachial nerves representing the mean level of distortion imposed on the sample of suckers in contact (Chapter 9).

The great advantage of this situation is that brain operations are unlikely to produce effects on touch learning attributable to interference with sensory analysing rather than learning mechanisms.

11.2.1 *Lesions to the vertical, superior frontal and optic lobes*

Research into the effects of brain lesions on touch learning began after Boycott and Young (1955a, b, 1957) had already established the importance of the superior frontal, vertical and optic lobes in visual learning. A first step was to discover whether removal of these parts produced corresponding defects in touch learning.

Fig. 9.2 indicates that the optic lobes play no part in touch learning, a remarkable result since the optic lobes together contain almost one half of all the nerve cells to be found in the octopus brain. Removal of the vertical lobe, or of the superior frontal lobe, which supplies most of the input to the vertical (Fig. 10.2) in contrast does affect touch learning, to an extent proportional to the amount of tissue removed (Fig. 11.2).

The deficiency shown by these octopuses is not attributable to an inability to distinguish between the objects. Once they have been trained, objects very similar to those used in training are not confused with them (Fig. 11.3).

Mistakes in training to make simple discriminations are predominantly due to acceptance of the 'negative' objects, which the animals are punished for taking. A similar situation is found if the vertical lobe

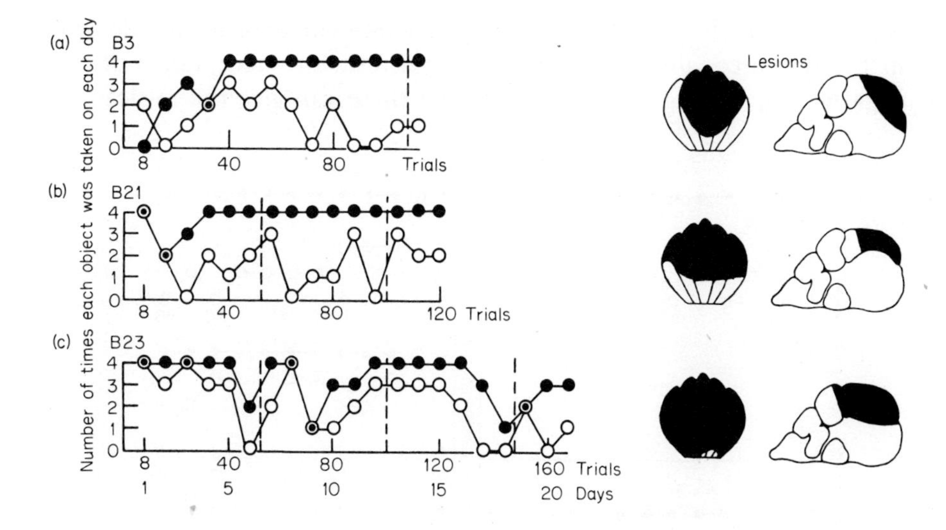

Fig. 11.2 Tactile training to discriminate between a rough and a smooth cylinder, showing the effect of removing progressively larger amounts of the vertical lobe. Lesions are mapped on diagrams representing a longitudinal section through the suboesophageal lobes and on a plan view of the vertical lobe (replotted from Wells and Wells, 1957b, where a number of further examples are given).

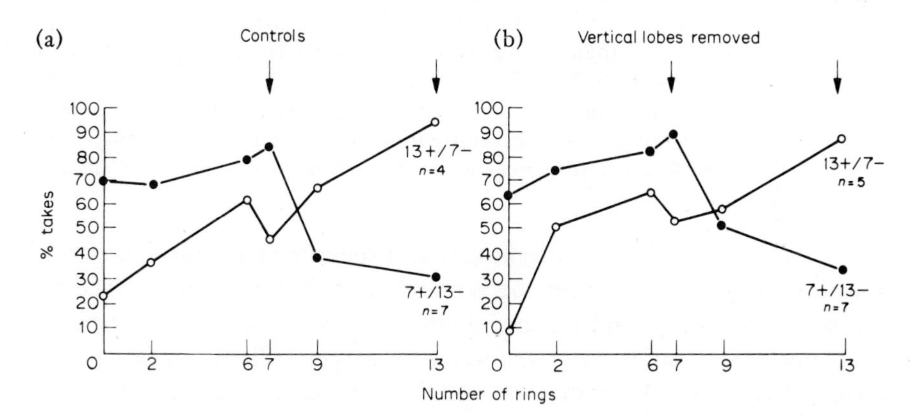

Fig. 11.3 Animals were trained to distinguish between spheres with 7 and 13 latitudinal grooves cut into them. After 10 sessions of 16 trials they were tested with further objects, 40 tests with each being interspersed between further training trials. The animals in (b) had the vertical lobes of their brains removed before training; they learned much more slowly than the octopuses in (a) but nevertheless showed a very similar pattern of transfer (from Wells and Young, 1970b).

is removed in the middle of training; the immediate result is a decline in performance, almost entirely due to a return to acceptance of the negative object (Fig. 11.4). Results like these can give the misleading

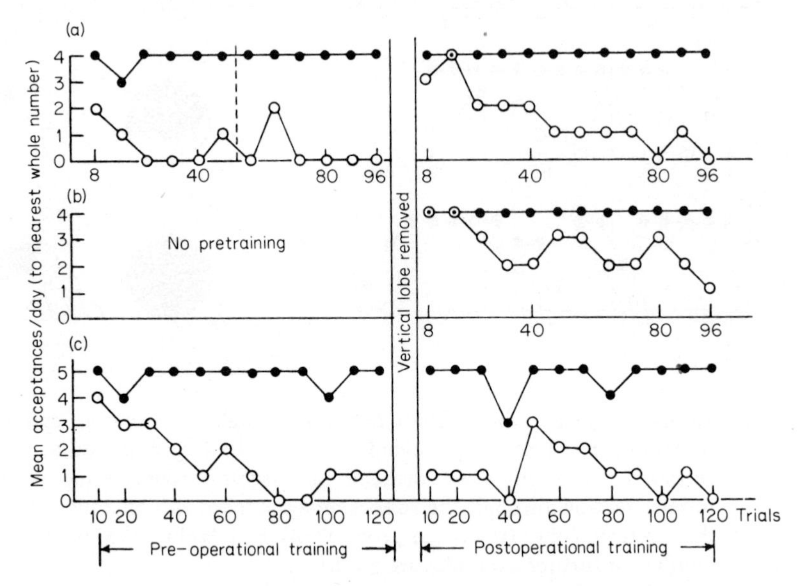

Fig. 11.4 Effects of a similar number of trials of preoperational training at 8 and 40 trials per day on the postoperational performance of animals following removal of their vertical lobes. (a) 3 animals: train-operate-train at 8 trials per day. (b) 6 animals; operate-train under the same conditions. (c) 3 animals; train-operate-train at 40 trials per day. ●, positive object; ○, negative object. Vertical dotted line indicates break in training of 36 h. There was a similar break after vertical lobe removal (from Wells and Wells, 1958a).

impression that vertical lobe removal has a specific effect upon negative learning. It should, however, be remembered that the animals feed by taking small objects and passing them inwards towards the mouth, so that any failure to learn or tendency to forget is liable to be characterized by a preponderance of mistakes in this direction. In fact, animals with their vertical lobes removed do not take an unusually high proportion of the 'negative' objects on occasions when their overall proportion of errors is the same as that of their controls (Table 11.3).

Animals with their vertical lobes removed require more trials than controls to achieve a given standard of accuracy of responses in discrimination experiments (Table 11.3). The difference between the two

Table 11.3 The proportion of errors due to acceptance of objects that the animals were punished for taking ('negative errors') in sessions when controls and animals without vertical lobes made the same total number of errors up to and including the session at which each animal made 75% correct responses (i.e. 5 errors) (after Wells, 1965).

		Total errors in a session of 20 trials										
		1	2	3	4	5	6	7	8	9	10	11
Controls, n=42 (analysis of 1920 trials = average of 2.3 groups to reach 75% correct)	−ve	1	12	11	49	48	69	44	75	36	80	36
	+ve	1	16	18	60	60	84	56	104	36	90	66
		=100%	75%	61%	82%	80%	82%	79%	72%	100%	89%	55%
Vertical and/or superior frontal lobes removed, n=35 (analysis of 2920 trials = average of 4.1 groups to reach 75% correct)	−ve	2	5	16	30	47	54	133	104	108	349	20
	+ve	2	6	27	36	60	72	168	128	153	360	33
		=100%	84%	59%	80%	79%	75%	79%	81%	71%	97%	61%

Controls and experimentals make about the same proportion of positive and negative errors, although the total number of errors made by the experimentals is much greater.

is reduced when the rate of training is increased from 8 to 40 trials per day (Wells and Wells, 1958a). The apparent narrowing of the gap between controls and experimentals with higher rates of training is, however, at least to some extent illusory. Because the animals without vertical lobes learn more slowly the proportion of trials at which they accept objects and receive shocks or rewards tends to be greater than that of controls, particularly at the higher rates of training which depress the overall level of take more rapidly in the controls. Since the animals can only gain information about the discrimination to be made by taking the objects, the animals without vertical lobes get more training in each group of trials than their controls – about 20% more with trials at 16 per session and 32 per day, for example. When their performance is reassessed in terms of takes not trials it is found that the experimental animals learn less from each action-reward sequence than their controls whatever the rate of training (Fig. 11.5, Wells and Young, 1969a). This difference can be traced in single-session learning experiments and even down to the level of single trials (Wells and Young, 1970a). Operated animals took about six times as long as their controls to

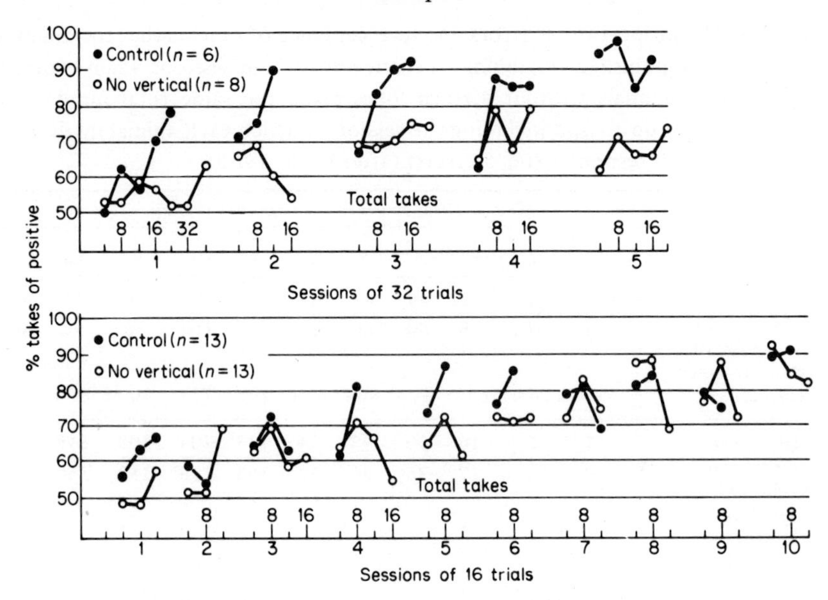

Fig. 11.5 Comparing the percentage of takes of the positive sphere in each 4 takes in each session of training at 16 and 32 trials per session. Animals with their vertical lobes removed take more often, but learn less from each take (from Wells and Young, 1969a).

examine and reject an inedible object when it was presented to them for the first time (Wells and Wells, 1957c).

Once a change has been produced by training, animals without their vertical lobes do not forget any more rapidly than their controls. In reversal experiments, both can relearn and both take longer to do so than in their original training (Wells, 1965). In retention tests lasting 5–15 days the performance of octopuses lacking their vertical lobes was not detectably different from that of their controls (Table 11.4). In subsequent studies involving much longer retention periods, Sanders (1970a, b, c) found that animals with their vertical lobes removed actually relearned faster than their controls when retrained 55–60 days after their initial training (Fig. 11.6).

Sanders (1970c) followed this rather surprising result with a demonstration that the learning defect produced by vertical lobe removal disappears altogether if a sufficient time is allowed to elapse between operation and training. 10 such octopuses kept for 3 months before their initial training learned as rapidly as their 19 controls.

This last result is important because it implies that there may be an

Table 11.4 Retention tests and removal of the vertical lobe. Each individual was trained until it reached a criterion of 85% (17) or 75% (15) correct responses in a group of 20 trials, and was then overtrained for an equal number of groups before the beginning of a retention period lasting 5 or 10 days. (after Wells and Wells, 1958b).

	Mean errors				*Increase in mean errors between last 20 training trials and tests*	
	At criterion	*In the last 20 training trials*	*In 20 unrewarded tests 5 days later*	*In 20 unrewarded tests 10 days after training or tests at 5 days*	*At 5 days*	*At 10 days (total = 15 days for the ⩽3 errors standard group)*
Controls	⩽3	1.4 (*n*=6)	3.2 (*n*=6)	5.4 (*n*=5)	1.8	3.8
	⩽5	2.3 (*n*=4)	—	6.8 (*n*=4)	—	4.5
Vertical lobe removed	⩽3	2.8 (*n*=11)	4.7 (*n*=11)	5.4 (*n*=6)	1.9	2.6
	⩽5	2.9 (*n*=7)	—	7.4 (*n*=7)	—	4.5

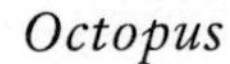

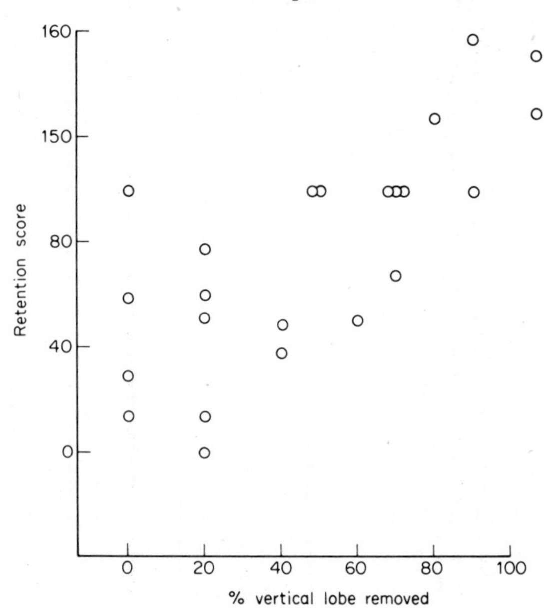

Fig. 11.6 The correlation between long-term tactile retention, as measured by savings on retraining after 55 to 60 days, and the percentage of vertical lobe removed. (Data from Sanders, 1970b and c.)

alternative explanation for the quantitative defects found in the very much larger number of experiments run with octopuses operated only days or weeks before training. If these defects really disappear with time, it is possible that they are all attributable to post-operational trauma? Perhaps the vertical lobe plays no part in tactile learning; but damage to it produces, directly or indirectly, a sustained chatter of irrelevant signals that interfere with the proper operation of mechanisms located elsewhere. A difficulty with any such hypothesis is that it yields the same predictions as others that envisage the vertical lobe as an additional memory store, or as a 'read-in' or 'read-out' device, responsible for channelling information in or out of stores located elsewhere. A further difficulty is that it cannot account for results of the sort summarized in Fig. 11.6, which shows a positive correlation between performance and the amount of vertical lobe tissue removed. Sanders (1970a, b) suggestion is that the better performance of the lesioned animals can be attributed to slower learning during the train-retrain interval; the old memory traces have not been overlaid or disorganized by the addition of subsequent material to the same extent as in controls. Which returns us to the propostion that the vertical lobe *is* implicated in touch learning after all. The argument remains unresolved.

Fortunately, the matter is not of overwhelming importance. Either way it is now clear that the vertical lobe is not one of the parts of the brain essential to touch learning. Under suitable conditions its contribution is slight and, so far as we can tell, entirely quantitative.

It must be pointed out that we know very little about the relation between tactile and visual learning. Investigations involving brain lesions have always required the animals to learn a visual *or* a tactile discrimination, never a sequential task necessitating the use of both sorts of cue. Although the two learning systems are based upon anatomically distinct parts of the brain, as we shall see below, there are extensive tracts connecting, for example, the inferior and superior frontal lobes. Perhaps these connexions serve to relate what the animal learns by touch to what it learns by sight, but we have no experimental evidence for this. It is one of many areas so far neglected in what appears, at first glance, to have been quite an extensive series of investigations on brain function in learning.

The function of the vertical lobe in visual learning, and the relation of the touch learning evidence to the interpretation of its function in vision are discussed in Chapter 12.

11.2.2 *The inferior frontal system*

Further extension of brain damage beyond the superior frontal and vertical lobes produces no additional decline in touch learning performance until the lesion begins to encroach upon the median inferior frontal, subfrontal and/or posterior buccal lobes. Removal of the basal lobes stops the animals from moving about in a well-integrated manner (see Chapter 10) but it does not prevent the individual arms from examining, taking and rejecting objects that they touch. The no-basal-lobe animal, lying with its arms in a tangled heap on the floor of its tank, learns as well as animals with only the vertical lobe removed (Fig. 11.7, Table 11.5).

Quite a different state of affairs is found after removal of the 'inferior frontal system' (the inferior frontal/subfrontal/buccal lobe complex). It is debatable whether these animals can learn to discriminate at all.

Like untrained octopuses, they err mainly by taking objects that

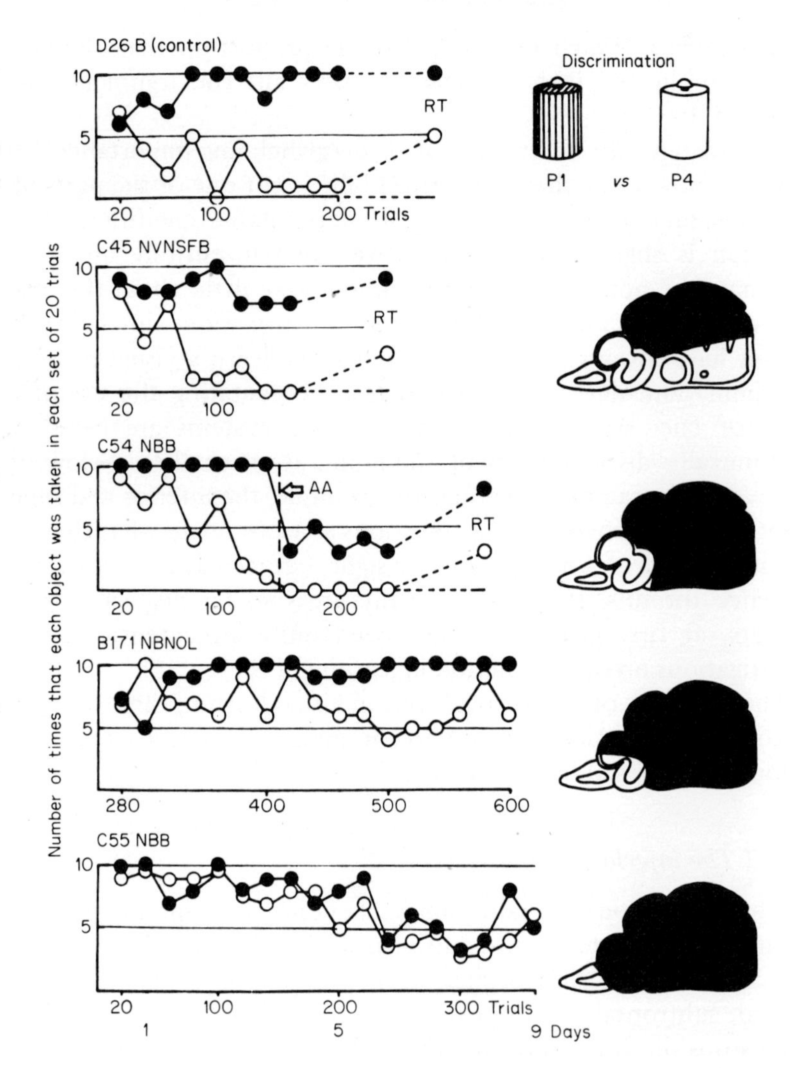

Fig. 11.7 The effect of lesions on the performance of octopuses in discrimination experiments, showing that while removal of the basal lobes does not seriously upset the performance of a tactile discrimination, damage to the inferior frontal system does. ●, Positive object; ○, Negative object. The plots show the number of times that each object was taken in each group of 20 trials (10+, 10−), there being two such groups per day. Where two points coincide, because the number of positive and negative objects taken was the same, these are displaced up- and downwards respectively. RT = Retention test, carried out after a break in training of 5 days; AA (in the plot of C54NBB) = arm amputated (not that used in training) an operation that reduced the level of take, changing the pattern of errors without disrupting discrimination (from Wells, 1959b).

Table 11.5 Trials taken to learn a tactile discrimination after brain lesions (after Wells, 1959b).

	Controls	*Vertical or vertical and superior frontal lobes removed (lesions like than shown for C45NVNSFB in Fig. 11.7)*	*Basal lobes removed as well as vertical and superior frontal (lesions like that shown for C54NBB in Fig. 11.7)*
Number of groups of 20 trials to reach a standard of 75% correct responses (figures in brackets show the range of results)	2.4 (1–4) n = 26	4.0 (1–8) n = 13	4.3 (4–5) n = 4
85% correct (same animals)	3.5 (1–7) n = 26	5.6 (3–10) n = 10	5.8 (4–8) n = 4

they should reject. Their overall level of response is often, though not invariably, high and may remain so despite a frequency of punishment that would almost entirely inhibit positive responses in normal octopuses (Wells, 1959b, 1961b – Table 11.6 summarizes the performance of 21 such animals). But it is not unalterable; repeated presentations eventually lead to a modest reduction in the proportion of objects taken (Wells and Young, 1968a) and the decline can be accelerated by giving shocks. A corresponding increase can be produced by feeding the animals (Table 11.7). What seems to be missing in these octopuses is any capacity for learning to reject one object of a pair while continuing to take the other.

11.3 Subdivision of function within the inferior frontal system

Once it became clear that some critical part of the touch learning machinery was located within the inferior frontal system, attempts were made to dissect the situation further. The inferior frontal system includes regions of very different structure, arranged as illustrated

Table 11.6 The effects of subfrontal removal on learning to discriminate by touch (from Wells, 1974).

		Scores in first 120 trials *Takes of*				*Scores in all trials* *Takes of*			
Animal	*No. of trials*	*Takes* +	—	% *correct*	% *takes*	*Takes* +	—	% *correct*	% *takes*
Animals trained smooth positive/rough negative									
1. *Operations including removal of the basal lobes**, *Wells (1959b)*									
C117	200	56	57	49	94	92	96	48	94
D43 RHS	200	42	30	60	60	59	48	56	54
D43 LHS	160	56	54	52	92	71	68	52	87
2. *Lesions to the inferior frontal system, basal lobes intact, Wells (1961b)*									
A112	60	14	30	37†	37				
B11	128	24	30	45	45	24	32	44	44
B13	128	33	45	40	73	36	48	41	66
B36	160	7	2	54	8	8	2	60	6
B37	140	7	1	55	7	7	1	54	6
3. *Split brains, basal lobes removed**, *Wells and Young (1965)*									
NKB	120	52	40	60	77				
NOR	120	9	10	49	16				
				50	51			51	51
$t = 0.034$: NS for scores in the first 120 trials									
Animals trained rough positive/smooth negative									
1. *Wells (1959b) as above*									
C55	360	53	57	47	92	126	127	50	70
D40	200	54	48	55	85	94	85	54	90
D50	80	21	5	70†	33				
2. *Wells (1961b)—as above*									
D35	200	17	14	53	26	20	16	52	18
3. *Wells and Young (1965) as above*									
NEY	240	46	23	69†	58	139	76	76	90
NPM	120	56	44	60	83				
NOQ	120	35	24	59	49				
NGF	120	55	33	59	73				
NNM	120	41	35	55	63				
NPL	120	20	16	53	30				
NGD	120	54	52	52	88				
				57	62			56	67
$t = 3.58$ $p < 0.01$ for scores in the first 120 trials									

* Removal of the basal lobes always included removal of the vertical, superior frontal, and subvertical lobes as well.

† Individual score significant at the $p < 0.05$ level (X^2).

Table 11.7 The effects of positive and negative training on the level of response. (After Wells, 1974).

Lesion	*Number of animals*	*Total takes in 32 unrewarded tests before (and after) training*	
1. *Positive training* (256 trials at 16 trials a day, at each of which the animal was fed)			
Controls (R.H.S. of split brains)	8	130	(197)*
Subfrontal lobe destroyed (R.H.S. of split brains)	6	68	(139)*
2. *Negative training* (256 trials at 16 trials a day, animal punished with a shock at each trial)			
Controls (R.H.S. of split brains)	5	95	(37)*
Subfrontal lobe destroyed (R.H.S. of split brains)	3	28	(17)*
3. *Habituation tests* (160 unrewarded tests at 32 trials per day in 2 sessions of 16)			
Controls	12	165	(115)*
Inferior frontal system destroyed bilaterally (R.H.S. tested)	5	105	(84) NS

* Difference significant at $p = 0.05$; NS, not significant (matched pairs t-test, one tailed).

diagramatically in Fig. 11.8. Tactile input comes from below, through the brachiocerebral tracts to the median and lateral inferior frontal lobes. From these, further neurones form two main circuits through the inferior frontal system. The more extended of these passes from the median inferior frontal into the subfrontal, and from there via the posterior buccal to descending pathways in the brachiocerebral tracts. A shorter circuit runs directly from lateral inferior frontal to the posterior buccal. In addition there are fibres running directly to the subfrontal from below including nerves from the palliovisceral lobe, which may carry signals indicating events at the hind end of the animal. The two sides of the inferior frontal system are linked in three places;

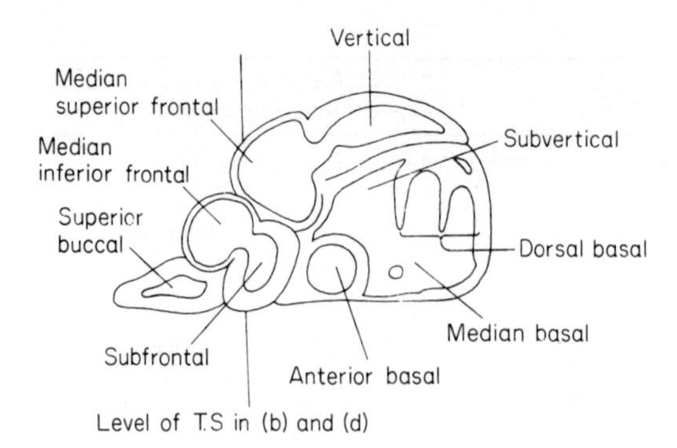
(a)
Vertical
Median
superior frontal
Subvertical
Median
inferior frontal
Superior
buccal
Dorsal basal
Median basal
Subfrontal
Anterior basal
Level of T.S in (b) and (d)

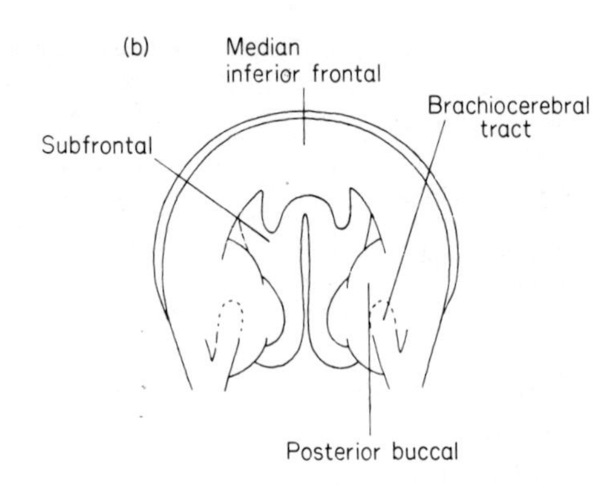
(b)
Median
inferior frontal
Brachiocerebral
tract
Subfrontal
Posterior buccal

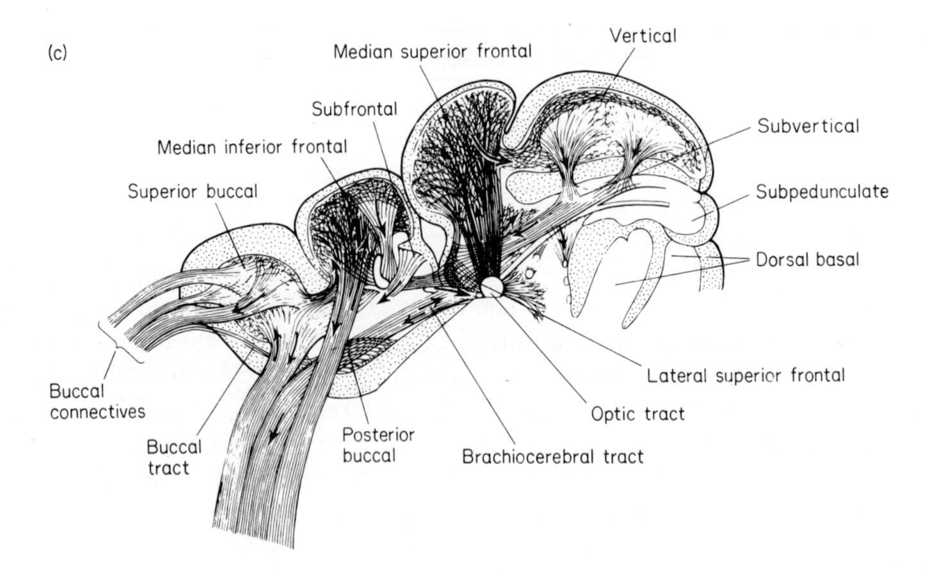
(c)
Vertical
Median superior frontal
Subfrontal
Subvertical
Median inferior frontal
Superior buccal
Subpedunculate
Dorsal basal
Buccal
connectives
Lateral superior frontal
Optic tract
Buccal
tract
Posterior
buccal
Brachiocerebral tract

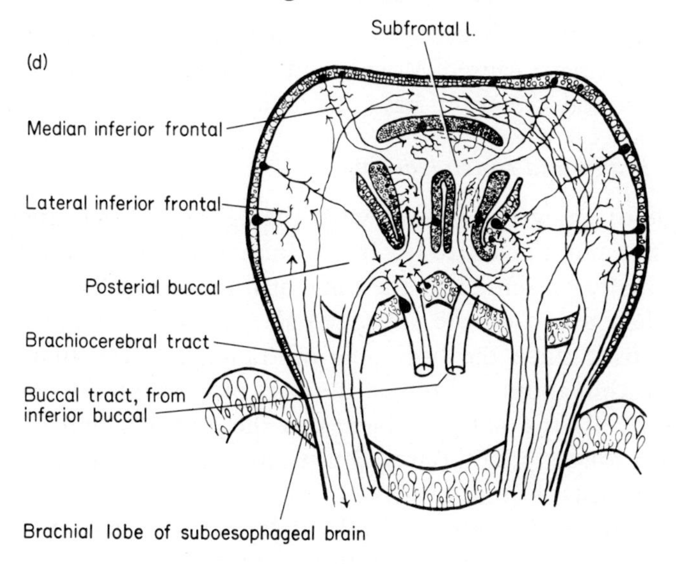

Fig. 11.8 Connexions within the inferior frontal region. (a) represents a longitudinal section through the supraoesophageal lobes. The two are used as standard diagrams on which lesions are plotted in Figs. 11.2, 11.7 and 11.11. (b) is a transverse section through this at the level indicated. (c) and (d) give further details of cell types and connexions within the inferior frontal system. Fig. 12.1 shows transverse sections corresponding to (c). (a) and (b) from Wells, 1959a, (c) after and (d) from Young, 1971.

through the median inferior frontal lobe, across the upper part of the subfrontal, where the neuropil of the two sides of the lobe is confluent, and by means of a commissure joining the two halves of the posterior buccal across the ventral part of the subfrontal.

The lateral inferior frontal lobes enclose the paired posterior buccal lobes* which in turn lie to either side of the paired subfrontal lobes. The subfrontal and posterior buccal cannot be removed without either destroying the median inferior frontal or the basal lobes or splitting the brain and approaching from within. Attempts to cauterize the subfrontal by means of a high frequency probe (Wells, unpublished)

* The distinction between the lateral inferior frontal and the posterial buccal lobes was not at first recognized, and some early accounts (Wells, 1959b; Young, 1963a) have described the whole region as 'lateral inferior frontal'.

proved unsatisfactory, because of the difficulty of estimating the extent of the damage caused. Most recent approaches to the problems of inferior frontal function have made use of split-brain surgery.

11.3.1 *Split brains and touch learning*

The left side of an octopus knows what the right side has learned, and vice-versa. But not immediately. If the animal is trained to reject an object repeatedly presented to the same arm on the same side of the body, and then tested on the untrained side, the result depends upon the time elapsed since the beginning of training. If the first test is carried out within about an hour of the start of training, the untrained side will nearly always take the object that the trained side has learned to reject. If a longer period (circa. 3 h) elapses, the untrained side will not make this mistake (Fig. 11.9).

There is some evidence of a similar delay when different arms on the same side of the body are trained and tested. These 'transfer' experiments have been interpreted as suggesting that each arm is represented by a separate pool of neurones in the inferior frontal system and that it takes time - of the order of 1–3 h - for the effect of training to spread from one pool to the rest (Wells, 1959a).

Experiments of this sort do not of course tell us whether the spread involves the establishment of memory traces in both sides of the brain. To prove this, the brain must be split after training.

Fig. 11.10 compares the results of train-split-test and split-train-test experiments. In these, the animals were trained to discriminate between rough and smooth spheres before or after having their supraoesophageal brains split by a longitudinal vertical cut. Training was restricted to the arms on one side of the body. Unrewarded tests were given to the other. The time scale, with training spread out over several days, allowed more than enough time for the spread of the effects of training to all parts of the system.

These experiments show (1) that only the trained side learns if the supraoesophageal brain is split before training; (2) that the untrained side performs as well as the trained, if the split is made after training and (3) that learning is quicker in animals with both halves of the brain intact.

The experiments indicate that the memory trace is established bilaterally and that both sides of the store contribute to the decision made when an object is picked up and taken or rejected (Wells and Young, 1966).

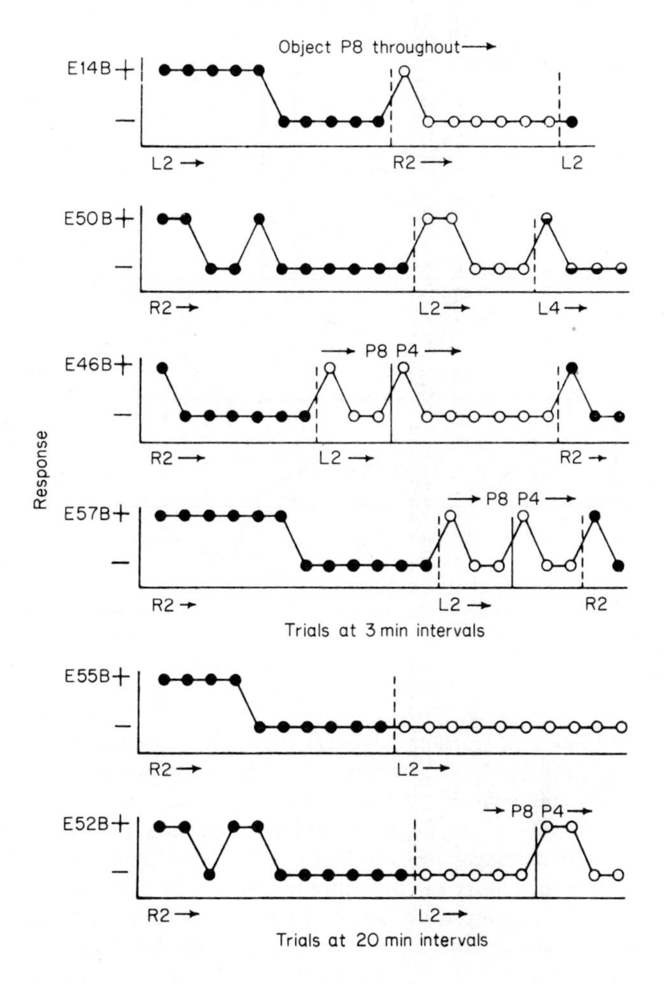

Fig. 11.9 Experiments on the time required for the effect of training to spread across the inferior frontal system. Each plot summarizes the performance of one animal, trained not to take the same object (P8 or P4), repeatedly presented to the second left, or the second right arm. After six successive rejections with the trained arm, the test object was presented to the corresponding arm on the other side of the body, and training continued. After training at short intervals, the object was generally taken in the trial immediately following the switch. With training spread out over a longer period, it was not. The full series included 24 animals trained with trials at 3 or 5 min intervals; 16 of these took the test object in the first trial with the untrained side. 12 animals had trials at 20 min intervals; only 2 of these took the test object in the trial following transfer (from Wells, 1959a).

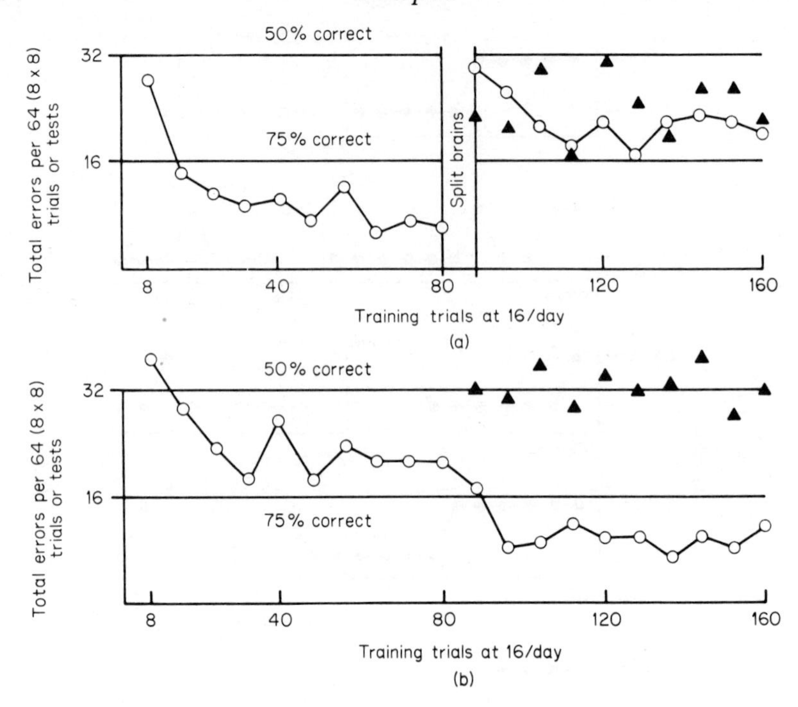

Fig. 11.10 Training and transfer tests: (a) train-split-test ($n = 8$); (b) split-train-test ($n = 8$). In experiment (a), eight octopuses were trained to discriminate between rough and smooth spheres. After 80 trials, restricted to one side of the body, the animals had the supraoesophageal brains split by a vertical longitudinal cut. Training continued together with unrewarded tests on the untrained side. The plots show total errors, ○ in the training trials and ▲ in the tests. Experiment (b) was carried out in exactly the same way, except that the brains of the animals were split before training was begun. These octopuses learned more slowly, but finally attained a standard similar to that of the unsplit octopuses in (a). In transfer tests, however, the untrained sides of their bodies failed to discriminate altogether (from Wells, 1966a).

11.3.2 *Untrained preferences and the effect of brain lesions*

As more and more brain lesion experiments were made, it became apparent that it was often easier to train animals in the rough^{+}/smooth^{-} direction than vice-versa. This effect can be seen, for example, in Table 11.6. It is particularly noticeable in animals with their median inferior frontal lobes removed, and in split-brain preparations. A possible explanation is that rough objects are more stimulating. Most of the mechanoreceptors in the suckers are rapidly adapting and the repeated changes of grip made when an octopus grasps and examines an object with the suckers might be expected to yield stronger signals from an irregular than from a smooth surface. Isolated arms and preparations with only the suboesophageal part of the brain remaining tend to take all objects presented to them although in fact capable of rejection as well (Section 10.2.6 - 10.2.8). The supraoesophageal brain is thus mainly inhibitory so far as tactile reflexes are concerned. Any damage might be expected to weaken this inhibition and allow rough, stimulating objects to become relatively more effective in evoking takes.

In the intact animal, some compensatory mechanism must operate, since it is equally easy to train normal (blinded but otherwise unoperated) animals in the rough^{+}/smooth^{-} and the smooth^{+}/rough^{-} directions (Wells and Young 1965). When the possibility of untrained preference was investigated in a long series of unrewarded tests, normal animals in fact showed a significant preference for *smooth* objects, that was absent in octopuses with brain lesions (Table 11.8). This preference could, of course, have been learned or innate.

Table 11.8 *Untrained preferences.* The animals were given 160 tests in groups of 16, 8 with rough and 8 with a smooth sphere presented alternately at intervals of 5 minutes. There were 2 groups of tests per day (after Wells and Young, 1968a).

	Number of smooth spheres taken	*Number of rough spheres taken*	*Total takes/ tests (%)*	*Preference: takes of smooth/ total takes (%)*
Normal animals (no lesion) $n = 10$	353	229	36	61*
Vertical lobes removed $n = 8$	397	366	60	52
Split brain animals $n = 12$	264	293	29	47
Inferior frontal system damaged $n = 6$	131	145	29	47
Inferior frontal system removed $n = 5$	272	231	63	54

* *t*-tests show that only the score of the normal animals is significantly different from chance at the $p = 0.05$ level.

11.3.3 *The subfrontal lobe*

The subfrontal lobe includes more than 90% of the six and a half million neurones in the inferior frontal system. On these grounds alone one would expect it to contain some vital parts of the touch learning mechanism, and this indeed proves to be the case. Early experiments involved an approach to the subfrontal from behind or above, eliminating the basal lobes or the median inferior frontal as well as the subfrontal. Some of these animals learned; others failed to do so. Careful mapping of the lesions concerned showed that every one of the successful octopuses had some fraction of the subfrontal lobe, or of the anterior part of the inner wall of the posterior buccal lobe remaining. The walls of the two lobes run together in this region and since both include very large numbers of very small cells the boundary becomes a matter of definition. In some instances animals learned successfully with only a few tens of thousands of these 'subfrontal' cells remaining, out of a total of more than five million present in the intact lobe. Octopuses (or sides of octopuses) with no trace of this tissue did not learn at all (Fig. 11.11; and Table 11.9) (Wells, 1966a). The continued presence of quite large parts of the lateral inferior frontal and/or of the lateral walls of the posterior buccal lobes in some of the animals did not alter this state of affairs, an observation consistent with the conclusion from structure that these are relay and distributive centres, unable by themselves to support learning.

Table 11.9 Minimal cell numbers and tactile discrimination (after Wells, 1966a).

Animal	*Side trained*	*No. of small cells remaining in the subfrontal lobe**	*Performance*		
			Percentage correct	*Positive object*	*No. of trials*
		A. *Discrimination experiments*			
Basal lobes removed (from Wells, 1959b)					
B171	R	Intact (2.6×10^6)	76	R	160
D47	R	342 000	64	R	200
	L	901 000	84	R	120
D50	R	926 000	66	R	100
	L	Nil	46	R	80
D49	R	15 000	53	S	220
C55	R	Nil	52	R	360
C117	R	Nil	47	S	200
D40	R	Nil	56	R	200
D43	R	Nil	58	S	200
	L	Nil	51	S	160

Animal	*Side trained*	*No. of small cells remaining in the subfrontal lobe**	*Performance*		
			Percentage correct	*Positive object*	*No. of trials*
Split brains (from Wells and Young, 1965)					
NPL	L	Intact	80	S	120
	R	Nil	54	R	120
NOQ	L	Intact	76	S	120
	R	Nil	60	R	120
NLP	L	Intact	71	S	120
	R	5000	50	R	120
NNM	L	Intact	71	R	120
	R	Nil	55	R	120
NKB	R	Intact	68	R	120
	L	Nil	60	S	120
NOR	R	Intact	63	R	96
	L	Nil	46	S	96
NGF	R	Intact	63	R	120
	L	Nil	60	R	120
NDO	R	45 000	80	R	180
NDL	L	16 500	74	R	280
NEF	L	10 600	74	R	180
NEY	L	Nil	70	R	280
NPM	L	Nil	60	R	120
NGD	L	Nil	53	R	140
B. *Repeated presentation experiments (from Wells, 1959a, b)*					
(Column 4 shows the number of times that the test object was taken and the animal punished in each group of trials.)					
B131	R	1.5×10^6	1	R	7
B126	R	12 800	3	R	9
	L	Nil	13	R	13
B132	L	41 000	11 (1, 2, 2)‡	R	13 (6, 6, 6)‡
	R	Nil	7 (6, 6, 5)‡	R	7 (6, 6, 6)‡
B128	L	21 700	13 (0, 0, 0)‡	R	13 (6, 6, 6)‡
	R	Nil	7 (4, 2, 6)‡	R	7 (6, 6, 6)‡
B127	R	17 000	6 (1, 0, 1)‡	R	8 (6, 6, 6)‡
	L	1 300	8 (2, 6, 3)‡	R	8 (6, 6, 6)‡
B122	R	12 800	14	R	15
B124	R	Nil	12	R	15
C117	R	Nil	20	R	20

‡ Three groups of trials at 6 per day per side following the first run of 7–13 trials. *Further details of the lesions in these octopuses are given in Wells (1959b) and Wells and Young (1965); it should be noted that many of the animals were trained in the R+/S– direction, which favours 'discrimination' by animals with large brain lesions.

More recent experiments have taken advantage of the possibility of removing the subfrontal alone from split-brain animals. These octopuses show fewer motor defects than octopuses with the basal lobes destroyed, and they can be prepared leaving the median inferior

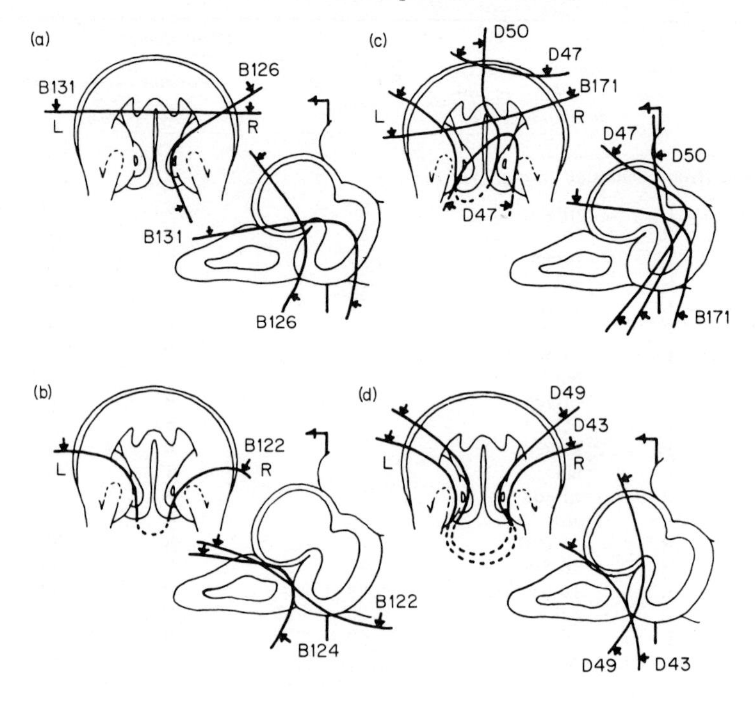

Fig. 11.11 Lesions to the subfrontal region of octopuses with their basal lobes removed, plotted on standard diagrams explained in Fig. 11.8. (a) shows the extent of removals in two animals that learned to recognize an object repeatedly presented (B126 learned with the right, but not with the left side). (b) Lesions in two that failed to learn (RHS tested) under the same conditions. (c) Lesions in three animals that were trained to discriminate between rough and smooth cylinders (B171 and D50 RHS learned, D50 LHS did not; D47 LHS learned, RHS learned, but very slowly. (d) Lesions in two animals that failed to learn (D43, both sides, D49 RHS only tested. Individual scores and cell counts, see Table 11.9 (from Wells, 1959b).

frontal intact. One series of these experiments is considered in some detail below, since it effectively summarizes the present state of our knowledge about the function of parts of the inferior frontal system.

In these experiments (analysed in full in Wells and Young, 1975) three measures of learning were made: (1) the response to positive training to take a smooth sphere (a necessary preliminary to discrimination training, because of the rough preference shown by animals with extensive brain lesions, see Section 11.3.2 above); (2) discrimination training in the 'difficult' direction smooth^{+}ve/rough^{-}ve; and (3) extinction and transfer tests in which the trained animals were given unrewarded tests with the spheres used in training and others intermediate in texture. The same animals were used throughout. Those with parts of the brain removed had these operations on the right side only, so that two sorts of control were available. The left, unoperated, sides of all the animals could be compared with the corresponding lesioned right sides, and with each other as well as with the left sides of the split but otherwise intact octopuses.

Fig. 11.12 compares the two sides of the split but otherwise intact

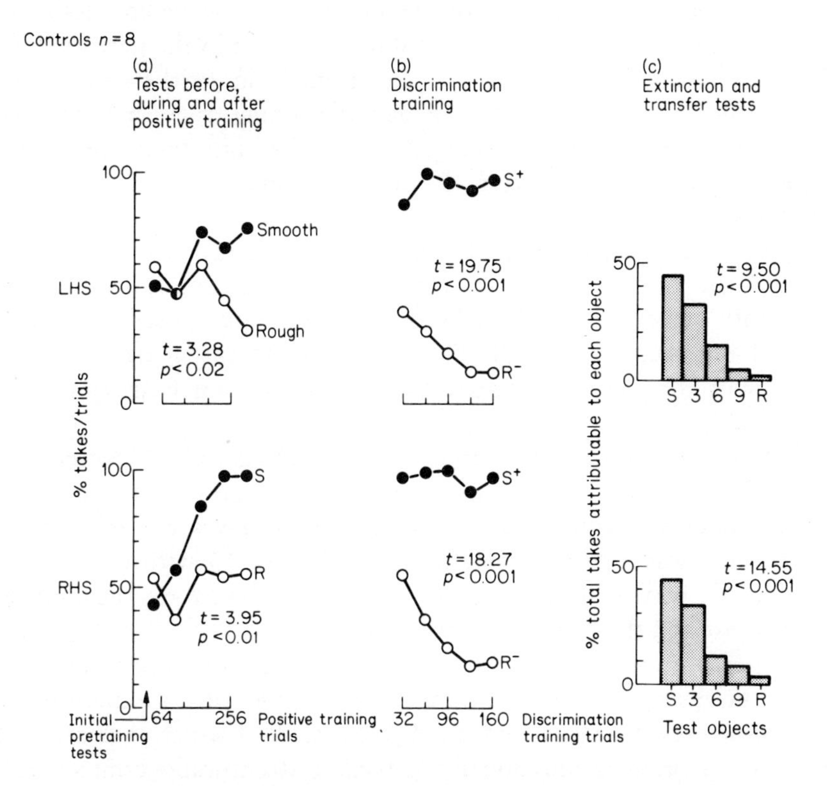

Fig. 11.12 Tactile learning by split-brain animals. Column (a) shows the scores made in groups of 32 unrewarded tests made before, during and after training to take a smooth sphere (see text). (b) records the scores in subsequent discrimination training at 32 trails per day. (c) shows the scores made in unrewarded tests run after the training; in these the original smooth and rough (13 grooved) spheres were presented, as well as further spheres with intermediate numbers of grooves. t-tests show the significance of differences in the number of takes of S and R in tests at the end of positive training, in the last 80 trials of discrimination training and between takes of S and R in the final tests (data from Wells and Young, (1975) where a full analysis is made).

animals. In the first part of the experiment only the right sides were trained. Before training the animals were checked for untrained preferences by means of 16 unrewarded tests with each object to each side. They were then trained presenting the smooth sphere to the arms on the right hand side of the body; if it was taken they were rewarded, if not the object was re-presented together with a piece of fish. Although this positive training was restricted to the arms on the right side, the animals were allowed to pass the objects under the interbrachial web to the mouth so that the bases of the left arms also made contact from time to time. The effect of these contacts is shown in the performance of the 'untrained' left sides of the animals in further groups of unrewarded tests given at intervals during training; they did not learn as rapidly as the more experienced right sides, but there were clear changes in the same direction. Both sides came to take smooth and reject rough.

After a marathon training of 256 positive trials (16 days at 16 trials per day) the animals were transferred to discrimination training; takes of smooth were rewarded as before, but the octopuses were now shocked for accepting rough. There were 32 trials per day, 16 with each side of the body. It was no surprise when the animals discriminated very capably.

After 160 trials, training ceased and the animals were subjected to a series of tests with the original smooth and 13-grooved spheres and further spheres having 2, 3 or 9 latitudinal grooves cut into them. Again, there was a considerable number of tests, 15 with each object to each side of the body. The animals, predictably, took the smooth sphere most and the roughest sphere least often.

The performance of the two sides of the body was almost identical during the discrimination and transfer tests phases of the experiments, confirming previous visual and tactile results, all of which indicate that the octopus brain is functionally as well as structurally symmetrical. Moreover it was found that the performance of the left side of the animals with brain lesions in the right could not be distinguished from the performance of the left sides of controls, a useful result since it not only confirms a lack of communication between the two sides of split-brain animals but also allows comparison of the performances of the right hand, lesioned sides of the different groups of octopuses in these experiments.

The right side of animals with the median inferior frontal lobe (m.i.f.) removed learned to discriminate more slowly than the right

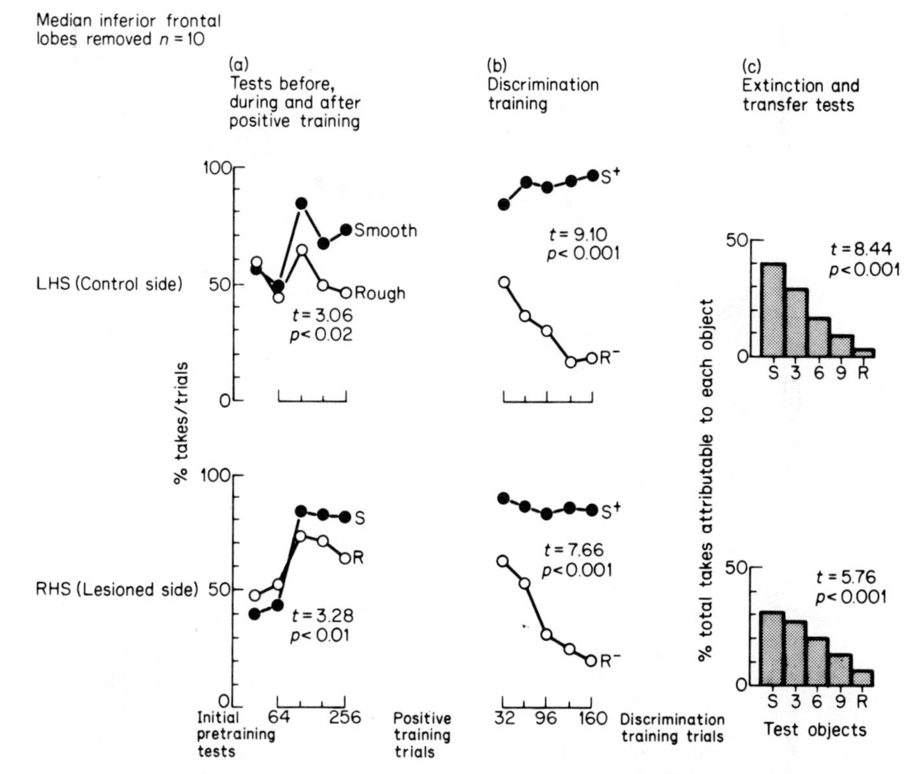

Fig. 11.13 The effect of removal of the median inferior frontal lobe from the RHS of split brain animals. Plotted as Fig. 11.12.

sides of unoperated controls or their own left sides (Fig. 11.13). They also performed less reliably in transfer tests. The differences from controls were not large; indeed the results of transfer tests were not significantly different (at the $p < 0.05$ level) on the left and right sides of these animals. These results are similar to those obtained using unsplit animals with the m.i.f. removed (see Section 11.3.6 below).

Damage to the subfrontal lobe, in contrast, produced large differences in tactile performance (Fig.s 11.14 and 11.15). In tests run before, during and after positive training, separation of the test objects was poor; animals with complete lesions failed to discriminate at all. In the discrimination training phase those with partial lesions showed a progressive increase in the capacity to discriminate, without, however, approaching the levels achieved by controls or m.i.f. animals. Of the 6 animals with none or very few of the subfrontal cells remaining (lesion 95% or more complete) only one achieved a significant score

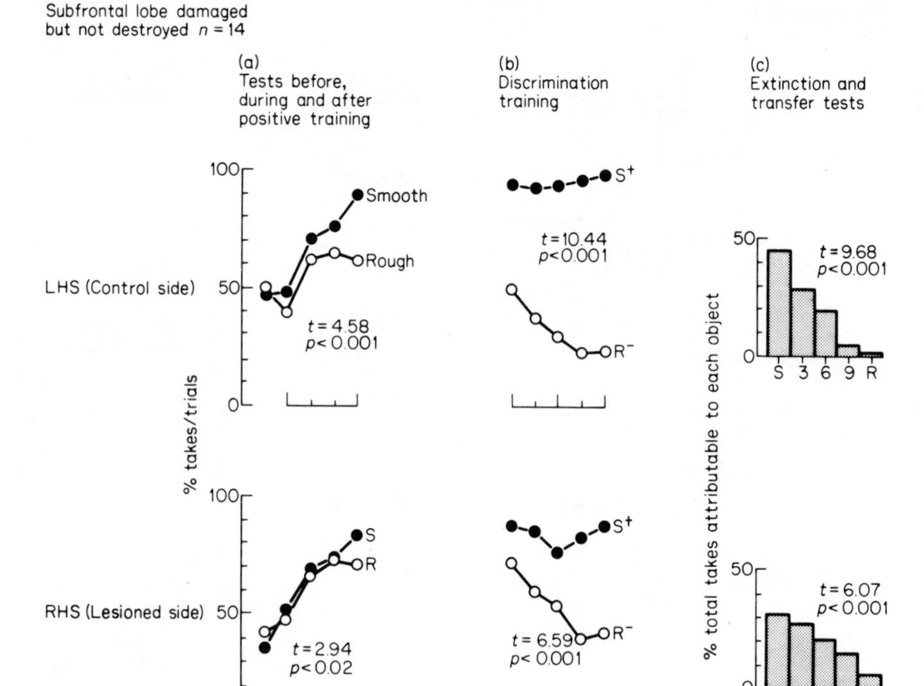

Fig. 11.14 The effect of damage to the subfrontal lobe. These split brain animals had between 50 and 90% of the subfrontal removed on the RHS; the LHS was intact. The lesioned sides learned more slowly than the RHS of controls (Fig. 11.12) or their own undamaged LHS. Plotted as Fig. 11.12.

in the second half of its discrimination training, and even this animal failed to discriminate in the subsequent extinction and transfer tests. Learning by these animals, is, at best, exceedingly slow. It should be noted that they all had the vertical lobe circuits intact, which suggests that any contribution to touch learning arising from the uppermost parts of the brain must be channelled through the inferior frontal system; by itself the superior frontal-vertical lobe system is unable to control tactile discrimination (Wells and Young, 1975).

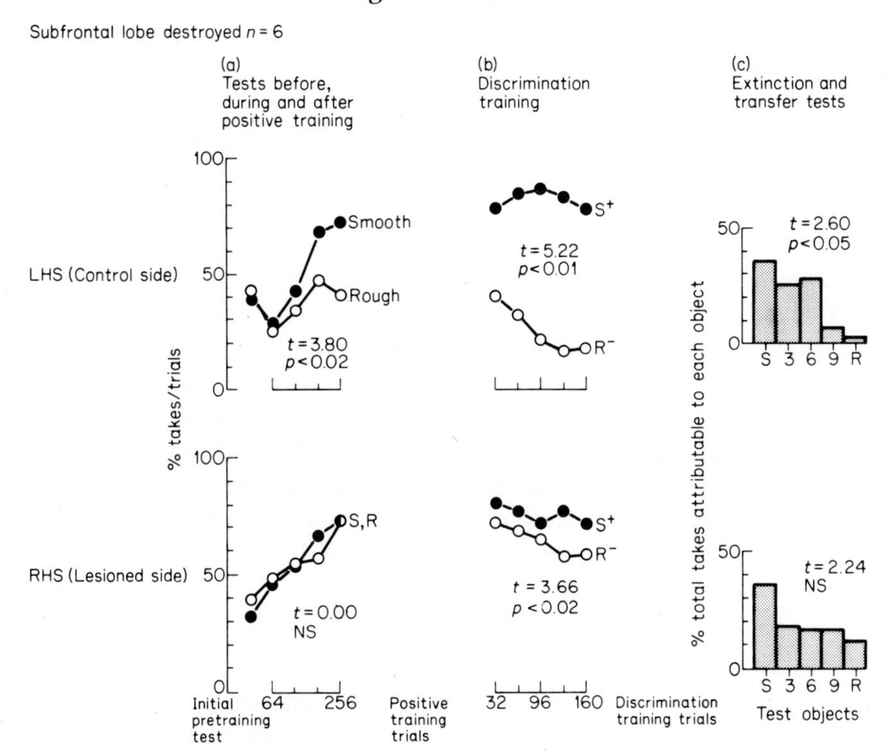

Fig. 11.15 The effect of removal of the subfrontal lobe from the RHS of split-brain octopuses. These lesions were 95% or more complete, very little larger than those in most of the animals concerned in the experiments plotted in Fig. 11.14. As usual, the unlesioned left sides performed well. Plotted as Fig. 11.12.

11.3.4 *The nature of the defects produced by subfrontal lobe removal*

Removing the subfrontal lobe has the same effects on touch learning as removal of the whole inferior frontal system. One can only conclude that the millions of small cells in the subfrontal form a vital part of the machinery for learning to discriminate. This is a matter of very considerable interest since the lobe is structurally homogeneous. Nearly all of the cells in it are tiny amacrines, without an obvious axon, and nearly all the connexions between the cells are limited to the subfrontal. The output of the lobe is provided by a few much larger cells, forming less than 1% of the total number of cells present (Table 10.1). If this is the place where the changes associated with learning take place, the mechanism is neatly boxed in and should repay a more detailed analysis of its structure.

The outstanding defect caused by subfrontal removal is an inability to learn to reject one object while continuing to take another. It is a discrimination failure, not a failure to regulate response levels. Comparison of Figs. 11.12 to 11.15 reveals no detectable differences in the rise in the level of takes that accompanies positive training. A corresponding drop as a result of negative training can be traced in Table 11.7.

A failure to learn to discriminate could arise because the animals are incapable of distinguishing between the objects used in training. It is not possible to distinguish this sort of failure from a failure to learn to discriminate between objects that the animal *is* capable of separating unless one can demonstrate general or individual preferences in untrained octopuses. When this was specifically tested in a group of 5 animals with the inferior frontal system removed, no significant preference was found (Wells and Young, 1968a; Table 11.8). Examination of the scores made by the 6 individuals in the tests summarized in Fig. 11.15 reveals no consistent individual preferences (Table 11.10).

Table 11.10 Individual scores in tests for untrained preference and in further tests after positive training to take a smooth sphere (S). R shows takes of a rough sphere which the animals were never rewarded for taking. Each object was presented 16 times in each set of tests. The collective performance of these animals, all of which had the subfrontal lobes removed, is summarized in Fig. 11.15.

Animal	*Before training*		*After 64 trials*		*After 128 trials*		*After 192 trials*		*After 256 trials*	
	S	R	S	R	S	R	S	R	S	R
RKA	9	7	9	7	3	6	5	2	4	5
RKD	12	16	16	16	11	12	16	15	15	16
RKL	1	2	5	13	7	7	0	0	13	15
RLE	0	2	9	6	16	15	16	15	16	16
RLG	1	1	1	3	1	1	13	10	10	7
RLH	7	10	4	1	13	12	13	14	12	11
Totals	30	38	44	46	51	53	63	56	70	70

Either the animals were unable to tell the objects apart, or they did not choose to do so in these experiments. Under discrimination training that is not preceded by prolonged smooth$^+$ve pretraining a slight preference for rough is revealed by animals with extensive lesions to the inferior frontal system. Such animals trained rough$^+$/smooth$^-$ often 'discriminated' while about half of those trained with smooth$^+$ve made 'perverse' scores (Table 11.6). The significance of this rough preference has been discussed in Section 11.3.2 above; it could arise at several levels, from those regulating sucker reflexes upwards, because rough objects stimulate the mechanoreceptors more readily.

11.3.5 *Classifying cells and a possible structure for the subfrontal lobe.*

The equivocal nature of the evidence on the ability of octopuses to distinguish between objects that they fail to learn to discriminate between, as a result of training, means that one cannot be absolutely certain that subfrontal removal interferes with learning rather than with some part of a stimulus analysing mechanism. At present the most economical hypothesis is to suppose that it does both, as would be the case if touch learning were based on 'classifying' cells of the sort that Young (1960a, 1965e) has postulated as the basis first for visual learning and subsequently for touch learning in the octopus. A 'classifying' cell is a neurone so connected as to respond to a specific pattern of stimulation (or, perhaps, the frequency of nerve impulses in the case of touch). It is seen as genetically determined, with outputs potentially capable of triggering positive or negative responses (take or reject in the case of touch). In the first instance, the stimulated classifying cell might do either, but the response can be switched by learning. In its original form (Fig. 12.21) Young's (1965e) 'mnemon' model is clearly imperfect (see Horridge, 1968; Wells, 1975, and Sections 12.2.1 and 12.2.2 below) but it is useful nevertheless because it draws attention to the likelihood that some sort of classifying cell plays a central role in learning. There is plenty of electrophysiological evidence that such cells exist in mammals (though the part they play in learning is not known) and cells that ought to behave in the same manner are abundant in the *Octopus* visual system (see Section 8.1.8). If touch learning depends upon classifying cells, these are presumably located in the subfrontal region, and it would not be suprising if lesions here were to affect both memory and the capacity to distinguish between objects. One can, moreover, make some predictions about the layout of these elements within the subfrontal. There must be many repeats of the same sort of cell, since partial lesions always produce a quantitative decline in the accuracy of response, never abolition of responses to specific stimuli. Elements responding to inputs representing objects of differing roughness must be so connected as to form a series of linear arrays, as experiments on stimulus generalization (Section 9.1.5)

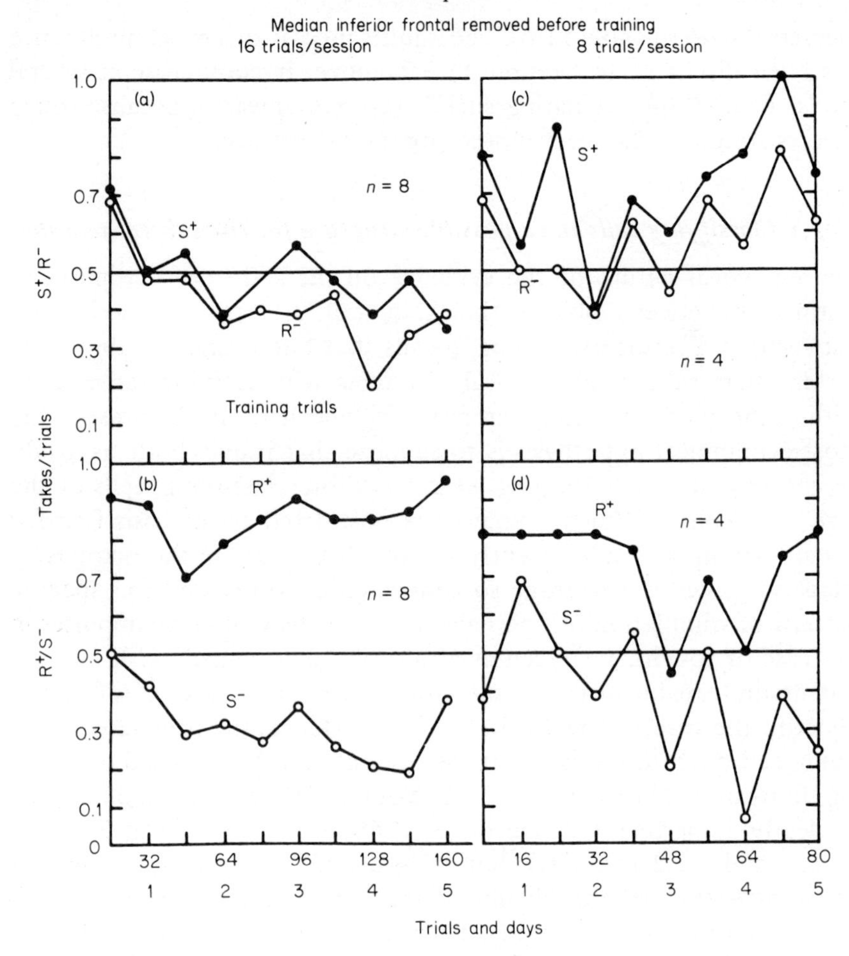

Fig. 11.16 Trained after median inferior frontal removal. The animals were trained to distinguish between a sphere with 13 latitudinal grooves cut into it (R) and a smooth sphere (S). (a) and (b) show the results of training at 16 trials per session, (c) and (d) at eight trials per session. Learning is always slower in the S^+/R^- direction, against rough preference, (from Wells and Young, 1973).

would suggest. Quite possibly they are stacked in columns, within the walls of the subfrontal, much as the cells responding to orientation are stacked in the visual cortex of mammals.

It is unfortunate that the millions of cells within the subfrontal are so small that they are unlikely to attract the attention of electrophysiologists for some time to come. The lobe is, however, compact and apparently homogeneous, and an analysis of the connexions within any small area of it would be of the very greatest interest. No such analysis is available at the present time.

11.3.6 *The median inferior frontal lobe*

The median inferior frontal provides most of the input to the subfrontal lobe (see Section 11.3 above). But its removal has surprisingly little effect on touch learning. Early experiments, reviewed in Wells and Young (1973), showed that damage to the m.i.f. produced a swing towards rough preference which tended to exaggerate differences between the operated animals and their controls. Fig. 11.16 shows how much more difficult it is to train octopuses with the m.i.f. removed in the smooth$^+$/rough$^-$ direction than vice-versa. As with vertical lobe removal, the effect of damaging the m.i.f. is quantitative. There is an immediate drop in the accuracy of discrimination due to increased errors in both directions; the positive objects are rejected, and negatives taken, more often than before the operation (Fig. 11.17). If rough preference is reduced by pretraining to take smooth, as in the experiments shown in Figs. 11.13 and 11.18, the difference between experimentals and controls is reduced. Whether, given sufficient training, the m.i.f.-removed animals could ever be brought to discriminate as well as their controls remains an open question. Fig. 11.18 would indicate that they cannot; results obtained with split-brain animals, summarized in Fig. 11.13, would suggest that they can.

The results of discrimination training experiments are very much what one would expect from the anatomy. The effects of removing the m.i.f. closely resemble the effects of damaging the subfrontal (cf. Figs. 11.13 and 11.14); one operation reduces the input to the memory store, the other removes part of the store itself.

It is also possible that m.i.f. removal damages the system in a more specific way by upsetting the proper distribution of tactile inputs. The m.i.f. is a region of interweaving tracts. It receives inputs from the arms on both sides of the body, and would appear to be arranged so that the nerves arriving in any one bundle are distributed to all parts of the m.i.f. and thence relayed to all parts of the subfrontal. Such an arrangement would ensure continuity of response when an object was passed from one arm to another, and it might be expected to play a part in the spread of the effects of training limited to any one arm.

Tests designed to reveal any such defect have yielded only equivocal

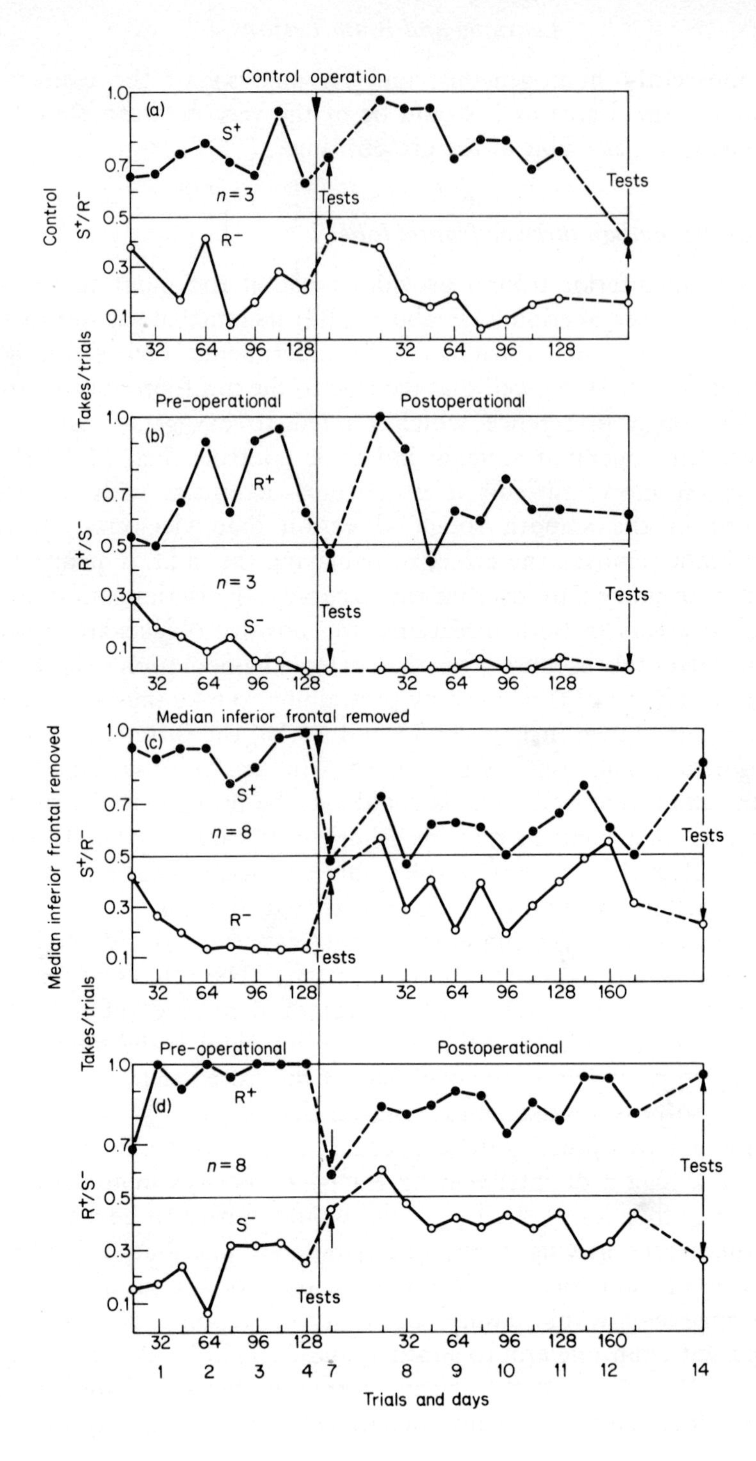

Control operation
(a)
S+
n = 3
Tests
R−
Control
S+/R−
Takes/trials
Pre-operational
Postoperational
(b)
R+
R+/S−
n = 3
Tests
S−
Median inferior frontal removed
(c)
S+
n = 8
S+/R−
R−
Tests
Pre-operational
Postoperational
(d)
R+
n = 8
R+/S−
S−
Tests
Trials and days

results. Animals with the m.i.f. split, trained on one side and tested on the other, perform equally well on the two sides. But then even octopuses with the whole of the inferior frontal system split seem to show some lateral communication. The small number of results available suggest that the two sides of the inferior frontal system can affect one another through the hind part of the supraoesophageal brain (Table 11.11). In an alternative type of experiment, attempts were made to train octopuses in different directions on the two sides of the body. *Ex hypothesis* this should be easier for m.i.f. animals than controls, which might be expected to suffer because inputs to either side of the inferior frontal system would be efficiently relayed to both. In the event, it proved quite easy to train controls to behave differently using the arms on the two sides of the body (Fig. 11.18). Animals with m.i.f. damage appeared to learn a little more readily (though the training on the 'reversal' side was in the 'easy' rough$^+$ direction). But those with the m.i.f. gone completely failed to learn at all; training and repeated shocks for taking the ex-positive object reduced these animals to a very low level of take at which further training was unlikely to be successful (Fig. 11.18).

These experiments do not prove that the m.i.f. has no specific function with regard to the spread of information through the touch learning system. They indicate only that its presence is not essential in a situation where plenty of time is allowed between training and testing. It is not improbable that the value of the m.i.f. lies in its capacity to ensure an immediate spread of tactile inputs to all parts of the tactile recognition system, a property that could be tested by repeated presentation experiments of the type made with animals lacking their vertical lobes by Wells and Wells (1957c). In these, the lesioned animals took several times as long as controls to recognize and reject an object touched and taken. Analysis of which arms were handling the object throughout the experiment (which would be easy enough to do through the glass floor of an aquarium) could be used to distinguish between delays attributable to learning by any one arm and delays attributable to a failure to share inputs.

Fig. 11.17 Training before and after removal of the median inferior frontal lobe. At sessions marked 'Tests' no rewards or punishment were given. The animals had a small number (up to 32) of training trials with the same objects before the start of this experiment. The control operation was a cut in the brain between the superior buccal and inferior frontal lobes (from Wells and Young, 1972).

Table 11.11 Lateral transfer after lesions to the inferior frontal system. Initial training to discriminate between smooth$^+$ and rough$^-$ (against untrained preference) was for 80 or 160 trials, all given to the same side of the body. Following this, training was continued for a further 80 trials, alternating with an equal number of unrewarded tests with the same objects presented to the arms on the untrained side. Details of these experiments, and individual scores, are given in Wells and Young (1966 and 1969b).

	% Correct in initial training on RHS	*% Correct in next 80 trials on RHS*	*% Correct in tests with untrained LHS*	*Difference between RHS and LHS (loss on transfer)*	*Reference*
1. *Controls: inferior frontal system intact, rest of supraoesophageal brain split*					
n = 6	66	70	67	2	Wells and Young, 1969
2. *Median inferior frontal lobe removed, rest intact*					
n = 6	66	67	55	12	Wells and Young, 1969
3. *Median inferior frontal split, rest intact*					
n = 2	81	86	82	4	Wells and Young, 1966
4. *Whole inferior frontal system split, rest intact*					
n = 6	72	78	57	21	Wells and Young, 1969
n = 5	72	88	73	15	Wells and Young, 1966
5. *Whole supraoesophageal brain split*					
n = 10	64	82	49	33	Wells and Young, 1966

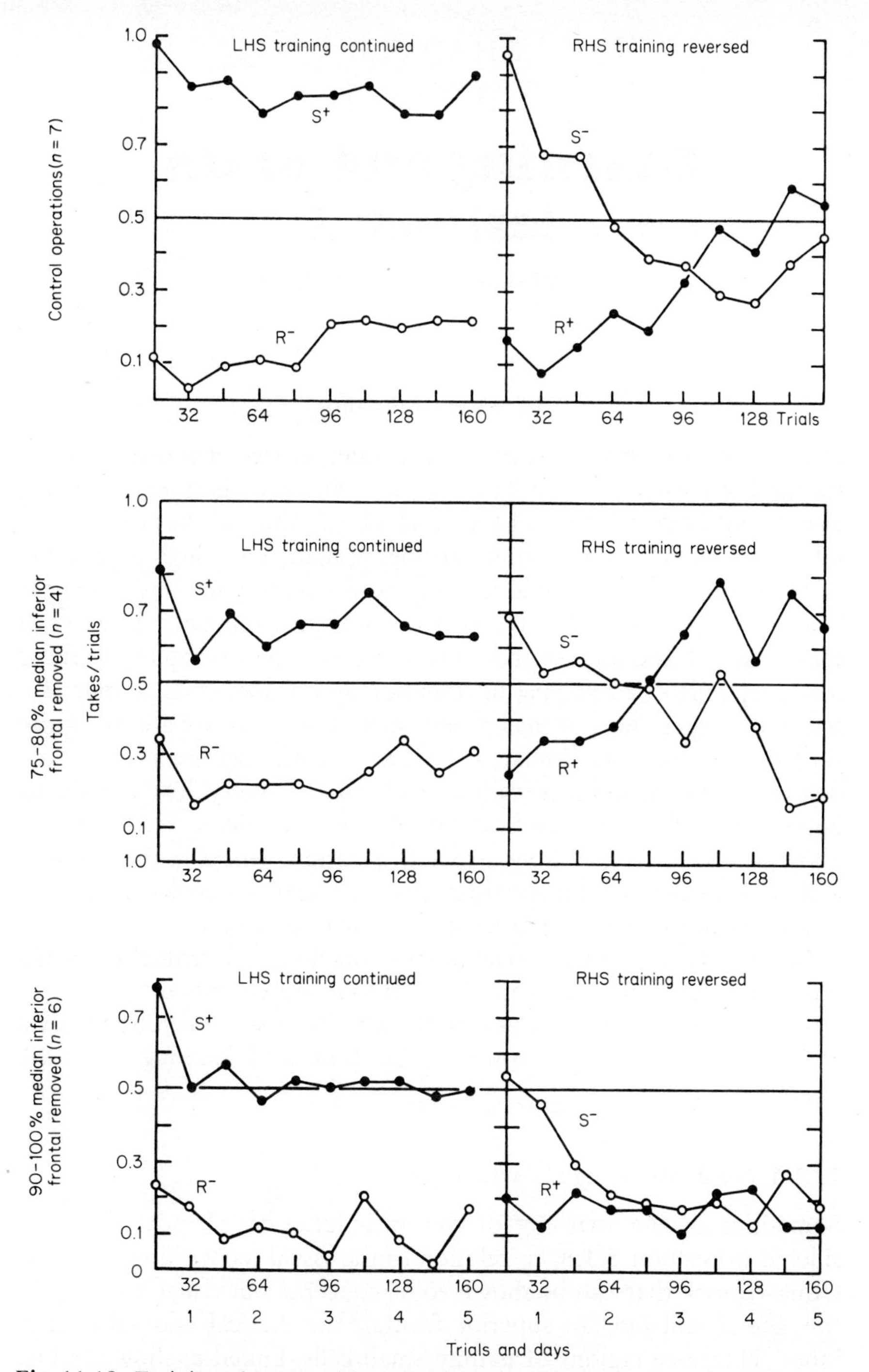

Fig. 11.18 Training the two sides of the same animals in opposite directions. These octopuses had had 256 positive training trials, to take a S^+, followed by 192 discrimination training trials (S^+/R^-) before the start of these experiments in which S^+/R^- training was continued on the LHS, simultaneously with R^+/S^- training on the RHS (from Wells and Young, 1972).

CHAPTER
TWELVE

Learning and brain lesions: 2

VISUAL LEARNING

12.1 Visual learning

Historically, experiments on visual learning preceded those on touch learning discussed in the last chapter. These earliest experiments, already outlined in Sections 11.1.1 and 11.1.2, showed that the vertical lobe plays an important part in learning by sight, and a high proportion of the brain lesion and visual training experiments made since then has been concerned with the effects of vertical lobe removal. As we shall see, these effects suggest that this part could be a complete learning machine, capable of storing information and changing its output as a result of experience. It has a very distinctive and apparently quite simple structure, very similar, at least at light microscope level, to that of the subfrontal and it is readily accessible to surgery. The tantalizing implication is that the vertical lobe has all the elements that could allow us to understand how associative learning works in the octopus, and perhaps all the information that we need to evolve a realistic neural hardware model for learning in animals generally.

In this last chapter, experiments on visual learning involving lesions to parts other than the vertical lobe are described first. There follows a rather longer survey of the effects of vertical lobe removal and a final discussion of some of the models of learning that have arisen from this.

12.1.1 *The anatomy of the visual system*

Something of the anatomy of the optic lobes has already been considered in Section 8.1.8, in relation to the visual recognition of shape. Other regions that can be shown to be concerned in visual learning are the lateral and median superior frontal, the vertical and subvertical lobes. These are regions of mainly small cells, linked as shown in Fig. 12.1 (Fig. 11.8c shows a corresponding longitudinal section). The relation between these parts, the optic lobes and other regions of the

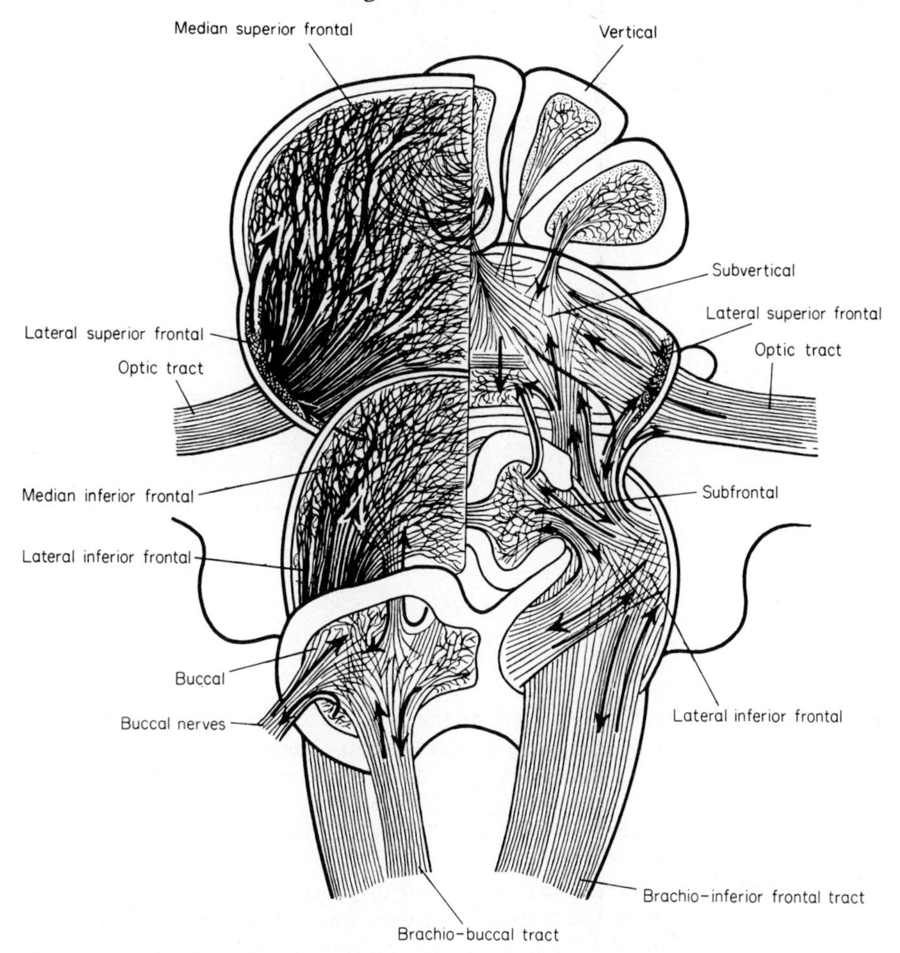

Fig. 12.1 Reconstruction of the front part of the brain from serial sections, to show tactile and visual pathways and the vertical lobe complex. Seen in transverse sections, viewed from in front. On the right buccal, medial inferior frontal and medial superior frontal lobes have been omitted (from Young, 1961a).

brain are summarized in the diagram, Fig. 12.2. There are obvious parallels between the visual and the tactile learning systems (illustrated in Fig. 11.8). Both include a pair of lobes, the superior frontal and vertical in the visual, the inferior frontal and subfrontal in the tactile, one of which is composed largely of interweaving tracts, the other of almost all of the cells. Messages pass one way only, from the tract lobe to the cell lobe, and thence only indirectly to regions concerned in

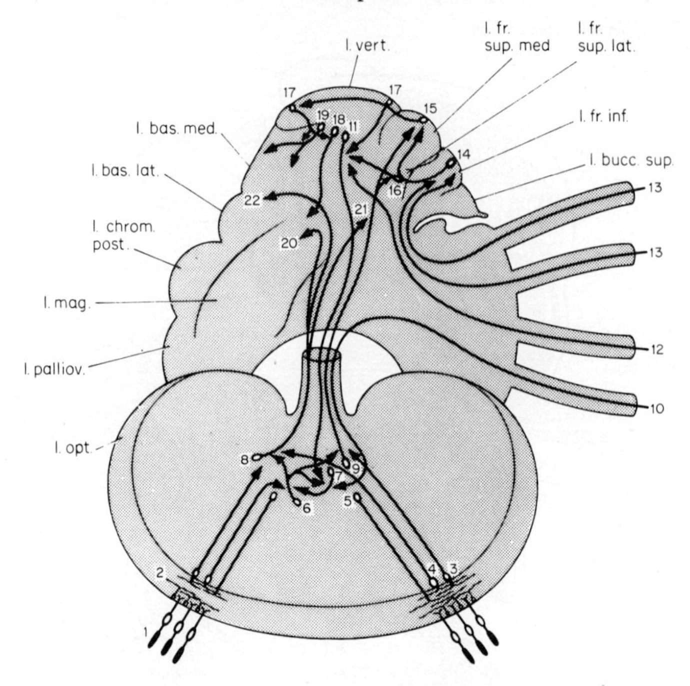

Fig. 12.2 Diagrammatic lateral view of the brain of octopus to show the main connexions of the vertical lobe system. The optic lobe is displaced downwards. 1. retinal element; 2. outer amacrine cell; 3 and 4. inner amacrine cells; 5. cell body in the inner part of the optic lobe, feeding back to the deep retina; 6 and 7. 'association cells' in the deep part of the optic lobe; 8. efferent tracts from the optic lobe to the anterior (21) and posterior (22) basal, and magnocellular (20) lobes; 9. optic-superior frontal tract; 10. brachio-optic tract; 11. subvertical-optic tract; 12. brachio-subvertical tract; 13. brachio-inferior frontal tracts; 14. inferior frontal-superior frontal tracts; 15. superior frontal-vertical tract; 16. lateral superior frontal-subvertical tract; 17. vertical-subvertical tracts; 18. subvertical-magnocellular tract; 19. tracts from subvertical 1 to subvertical 2 and 3. (from Young, 1961a).

motor control, via the subvertical or posterior buccal lobes. There is an alternative 'short circuit', parallel with this, running through the lateral superior frontal and subvertical (in the visual system) or the lateral inferior frontal and posterior buccal (in the tactile). The presence of two largely independent learning systems in one brain provides attractive possibilities for comparative work, since there are presumably common features in the manner in which the systems operate, despite the rather different nature of the information that they handle (Young, 1963a; 1964b, 1965e).

12.1.2 *Brain lesions and visual learning*

Early experiments on brain lesions and visual memory in the cuttlefish and some of the subsequent studies of the effect of vertical lobe removal on learning in *Octopus* have already been reviewed in Sections 11.1.1 and 11.1.2.

A more extensive report can be found in Boycott and Young (1955a). These authors confirmed that large parts of the brain of *Octopus* could be removed, producing considerable defects in visual learning without, at the same time, affecting the ability to make visual responses. They also showed that the animals could be anaesthetised, their craniums opened, brains prodded and stimulated faradically without disrupting visual memory traces.

Progressively larger lesions made to the top part of the brain showed that the capacity to learn a crab vs crab plus square discrimination was much reduced by removal of the vertical or superior frontal lobes or by section of the tract connecting the two (Table 12.1). Larger lesions did not noticeably increase the already very considerable defect (Table 12.2) until they began to encroach upon motor control areas in the basal lobes and destroy the capacity to react at all. Damage to the lateral part of the superior frontal lobe appeared to inhibit attacks on crabs seen at a distance (Table 12.1). The number of trials and animals used in these early experiments was, for the most part, small but the collective trend was clear enough. Even octopuses with the whole of the superior frontal, vertical and subvertical lobes removed showed signs of learning to discriminate. The signs were, however, slight, and there seemed (and still seems) to be little doubt that visual learning is a collective property of the uppermost part of the brain, the 'silent' areas with many small cells (Section 10.1.2), and the optic lobes. Lesions made to the only areas with comparable structure, in the inferior frontal system, have no detectable effect on visual learning (Fig. 11.1).

12.1.3 *Lesions to the optic lobes*

From their structure, it seems probable that the optic lobes include cells that will respond to particular patterns of visual input (Section

Table 12.1 Brain lesions and visual discrimination (from Boycott and Young, 1955a).

Vertical lobe removed

Octopus	Lesion (% removed)				Attacks			
	Vertical lobe	Subvertical lobes			Pre-operative		Post-operative	
		1	2	3	C	CS	C	CS
EX	100	0	0	0	29/33	12/30	6/8	9/9
PU	100	0	0	0	38/45	3/45	3/21	0/21
SO	100	0	0	0	15/15	6/15	9/9	10/10
TH	100	0	0	25	9/9	4/9	13/14	11/14
SW	100	25	25	25	12/13	2/15	4/9	2/9
HI	100	25	50	100	8/8	8/20	13/15	10/18
SP	100	25	75	100	15/15	6/15	16/24	18/24
PR	95	0	50	75	24/24	4/24	4/22	2/22
TG	95	25	50	50	9/9	3/9	8/8	7/8
SQ	95	0	0	0	9/9	4/9	21/24	16/24

Superior frontal to vertical lobe tract cut

Octopus	% of tract severed	Attacks			
		Pre-operative		Post-operative	
		C	CS	C	CS
HG	100	–	–	12/17	18/19
SK	100	–	–	19/21	17/21
HM	100	12/12	12/19	7/8	11/16
HJ	100	6/6	6/15	18/20	30/34
RC	100	31/33	5/33	8/21	1/21

Injury to lateral superior frontal lobe

Octopus	Lesion (% removed)							Attacks			
	Vertical lobe	Superior frontal lobe			Subvertical lobe			Full length of tank (80 cm)		Half length of tank (40 cm)	
		Medial	Lateral		1	2	3	C	CS	C	CS
HAB	100	50	25	0	25	0	0	7/18	9/18	8/9	7/9
HAM	100	100	50	100	50	50	50	7/18	4/18	6/9	5/9
HAL	100	50	75	0	50	25	25	5/15	5/15	9/9	8/9

Animals with lateral superior frontal lesions seldom attack when the crab or crab and square are far away but do so more often if they are placed half-way down the tank; no shocks were given to these octopuses.
C = crab; CS = crab plus 5 cm square.

Table 12.2 Extensive lesions to the supraoesophageal lobes and visual discrimination (from Boycott and Young, 1955a).

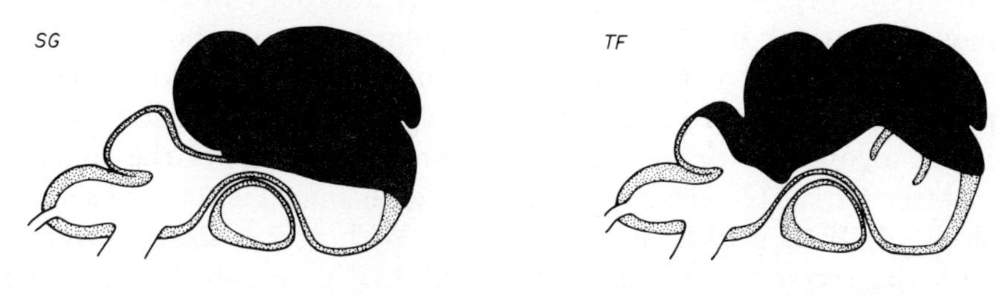

Octopus	Lesion (% removed): Vertical lobe	Superior frontal lobe: Medial	Superior frontal lobe: Lateral		Subvertical lobe: 1	2	3	Attacks: Pre-operative C	Pre-operative CS	Post-operative C	Post-operative CS
ST	100	100	75	0	25	0	0	8/8	4/10	6/18	5/18
RP	100	100	75	50	25	0	0	–	–	9/25	4/27
SG	100	100	100	75	100	100	100	–	–	35/36	23/36
TF	100	100	100	100	75	50	50	9/9	3/9	14/14	11/14
FT	100	100	100	100	100	100	100	–	–	14/20	19/29
PN	50	50	0	75	50	50	50	18/18	4/18	15/24	4/24
PL	50	50	0	50	50	50	50	16/21	4/21	5/21	1/21

The lesions in octopuses SG and TF are mapped on diagrams representing a median longitudinal section through the supraoesophageal part of the brain. In octopus RP the top part of the inferior frontal lobe was also damaged and there was slight damage to the subfrontal lobe. In octopuses FT and TF the inferior frontal lobe, subfrontal lobe and brachio-cerebral tract were damaged. *C* = crab, *CS* = crab plus 5 cm square.

8.1.8). They could also be the site of memory traces established when the animals learn to recognize objects by sight.

Very large lesions to the optic lobes are bound to disrupt the visual analysing system, and will partially or totally blind the animals. Small lesions (cuts at random in the optic lobes) had little or no effect on learning (Boycott and Young, 1955a). The view that visual learning as well as analysis could proceed within the optic lobes depended for a long time mainly on arguments from structure (Young, 1965a), on a report (Boycott, 1954) that habituation had been observed in an unspecified number of octopuses following removal of the rest of the supraoesophageal lobes, and on the rather scant evidence summarized in Table 12.2.

This situation was resolved in 1963, when two series of experiments were made, both indicating that the optic lobes are, indeed, a likely site for visual memory stores. Muntz (1963) trained octopuses to recognize horizontal or vertical 10 × 2 cm rectangles without moving from homes placed in the middle of their aquaria. The animals were shown one of the shapes and shocked after 5 seconds, or earlier if they moved to attack it. The fact that the animal was punished before (or as soon as) it had started to move meant that Muntz could arrange to present the stimulus to a specific retina or parts of the retinae by showing it to one eye or immediately behind or in front of the animals.

After 4 trials (2 each day, for 2 days) each animal was tested successively; 1. With a crab, shown alone; 2. With a crab shown together with the rectangle used in training and 3. With a crab shown with a 'transfer' shape (a horizontal rectangle when a vertical had been used in training, and vice-versa). The animals almost invariably attacked the crabs shown alone, but were more reluctant to attack crabs shown together with a rectangle and in particular any crab shown together with the rectangle used in training. Tests continued over a period of days until all three stimuli were attacked regularly.

Learning in this experiment can be expressed in terms of the number of occasions when no attack was made or in terms of the response latency at trials when an attack was made.

Fig. 12.2a summarizes the results with animals trained using one eye and tested using the other. The octopuses performed most reliably using the 'trained' eye, but there was clearly some interocular transfer (Fig. 12.3a), as was to be expected from Muntz's own previous work (Muntz, 1961a, b, c; Fig. 10.4 and Section 10.2.1).

When similar tests were carried out on octopuses that had been trained using only the backs or only the fronts of the retinae, almost perfect transfer was found (Fig. 12.3b); apparently the visual system includes a mechanism for spreading information so that receptor generalization is achieved, just as it is between arms in the tactile system (Section 11.3.1).

Receptor generalization between front and back of the retina did not occur if each optic lobe was partially divided by means of a tranverse vertical cut (Fig. 12.3c). This can only mean that the spread of information leading to receptor generalization proceeds from front to back (or vice-versa), across the optic lobes. It cannot be signalled through the optic tracts to some other part of the brain to establish changes elsewhere that then determine responses when the same stimulus falls upon some other part of the retina. Radial pathways would not be affected by the vertical cut. The simplest hypothesis is that any changes associated with the establishment of visual memory

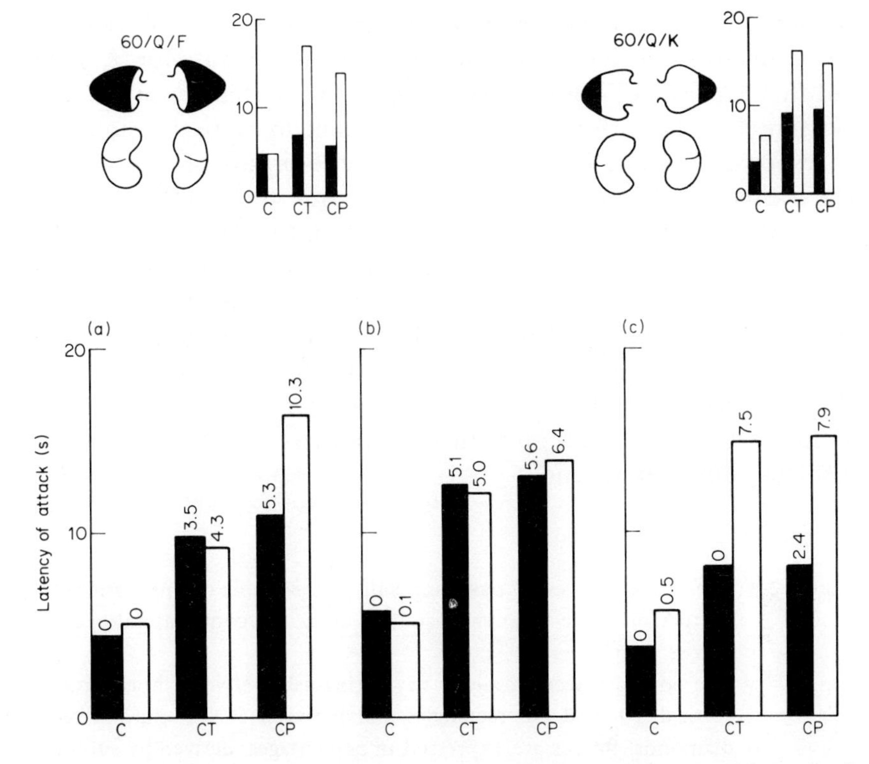

Fig. 12.3 Experiments on interocular and intra-ocular transfer. For (a) 4 animals were trained not to attack a rectangle, using one eye and then tested with this and the other eye. Test stimuli were C, crab presented alone; CT, crab plus a rectangle (not that used in training); CP, crab and the rectangle used in training presented together. Columns show mean attack latency and the mean number of presentations at which no attack was made, □ by the trained eye, ■ by the untrained eye. (b) shows a similar experiment in which the rectangle used in training was shown in front of or behind the octopus, testing intraocular transfer between the front and back halves of the retinae, n = 8. (c) as (b) but with a transverse vertical slit in the optic lobes, as shown for 60/Q/F and 60/Q/K (the largest and smallest lesions) above, n = 8 (from Muntz, 1963).

stores actually take place in the optic lobes (Muntz, 1963).

A second line of experimental evidence supporting the theory that the optic lobes are the site of visual memory stores was provided by Parriss (1963). She showed that octopuses trained after removal of part or the whole of one optic lobe performed less accurately than octopuses with the optic nerves cut on one side. The defect only became obvious after training on two successive discriminations

(Table 12.3). We know that interocular transfer survives removal of the 'trained' optic lobe (Boycott and Young, 1955a; Muntz, 1961c) and this experiment implies that in the visual system, as in the tactile (Section 11.3.1) both sides of the brain contribute to decisions made in response to stimuli received by the sense organs on any one side. The difference between intact animals and those with the optic nerves cut on one side is a further indication that lateral transfer may be incomplete. These animals were trained by presenting the figures to be distinguished successively at the far ends of their tanks. The intact animals could (and presumably did) view the figures with both eyes during training and the result obtained (intact animals learned better than those blinded in one eye) is just what one would expect if training establishes traces on both sides, but less effectively on the side that lacks direct experience.

Table 12.3 The consequences of removing half or the whole of one optic lobe on the performance of animals trained using the contralateral eye (from Parriss, 1963).
The animals were trained to distinguish between horizontal and vertical 10 × 2 cm rectangles and between 5 × 5 cm squares and diamonds. Scores are expressed in percentages, derived by subtracting the number of takes of the negative object from the number of takes of the positive and dividing this by half of the number of trials, × 100.

Group	*Initial training on horizontal versus vertical rectangles (55 trials) %*	*Subsequent training on square versus diamond (96 trials) %*	*Retraining on horizontal versus vertical rectangles (55 trials) %*
1. Optic nerves cut on one side, *n* =4	57.0	15.2	25.9
2. Half of one optic lobe removed and remaining nerves cut on that side, *n* = 4	41.4	11.3	11.9
3. One optic lobe removed, *n* =4	51.7	17.7	13.9
4. Intact animals, *n* = 7	59.7	15.4	70.9

Differences between groups are significant at $p = <0.05$ only in retraining on H versus V when groups 1 or 4 are compared with 2 or 3.

12.1.4 *The vertical lobe and reverberating circuits*

The first few experiments to be made on visual learning in *Octopus* involved crabs and learning not to attack crabs. The crabs were shown alone, or together with a 'neutral' stimulus (a white square or disc) and the animal was given a small electric shock if it attacked. Animals with their vertical lobes removed seemed to be capable of remembering not to attack crabs for a few minutes only, while unoperated animals could remember for hours or days. Forgetting could be delayed by leaving the crab and square in view of the octopus (Table 11.2, Section 11.1.2).

Young (1951) was impressed by the existence of the circuit of nervous connexions running from the optic lobes to the vertical lobe and back to the optic lobes (Figs. 12.2 and 11.8c). He pointed out that most of the results then available could be explained if one supposed that the function of the vertical lobe was to re-present from within, maintaining activity in the optic-vertical lobe circuit for long enough to establish permanent changes within the optic lobes. This idea has remained in the literature ever since. It is, for example, cited by Young (1970a) in accounting for the relatively short duration of memories promoting attack on rectangles after removal of the superior frontal lobe, and by Wells (1967) in discussing interocular transfer in detour experiments. Self re-exciting chains of neurones have remained the basis of Young's (1963c, 1964a, b) models for learning (discussed later, in Section 12.2.1), as a means of 'maintaining the address', so that the signals indicating the results of actions recently taken can be steered to the neurones that have initiated the action.

An apparent flaw in the reverberating circuit theory was the very considerable deficiency in performance shown by animals trained not to attack crabs before interruption of the vertical lobe circuit. Their performance was not noticeably better than that of animals with no preoperational training, although the presumed function of the vertical lobe was to establish effective memory traces within the optic lobes (Table 12.1). The vertical lobe, or some other element in the interrupted circuit (though the vertical lobe was the obvious target, since it contained most of the cells) must have some function other than re-presentation. In the case of crabs, an obvious possibility was that it provided a brake on attacks, which were clearly more liable to occur in its absence. This was the view taken in Boycott and Young (1950), which re-emerges later as a component of Young's (1963c, 1964b) 'paired centres' theory, outlined below.

It is important to recognize that there has never been any direct evidence for reverberating circuits in the octopus. The required anatomy is certainly present, and the system plainly could work in this manner. The view that it does so is compatible with the results of a wide range of training experiments which show (1) that octopuses learn more slowly if the vertical lobe circuit is interrupted and (2) that this deficiency can be partly remedied by packing trials more closely together. It is also compatible with the consequences of electro-convulsive shock treatments which indicate that the establishment of long-lasting memory traces can be severely impaired by stopping electrical activity in the CNS soon after an experience that the octopus would otherwise remember (Maldonado 1968, 1969 - see Table 11.1).

An attempt to demonstrate prolonged electrical activity following visual stimulation had, however, only very limited success. Stephens (1974) placed extracellular microelectrodes in the vertical lobe and observed evoked potentials 0.5 s after a 1 s light stimulus, followed after a further 1.2 s by a second round of potential changes - 'This suggests that impulses from the eyes do reverberate at least once round the loop'. But once is surely not enough? Removal of the vertical lobe produces an approximately five-fold increase in the number of trials required to produce a given behavioural change (Young, 1961b). While it remains possible that Stephens' results were a consequence of the non-physiological conditions of his experiments (the animals were immobilized by cooling them to 8 °C) the only possible conclusion from them at present is that there is no good electrophysiological evidence to support the reverberating circuit hypothesis.

12.1.5 *Memory traces present but ineffective*

As soon as training experiments were done without crabs, by presenting octopuses with unfamiliar figures which they would not normally be expected to associate with food, the apparently clear-cut difference between octopuses with and without vertical lobes began to disappear. Even animals with very large lesions could be taught to discriminate at eight or ten trials per day, a rate of training that would never, it seems, have produced successful discrimination in the crab/crab plus figure situation. Crabs apparently constitute an overwhelmingly attractive stimulus and if they are omitted in training trials that otherwise include both a crab and a figure, even octopuses that have shown

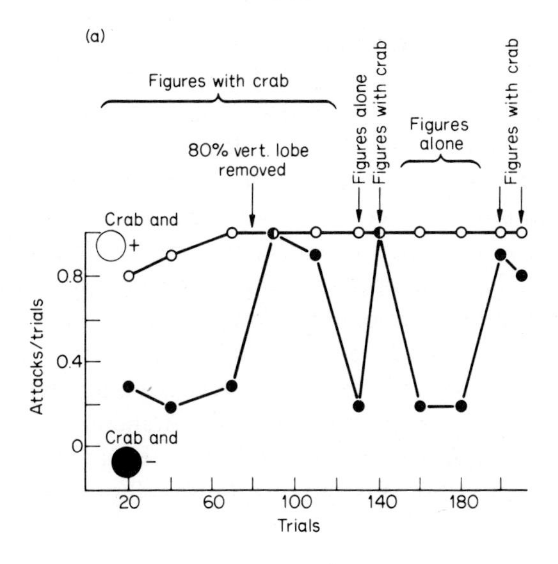

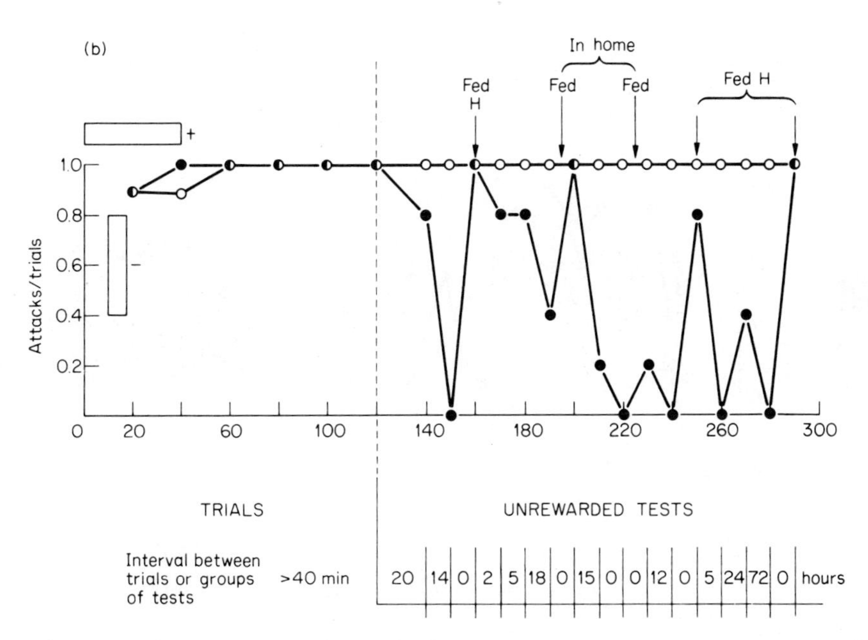

Fig. 12.4 (a) Train-operate-train, with vertical lobe removal. There were 16 trials per day, given in pairs (one $^+$, one $^-$) with 30 min or more between pairs (after Boycott and Young, 1957). (b) Operate-train following 85% vertical lobe removal. Trials were at 16 per day, as before (minimum interval between pairs, 40 min). Unrewarded tests were given in groups of 10 (5^+, 5^-) at the intervals shown, 5 min between each test within a group (after Young, 1958a).

no previous signs of discrimination may be shown to have learned (Fig. 12.4a, Boycott and Young, 1956b, 1957). A similar reversible masking of the ability to discriminate may sometimes be produced simply by feeding the animals (Fig. 12.4b).

The results of both sets of experiments imply that the vertical lobe has a mainly inhibitory effect. But both are concerned with food and feeding where any tendency to forget training would be characertized by a rise in the level of attacks. When 'neutral' figures rather than crabs are used in preoperational training, removal is as likely to result in a reduction as in an increase in the number of attacks made (Young, 1958b). If the problem of response levels is eliminated by showing both of the figures to be discriminated simultaneously, the difference in performance between controls and animals lacking their vertical lobes is very much reduced (Fig. 12.5, Muntz, Sutherland and Young, 1962).

Considered collectively, experiments of these sorts suggest that memory traces can be established in octopuses lacking their vertical lobes, but they are often ineffective. There is, at least, no longer any

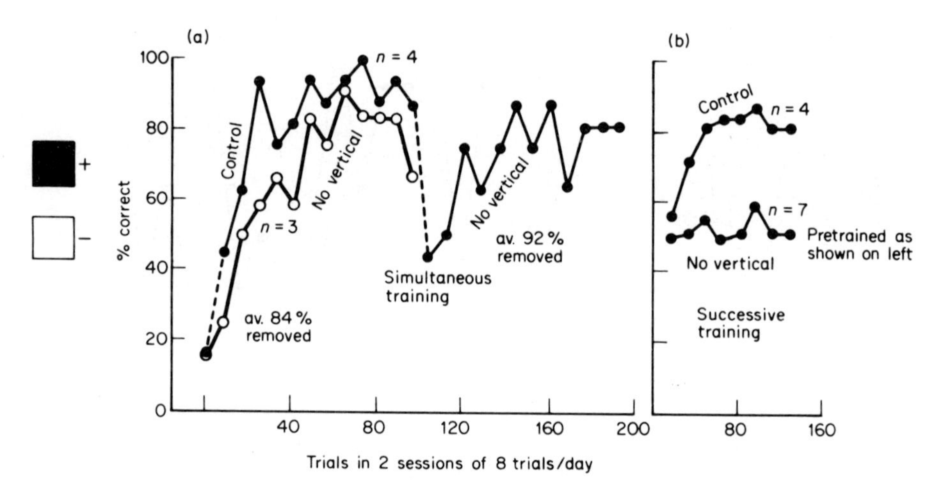

Fig. 12.5 Simultaneous and successive training and the effect of vertical lobe removal. The animals were trained to discriminate between 5 cm black and white squares, with white, the initially preferred figure, negative. In (a) the squares were presented simultaneously. The 4 control animals were operated upon after 12 days (96 trials) of training. In (b) the same figures were presented successively, one or the other at each trial. A new group of controls were used, but the no vertical animals were the same (after Muntz, Sutherland and Young, 1962).

doubt that memory traces can be set up. Indeed given sufficient training, they can be brought up to a standard of effectiveness approaching those in controls, at least for easy discriminations (Fig. 12.6). Even

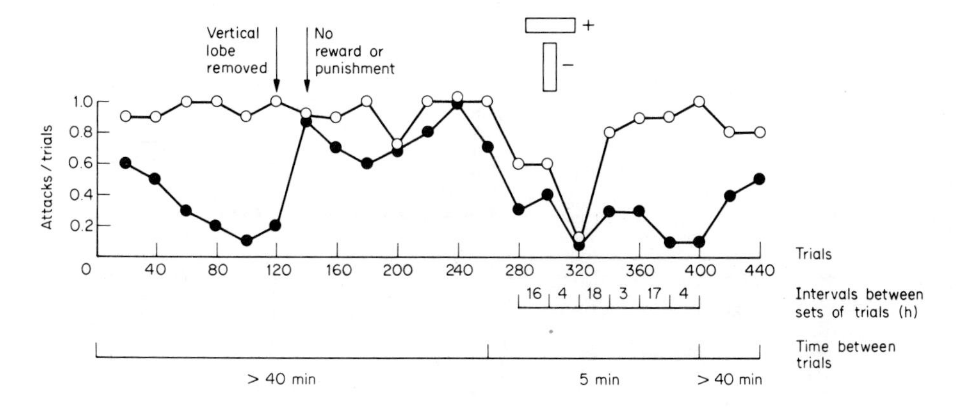

Fig. 12.6 Training before and after vertical lobe removal. The operation produces an immediate drop in the accuracy of response which can, however, be restored if the animal is given sufficient post-operational training. The first part of the training was with trials at not less than 40 min intervals. The rest was with groups of 20 trials (10^+, 10^-) at 5 min intervals; the times between the groups were as indicated (after Young, 1958b).

the 'impossible' task of learning not to attack crabs can be mastered and, if the animals never actually learn to leave crabs alone after an overnight break in training, at least very considerable savings can be observed on retraining (Fig. 12.7).

A scattering of experiments in which more difficult discriminations (V vs ≶ , ⊏ vs ∟ , ◇ vs □ etc.) have been attempted indicate that the effect of vertical lobe removal may be so serious in the performance of difficult tasks as to prevent discrimination altogether (Boycott and Young, 1957; Muntz, Sutherland and Young, 1962).

One particular class of 'difficult' discrimination is that involving reversal of a discrimination already learned. Control animals can manage to do this; those lacking vertical lobes only seem to succeed when the reversed direction coincides with apparently innate preferences. Thus they can, for example, be trained to take a vertical and reject a horizontal rectangle after previous training on H^+/V^-; but not *vice versa* (Young, 1962c; for a review of reversal and repeated-reversal experiments, see Section 8.1.15).

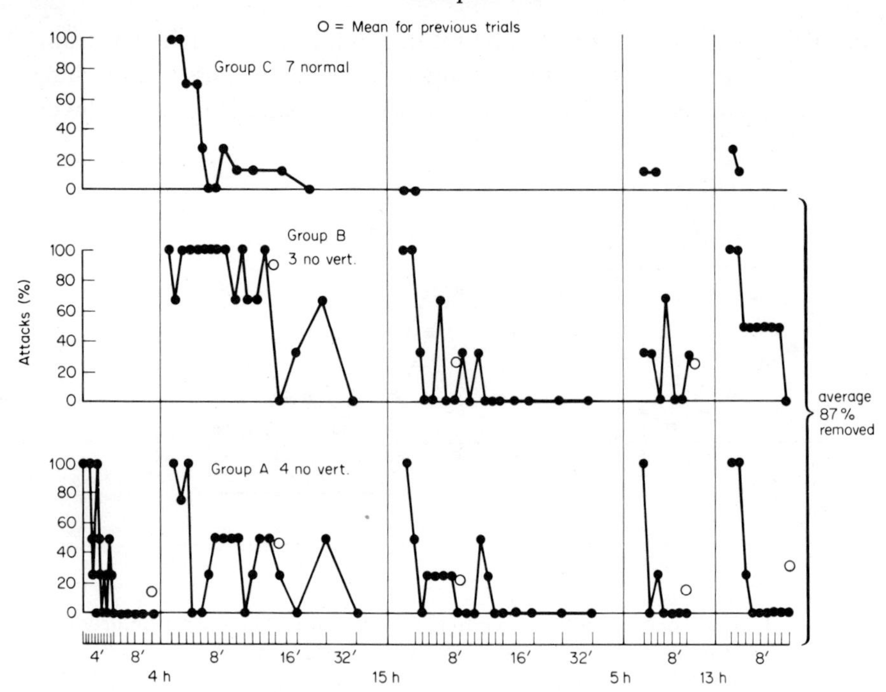

Fig. 12.7 Learning not to attack a crab. The vertical lobe has been removed in the animals in groups A and B; group C were control animals. The tests were made at 4 min intervals in group A, and then at greater intervals. Groups B and C were tested at 8 min intervals and subsequently at longer intervals (from Young, 1965e).

12.1.6 *Learning at different rates of training*

Increasing the number of trials within a given period tends to reduce the proportion of attacks. Whenever the errors made are mainly negative, as is usually the case, performance improves. It is difficult to disentangle this cause of improved performance from an increase in the capacity to discriminate independent of the level of response. Fig. 12.6 illustrates the problem; the improvement in result arising from the higher rate of training could have been due to the more ready establishment of memory traces, or to the unmasking of traces already present.

This problem was recognized almost from the beginning of studies of the effects of vertical lobe removal. Young (1960d) compared the rates at which discrimination developed when octopuses were trained

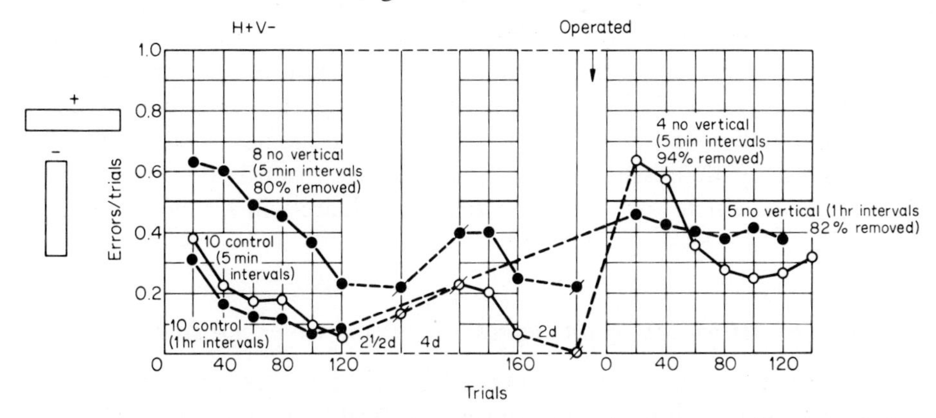

Fig. 12.8 Proportion of errors by control animals and octopuses with no vertical lobes. Horizontal rectangle as positive, vertical as negative figure. Trials at 5 min intervals, twenty daily in two groups of 10 (each 5^+, 5^-), or at 8 trials per day, (4^+, 4^-) with an approximately 1 h interval between pairs (1^+, 1^-) Scores indicated ⌀ or ● show the performance of the animals in unrewarded retention tests. (data from Young, 1960b).

with trials at five minute and at approximately one hour intervals. Some of his results are shown in Fig. 12.8. Learning was apparently more rapid, trial for trial, at the higher rate of training. But this is at least partly an illusion; where performance at the *first* trial only of each 10 trial, 5 minute interval session was considered, long-term cumulative learning was not significantly better than that shown with trials at hourly intervals; the improvement was all within sessions, and largely due to a drop in the proportion of attacks (Young, 1960d).

A further problem is that animals, presumably, can only learn by experience if they attack. Octopuses with their vertical lobes removed tended to attack more, so they got more training, trial for trial, than their controls. The matter of learning per experience, rather than per trial, has rarely been adequately analysed in visual experiments. But a consideration of tactile training experiments, where, too, there is an apparent relative improvement in the performance of no verticals with a rise in trial frequency, shows quite clearly that the narrowing of the performance gap is due to the greater amount of training received by the experimentals. A rise in the rate of training depresses response levels in controls more readily than in the no vertical octopuses, which get relatively more training with every increase in trial frequency (Wells and Young, 1969a; see Section 11.2.1).

There is therefore some doubt as to whether any improvement in long-term learning is achieved by packing trials closely together. All the evidence suggests that animals without vertical lobes learn less, take for take, than their controls, whatever the rate of training. The main advantages of a high rate of training are that it is convenient experimentally, and that otherwise ineffective memory traces can be unmasked by depressing the rate of response towards the 50% level, the rate at which any tendency to discriminate will be most readily revealed.

12.1.7 *Quantification of the defect; learning to do or not to do*

Subsequent to the realization that long-term learning was slow rather than absent in octopuses lacking their vertical lobes, a series of experiments was made in an effort to quantify the defect, the assumption by this time being that the effect of vertical lobe removal was quantitative rather than qualitative. Young (1960d) used white plastic 10 × 2 cm horizontal and vertical rectangles and concluded that octopuses lacking their vertical lobes needed between three and five times as many trials as controls to reach an asymtote of about 80% correct responses. (The rectangles were used as a standard test in most of the experiments on vertical lobe removal from 1958 onwards, usually in the form Horizontal$^+$/Vertical$^-$, since a vertical rectangle moved along its long axis is the 'preferred' figure (Section 8.1.2).) In further experiments, one rectangle was presented repeatedly and the octopuses trained to attack or not to attack it. The results show that octopuses lacking their vertical lobes are slower to learn in both directions - about five times as many trials were required to reach the same standards as controls (Figs. 12.9, 12.10; Young, 1961b). Studies of the effect of incomplete lesions showed a clear relationship between the numbers of positive and negative errors made and the proportion of the vertical lobe removed (Fig. 12.11; Young, 1958a).

12.1.8 *Removal of the vertical lobe after training*

Train - operate - train experiments generally show rather little retention of learned visual responses following vertical lobe removal - certainly a great deal less than is shown in equivalent touch learning experiments (Section 11.2.1).

The examples shown in Figs. 12.5, 12.6 and 12.11, and others given in Boycott and Young (1957) and Young (1960d) can be summarized as showing: (1) that some effect of pretraining is usually detectable, at

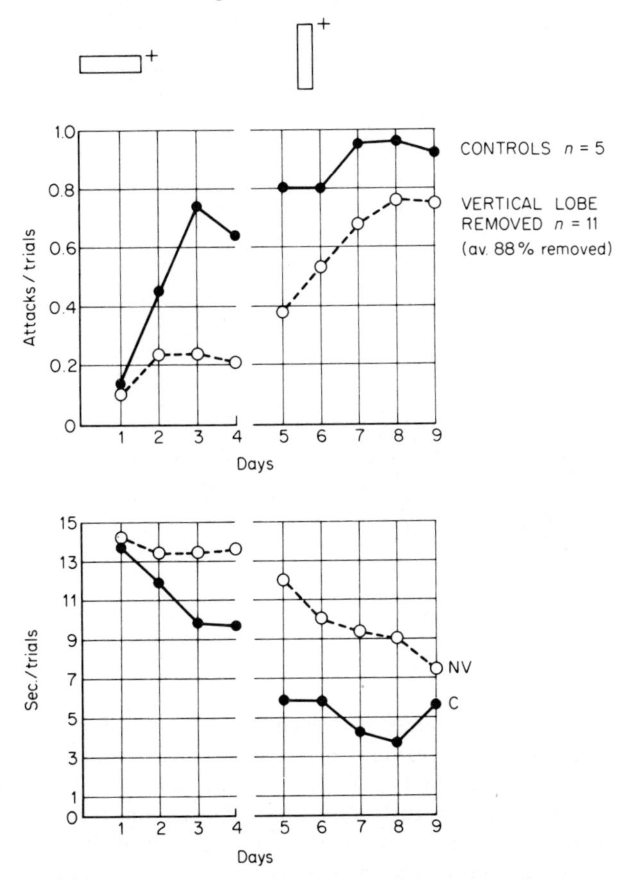

Fig. 12.9 Positive learning after vertical lobe removal. The animals were rewarded if they attacked, offered food together with the rectangle if they did not. The points show the proportion of attacks made and the mean times taken to launch attacks during the five trials of each day, given at 15 min intervals. Learning is much slower by the operated animals (from Young, 1961b).

least in the form of savings on retraining; (2) that the immediate drop in quality of performance cannot be attributed to forgetting during the delay (often only a few hours) between operation and the resumption of training (Fig. 12.8 shows some typical retention tests); and (3) that although the rise in errors is generally associated with an excess of attacks at the 'negative' figure, the failure to discriminate cannot be attributed simply to a change in the level of responses. The animals' performance is worse, even in experiments involving simultaneous presentation (Fig. 12.5).

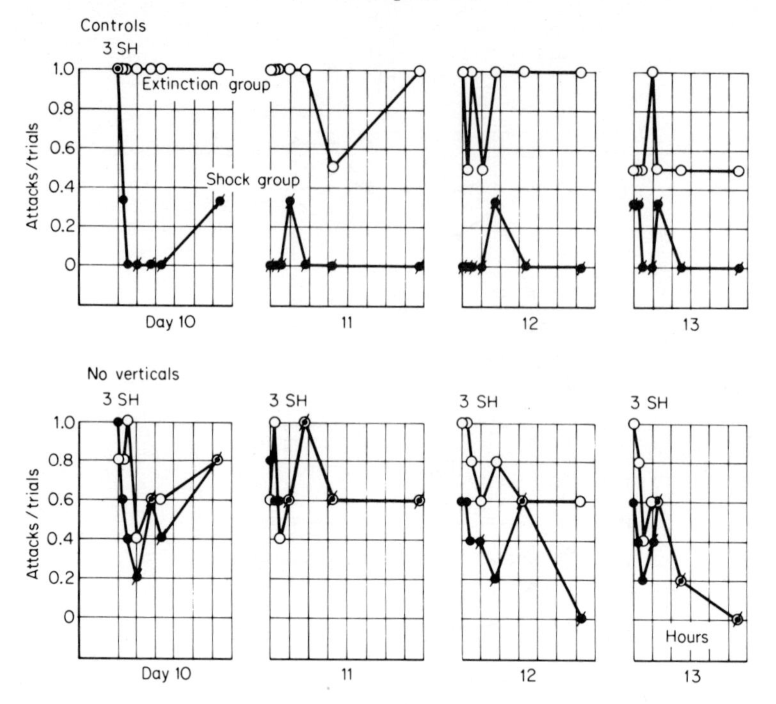

Fig. 12.10 The effect of shocks in preventing attacks at a 'positive' figure. The animals had all learned to attack a white vertical rectangle on days 1–9 as shown in Fig. 12.9. Then on day 10, three of the controls and five no-verticals were given shocks on three occasions after showing the rectangle, whether they attacked or not (filled circles). Their responses were then tested by no-reward trials (crossed circles). The control animals were given no further shocks but the operated ones received three further shocks at the beginning of each of the next 3 days. Two control and five no-vertical animals provided 'extinction' groups (open circles). These were shown the rectangle without reward or shock (from Young, 1961b).

12.1.9 *The vertical lobe as an additional memory store*

The immediate drop in the quality of performance of animals trained before the operation, the relative ineffectiveness of memory traces in lobeless animals and the quantitative effects of vertical lobe removal in both positive and negative learning are all consistent with the view that the vertical lobe is itself the visual memory store, or at least a large part of it (Young, 1958a, 1961a, b).

Such a view is consistent too with the findings from tactile training

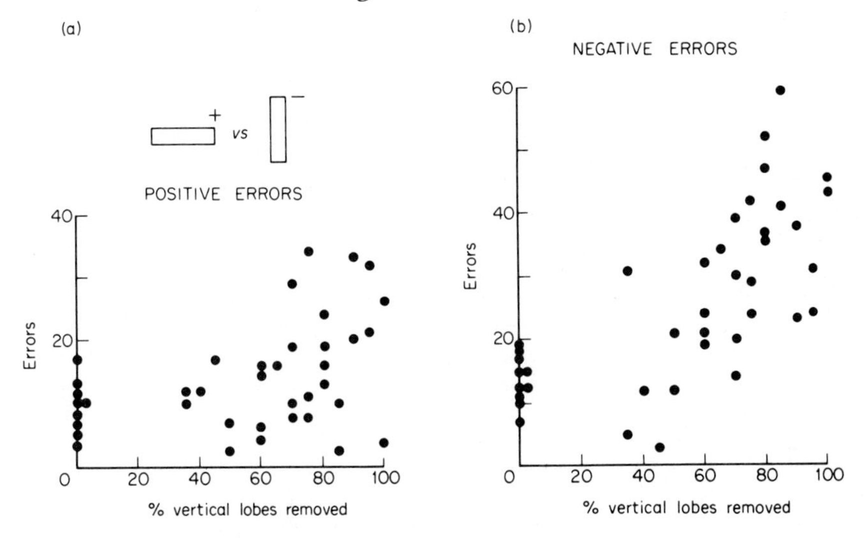

Fig. 12.11 Errors made in 120 trials by octopuses with their vertical lobes removed or damaged (from Young, 1958a).

experiments, where again the effect of removal is quantitative (Section 11.2.1). The vertical lobe has a structure closely resembling that of the subfrontal lobe, which is almost certainly a site of memory traces in tactile learning (Sections 11.3.3–11.3.5). The subfrontal lobe is absent in squids, which have little use for touch learning, and the vertical lobe is absent in *Nautilus,* which has poor eyes and, one would suppose, a rather limited capacity for learning by sight (Young, 1965c).

12.1.10 *The vertical lobe as a short-term memory system*

The conclusion that the vertical lobe is itself a substantial part of the long-term visual memory store is not incompatible with the view that it plays a part in the establishment of such stores elsewhere. An additional function could well be the maintenance of electrical activity for a period after each event, facilitating the establishment of long-term, presumably structural, traces elsewhere, as well as within the lobe itself. The two views of vertical lobe function are complementary, not contradictory; the only argument that arises is whether it is necessary to propose the second function at all.

Some of the effects of visual experience in discrimination training are cumulative. Others seem to disappear altogether in a few seconds

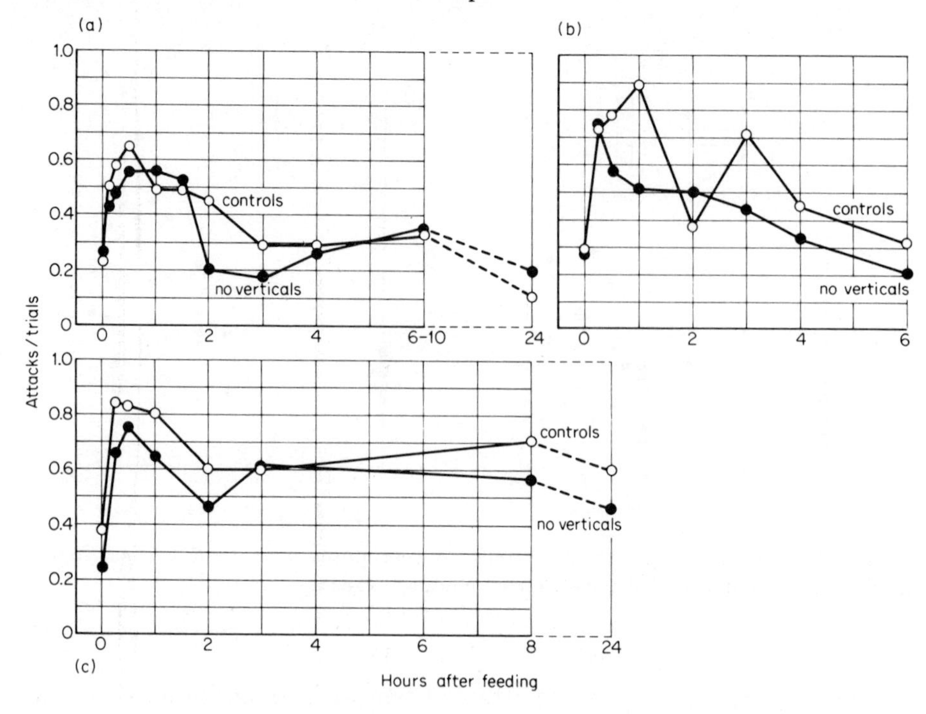

Fig. 12.12 Proportion of attacks made at a white vertical rectangle at various times after feeding with and without showing the figure. (a) shows the attacks by fifty-four control (○) and thirty-nine no vertical (●) animals tested at the times shown after feeding in the home and before any feeding in association with the rectangle. (b) shows the responses of the same animals on the first occasion after food had been given immediately following presentation of the figure. (c) shows the responses after a second such occasion, 24 h after (b). The zero points in (a) were obtained from tests at times several hours after any feeding on the day before the main experiments began (from Young, 1960c).

or hours. A partial or total failure to show cumulative effects is compatible with the view that the vertical lobe is a short-term memory store. But it does not prove that the view is correct. The same sort of result would be predicted by a simple reduction in the total number of units available in a long-term store, the effectiveness of which might depend upon the number of neurones involved, the degree to which the state of each was altered by the experience and the time course of the fade-away of its effect. Better evidence for a short-term store, qualitatively different from the long-term store, is obtained from experiments in which the effect of single trials is followed and from the performance of octopuses in delayed response and delayed reward learning experiments.

12.1.11 *Unit processes in learning*

If an octopus is fed, it becomes for a while more ready to attack unfamiliar objects than before. The effect, which follows the same time course as the filling and emptying of the crop (Section 4.2.2), lasts for 2–4 hours and can be quantified by feeding a number of animals and then testing groups of them at intervals. The tendency to attack otherwise fluctuates little, sometimes rising slightly towards evening. It is not noticeably increased by starvation (Young, 1958b).

In Fig. 12.12a the effect of feeding is compared for groups of 54 control and 39 no vertical animals. Both attacked more in the period after feeding despite the fact that the test object, a white vertical rectangle, had at no time been shown in association with food, which was given in the home 35 minutes before the first test. In Fig. 12.12b and c, the performance of the same octopuses is followed through two

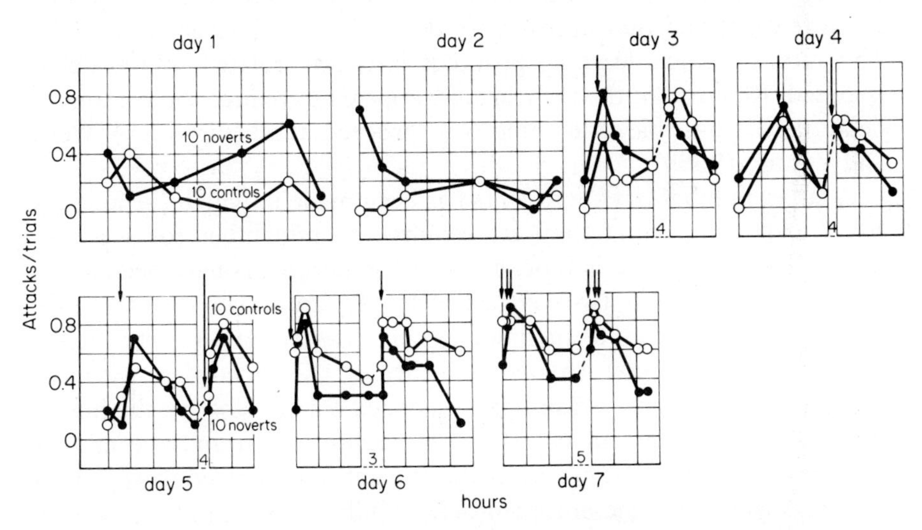

Fig. 12.13 Attacks made by ten control (○) and ten no vertical (•) animals (average 91% removed) at a white vertical rectangle. Food was given only on the occasions indicated by the arrows. In was given in the home 35 min *before* showing the rectangle on days 3 and 4, but immediately *after* showing it on days 5 to 7 (from Young, 1960c).

subsequent days, on each of which the octopuses were again fed once only, but this time immediately *after* showing them the rectangle for the first time on each day. Feeding in association with the rectangle has a similar effect to feeding *before* showing it, with the difference that the transient rise and fall in response is now accompanied by a cumulative and much more enduring effect (Fig. 12.12b, c and 12.13).

Two features of these experiments are relevant to any consideration of the defects arising from vertical lobe removal. The effect of each feeding on the transient change in probability of attack is less marked and dies away more rapidly in the lesioned animals (Fig. 12.12b, c). And its effect upon their cumulative change in behaviour is relatively slight (Fig. 12.13). The two effects are evidently linked; when short-term learning is poor, so is the long-term cumulative effect.

Corresponding effects can be shown in negative training; the effect of small numbers of shocks fades more rapidly and leaves less permanent traces in the lesioned octopuses (Young, 1960c, 1970).

12.1.12 *Learning with delayed rewards and performance in delayed response tasks*

Normal octopuses will learn to make visual and tactile discriminations when rewards and punishments are delayed for as long as 30 seconds. The rate at which learning proceeds is clearly related to the delay imposed. Removal of the vertical lobes appears to reduce the maximum delay that is possible if learning is to proceed at all (Fig. 12.14; Wells and Young, 1968b).

A larger number of experiments has been made in which the animals are shown figures that they are required to remember for a period during which response is blocked. These experiments have been made in two forms. In one, two vertical rectangles are shown on either side of a vertical partition, as in simultaneous discrimination experiments. A crab is shown next to one or other of the rectangles, with the octopus held back by a transparent perspex barrier close to the home. An opaque shutter replaces the transparent screen and the crab is removed; the shutter is removed and the octopus allowed to approach the rectangles after the enforced delay (Dilly, 1963; Sanders, 1970c). In a second type of experiment, an octopus is allowed to see a crab through a transparent barrier to one side of its home compartment. To get the crab it has to pass down an opaque corridor and make an appropriate left or right turn at the end of the corridor. The octopus

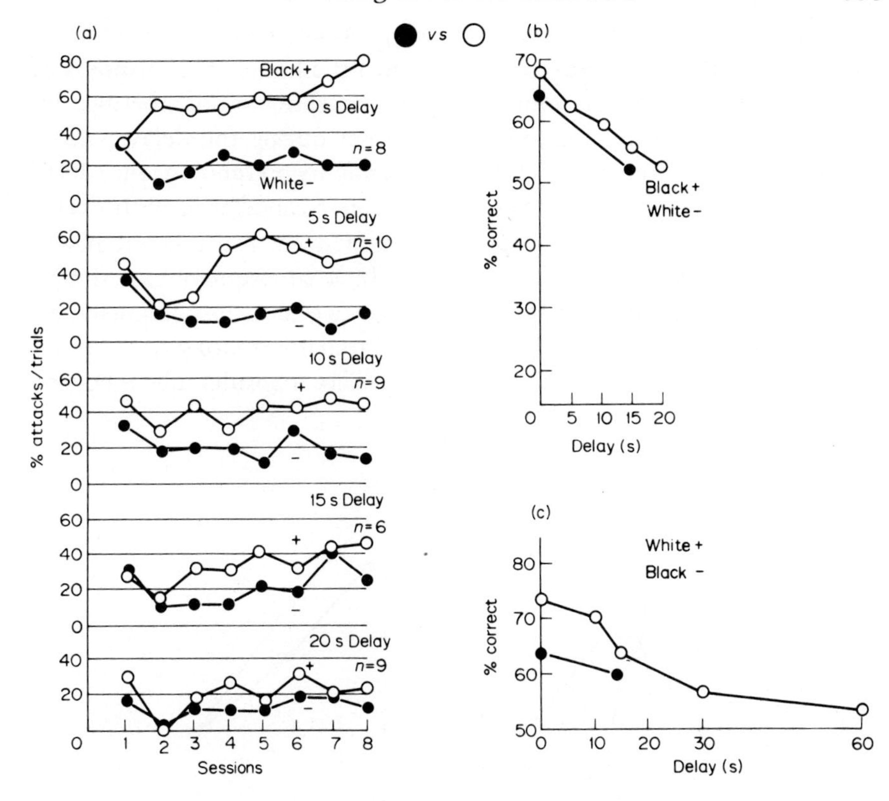

Fig. 12.14 Delayed rewards and the effect of vertical lobe removal (NV animals). (a) summarizes experiments with controls. At each trial an animal was shown a disc. If it attacked, the disc was removed and the animal fed or shocked after the stated delay. (b) replots these results and compares control performance (○) with that of 12 NV animals (●) (6 with 0 s delay, 6 with 15 s). (c) shows a comparable series; controls with 0, 10, 15, 20 and 60 s delay (respectively, n = 5, 5, 5, 9, 5) and NVs with 0 and 15 s (n = 3 and 3). In (a) and (b) there were 10 trials per session, 2 sessions per day; trials in semi-random sequence. (c) had 160 trials, with 16 per session, alternating + and −. Comparable experiments were made and results obtained using tactile stimuli (from Wells and Young, 1968b).

can be delayed by shutters at either end of the corridor. The apparatus used is shown in Fig. 9.13 (Schiller, 1949a; Wells, 1967).

In the first type of experiment, Dilly (1963) found no difference between controls and experimentals when delays of up to 30 seconds were imposed. In some of his experiments the animals were disturbed in various ways during the delay period, generally causing them to lose contact with the doorway and retreat into the home. This disposes of the possibility that the animals were somehow 'remembering' by maintaining a position relative to the target during the delay period. Sanders (1970c) repeated Dilly's experiments extending the delay period to as long as 4 minutes. Controls performed well with delays of up to 2 minutes, but octopuses with their vertical lobes removed failed if held for more than 1 minute. In a subsequent experiment Sanders used animals operated upon up to two months previously. As in his tactile experiments with animals operated upon a long time before training (see Section 11.2.1) he found a very considerable recovery of learning capacity (Fig. 12.15).

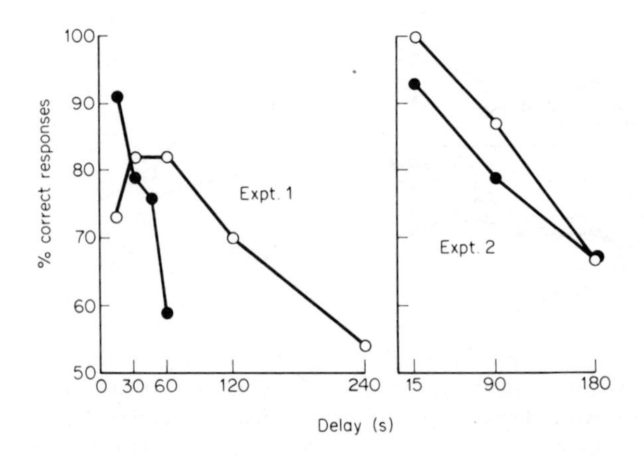

Fig. 12.15 A comparison of the performances on a delayed response task of octopuses with and without the vertical lobe. ○, controls; ●, octopuses with the vertical lobe removed. The performance of the controls with a 15 s delay in experiment 1 is atypically low. Note the marked effect on the performance of the lesioned octopuses of the length of the interval between vertical removal and testing. This interval was 27 days for experiment 1 and 54 days for experiment 2 (data from Sanders, 1975).

Very similar time-scales were established from detour experiments. Controls can perform correctly with delays of up to 2 minutes while octopuses with their vertical lobes removed appear to forget within about 1 minute regardless of whether the delay is enforced or self-imposed (Fig. 12.16).

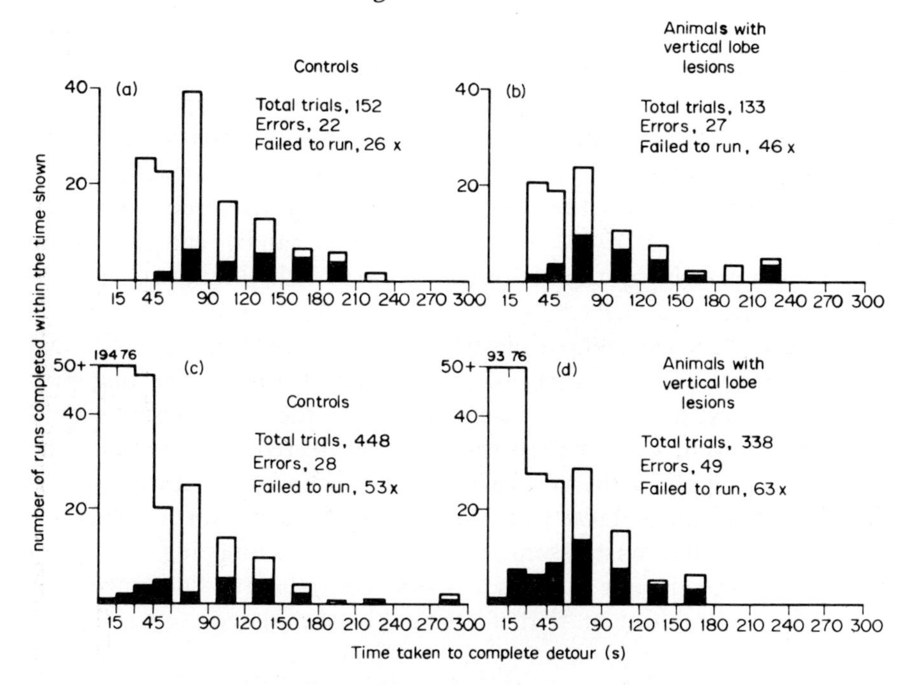

Fig. 12.16 Detour experiments and the effect of vertical lobe removal. (a) and (b) show the results of trials including delays imposed by shutting the animals into the corridor in the course of their detours. (c) and (d) show the results of undelayed trials. Errors are indicated in black on the histograms. 'Failed to run' means the animal returned home without completing the detour. If undelayed, octopuses lacking the vertical lobes complete their detours just about as rapidly as unoperated animals. But they still make twice as many mistakes (from Wells, 1967).

12.1.13 *Interocular transfer and the vertical lobe*

Muntz (1961a) was able to show that in control animals, trained using one eye and tested using the other, interocular transfer is complete, or very nearly complete for simple visual discriminations (Sections 10.2.1, Fig. 10.4). He also showed that transfer is prevented by removing the vertical lobe, or by splitting the lobe down the mid-line; the minimal lesion that blocked transfer was a vertical cut, preventing any cross-over of the tracts running from the superior frontal to the vertical lobe (Muntz, 1961b). Apparently, the optic commissures are not by themselves sufficient to ensure that a long-term learning trace is established in, or available for read-out by, both sides of the brain.

In a short-term situation the optic commissures evidently suffice. Octopuses with their vertical lobes missing are not visibly confused when their prey (or any other object seen) moves from one visual field to another, and they quite readily switch from leading with one eye to leading with the other in the course of detour experiments (Wells, 1967), something that they are not able to do after section of the commissures (Wells, 1970). A short-term trace can be read out via the optic commissures by the contralateral part of the brain, a long-term trace cannot.

These results are compatible with either or both of the principal views of vertical lobe function. If the vertical lobe is the long-term store, severing it from the visual input on the trained side would naturally prevent correct performance when tested on the other. The same would apply if the vertical lobe were a short-term store responsible for establishing traces in the optic lobes by some sort of reverberating circuit. One must assume in both cases that ongoing activity in one optic lobe is not mirrored on the contralateral side since if it were, the vertical lobe on that side would be stimulated.

12.2 Models of learning

12.2.1 *Paired centres for the control of attack*

In the visual (and tactile) learning systems the lobes are arranged in pairs, connected in the manner that Young (1963c) has summarized in Fig. 12.17 (for more anatomical detail, see Fig. 12.1). He has suggested (Young, 1963c, 1964b) that the members of each pair exert opposing influences on the probability of a positive or negative response. Thus, for example, the median superior frontal will increase the probability of attack unless this action is vetoed by the vertical lobe, through which the output from the median superior frontal must pass.

It is not easy to test this proposition experimentally, because the output from the lobe enhancing positive responses runs through the lobe responsible for repressing these. Damage to one cuts the input or the output from the other. Some experiments on the superior frontal/vertical lobe pair show clear effects in the predicted directions (Fig. 12.18). Others, in which the animals were not preselected because they attacked vigorously, have yielded more equivocal results (Fig. 12.19). All that can be said with certainty is that such operations often produce gross changes in the response level of individuals (Fig. 12.19) and that they seem to make the animals more susceptible to

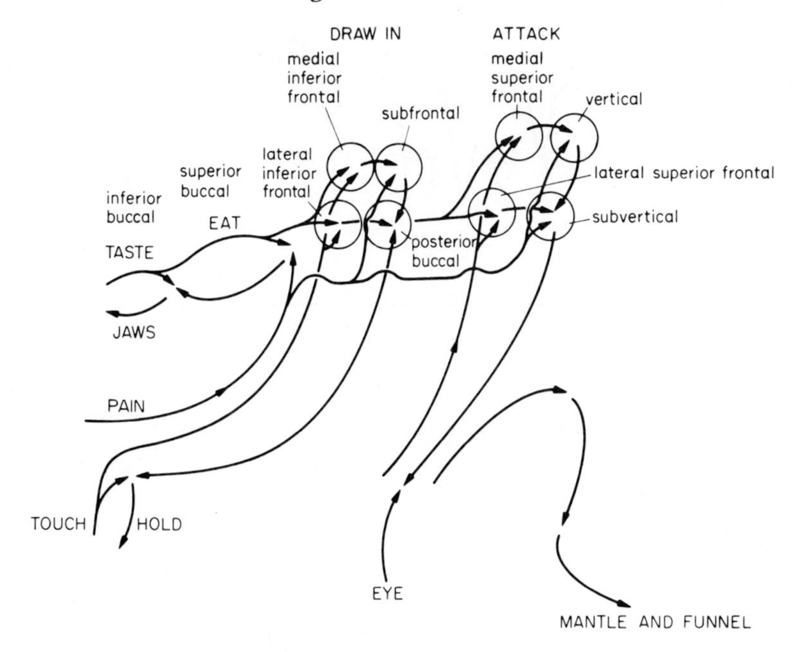

Fig. 12.17 Diagram of the centres concerned with eating and obtaining food in an octopus (from Young, 1963a).

other factors that can shift response levels; the lesioned octopuses in Fig. 12.20 for example, show a clear diurnal rhythm, attacking more in the mornings, which is overridden in the rapidly-learning controls.

The paired centres proposition forms part of a comprehensive scheme for learning in *Octopus* which is developed in Young, 1963c, 1964b, and 1965e. He has pointed out that as in the course of evolution animals become more sophisticated, so there is an increasing delay between action taken and the arrival of signals that show the consequences of the action. Food is bitten and tasted, but if it is to be recognized from contacts made with the arms there is a delay, during which the object must be passed up to the mouth, before the result of this course of action is known. With visual stimuli the delay will be even longer, while the prey is recognized, attacked and drawn in under the web. If the animal is to learn by experience, the neurones excited by contact or sight must somehow be held ready to receive the signals of results until these arrive. Young postulates that the circuits running through the paired centres are responsible for 'holding the address' by maintaining activity in the 'classifying' cells that recognized the original signals and ordered the responses (Fig. 12.21). Classifying

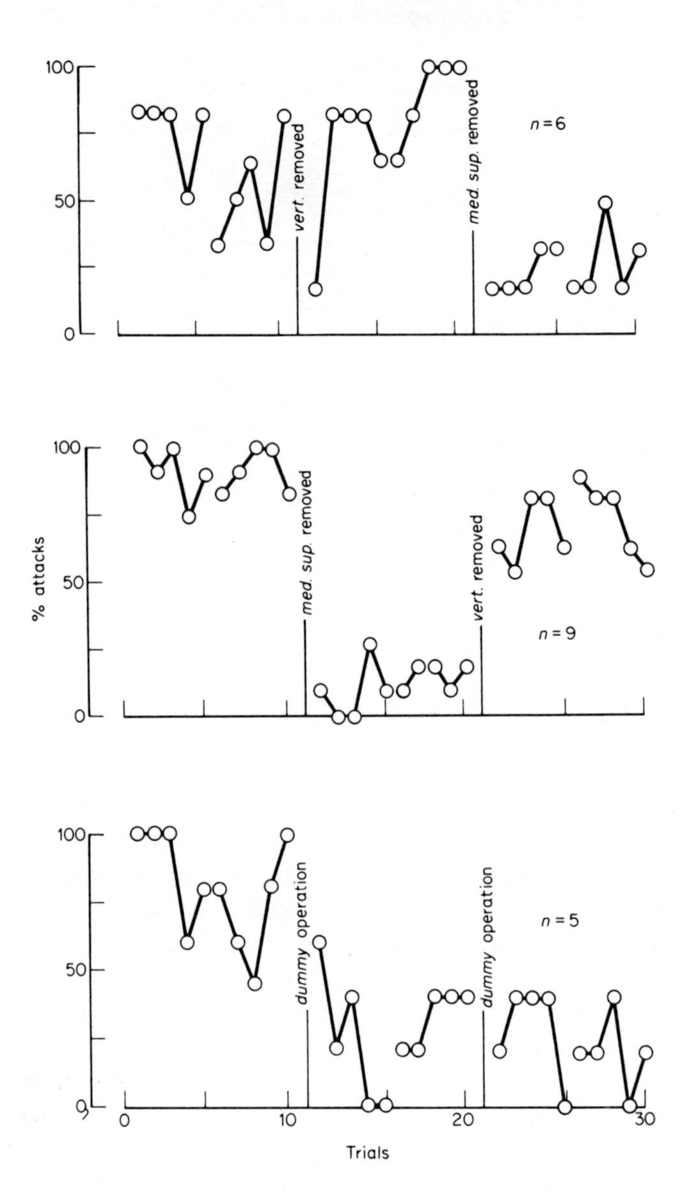

Fig. 12.18 Changes in tendency to attack crabs after various operations. The crabs were shown attached to a thread and were withdrawn when attacked or after 60 s. After ten initial trials, given in groups of 5, morning and evening, operations were performed as shown. The next day, ten further trials were given and the second operations performed. The final ten trials were given on the following day (from Young, 1964b).

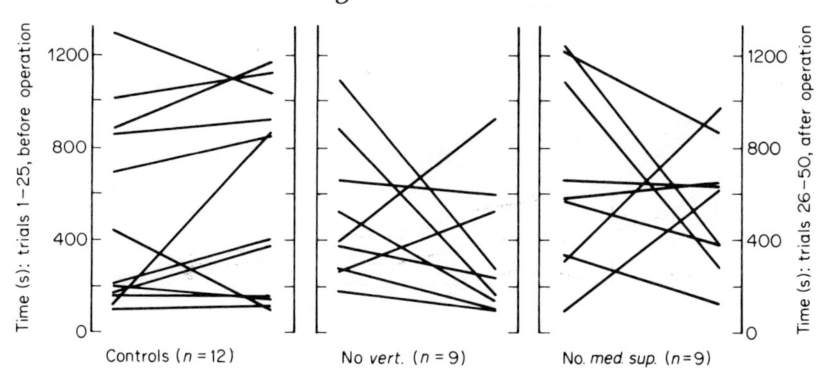

Fig. 12.19 Changes in the tendency to attack crabs after various operations, 25 trials before and 25 after. Each line links the mean time taken to attack before operation to the mean time taken to attack afterwards by the same animal (from Young, 1964b).

cells are seen as being within the optic lobes or (for touch) the posterior buccal lobes (see Sections 8.1.8 and 11.3.5).

There are two major difficulties about this scheme. One is the nature of the signal that is supposed to return to the classifying cells. If the pairs function as Young suggests, the first member enhances the signal, increasing feedback to the unit that ordered the response. This keeps the classifying cell active, so that when the signals of results arrive they can affect one or other of its output pathways. But what of the supressor effect of the second member of the pair? If this cuts the feedback, the 'address' is lost. This could account for the non-enhancement of a positive response, indeed it might allow the response to extinguish if one assumes that each pathway is liable to habituate in the absence of feedback. But it could not very well account for the more rapid suppression of a response that takes place when shocks are given. The issue cannot be evaded by supposing that suppression is automatic and rapid in the absence of a feedback from the paired centres, since we know that animals will continue to attack regularly and rapidly when the subvertical lobe is entirely destroyed (Table 12.1) and that they will continue to take objects that they touch when the whole of the supraoesophageal part of the brain has been eliminated (Sections 10.2.6 to 10.2.8).

A second difficulty, as Horridge (1968) has pointed out, is the degree of detail of the genetic specification implied by Young's model.

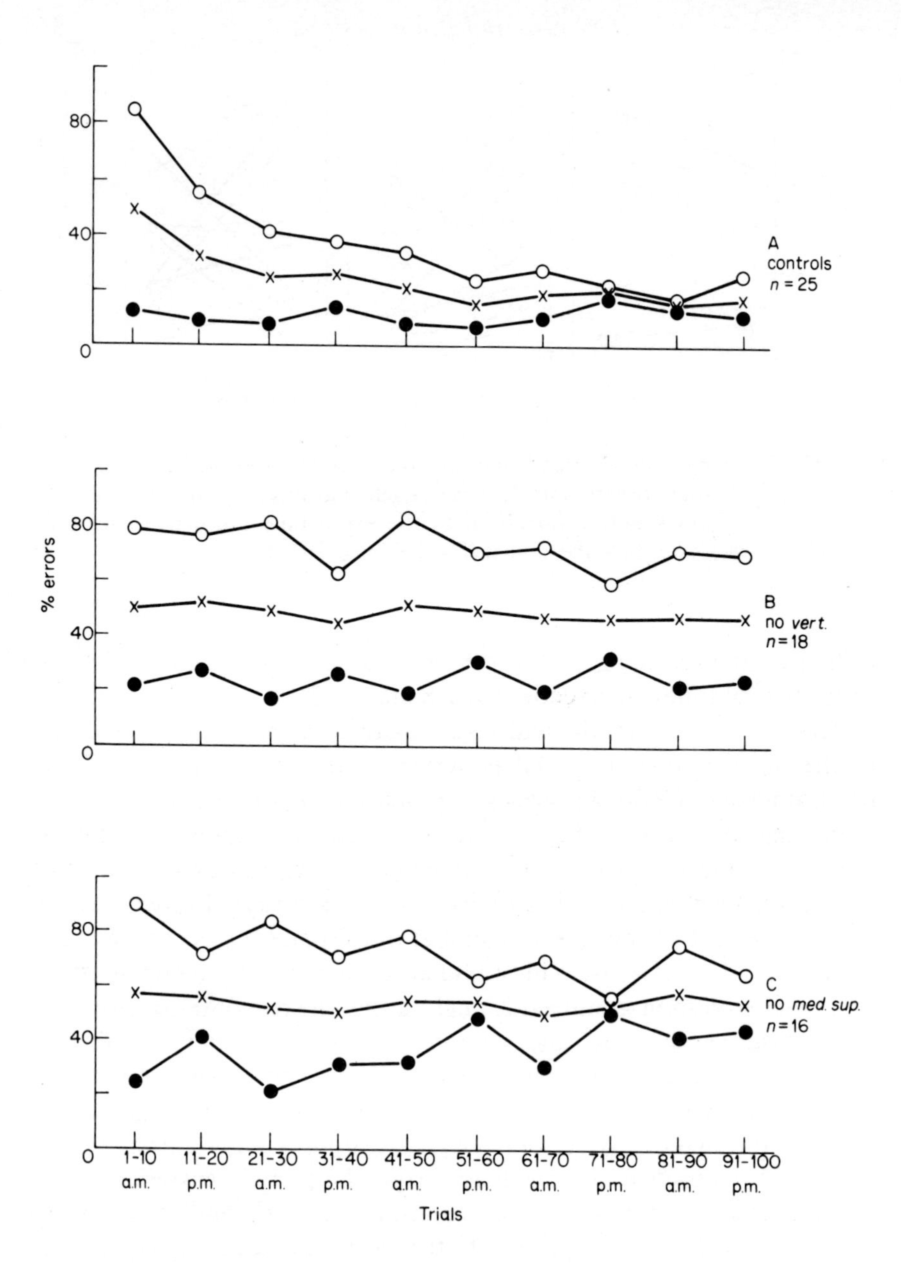

Fig. 12.20 Errors made in training to discriminate between a horizontal rectangle, the positive object (errors plotted ●) and a vertical rectangle, that the animals were shocked for taking (○). Before this, each had 45 trials in which it was trained to take the horizontal rectangle. Removal of the vertical lobe produces a higher proportion of negative errors than removal of the superior frontal (from Young, 1964b).

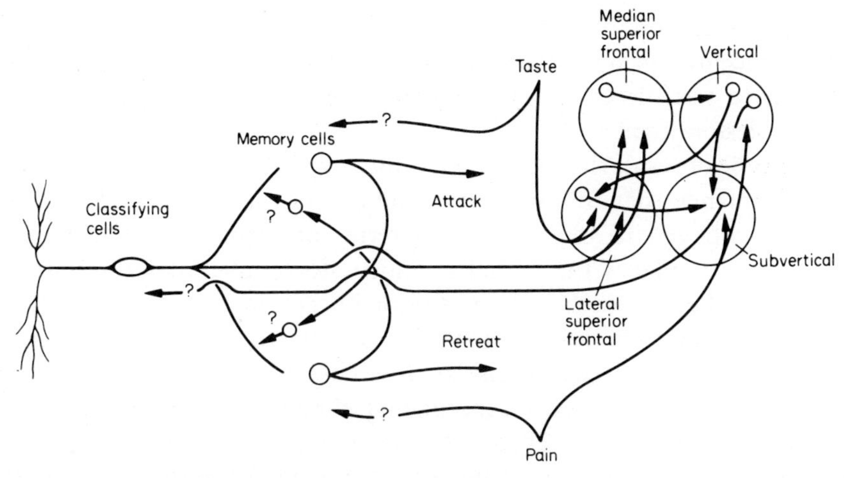

Fig. 12.21 Young's model for learning in the octopus visual system. The equivalent lobes in the tactile system would be the median and lateral inferior frontals, the subfrontal and posterior buccal. ? indicates regions where the pattern of connexions is uncertain (from Young, 1965e).

Each classifying cell, of which one must suppose several millions within the visual system, must have its own circuit through the complexities of the two layers of paired centres. If receptor generalization is to occur, so that all classifying cells of a similar type are trained at once, despite the fact that only some of these receive the original signal, each cell must not only be wired to feed back upon itself, but also linked to all other classifying cells with the same properties. Young (1965e) in fact suggests that this could be the basis for progressive enhancement of the response, as more and more cells are 'switched' to produce the appropriate response. If one grants that the necessary degree of specification could occur for all the elemental classifying cells, is one further to suppose that all the visual patterns that an octopus might learn to recognize are foreseen genetically and represented by second and third order classifying cells, each with its own feedback loop? The whole operation implies some formidable problems in morphogenesis.

12.2.2 *An alternative model for learning*

The paired centres hypothesis, outlined above, is a mainly 'anatomical' theory. It fits the anatomical facts, so far as we know them. It accounts for the enormous numbers of cells in the vertical lobe by supposing that one function of the superior-frontal/vertical lobe circuit is to ensure receptor generalization (Young, 1963c, 1965e). The theory is difficult to test directly, but it is compatible with most of the results of brain lesion work.

One type of result that it cannot readily explain is the abrupt decline in performance that follows vertical lobe removal. The paired centres theory is obliged to explain this in terms of regulation of response levels since it places the actual memory stores unequivocally within the optic lobes (or within the posterior buccal, in the case of touch). Since we know that the decline in the capacity to discriminate cannot be attributed to an overall rise or lowering of the rate of response (simultaneous presentation experiments eliminate this possibility) an explanation in terms of the paired centres theory is obliged to postulate complex effects of lesions, raising the proportion of positive responses to the negative object at the same time as lowering the rate of response to the positive.

It is more economical to assume that destruction of the vertical lobe (or elimination of its input by destruction of the median superior frontal) drastically reduces the memory stores available to the animal, weakening any control that the effects of past experience can have on behaviour in the present.

Seen in these terms, the vertical lobe (and its parallel in the touch learning system, the subfrontal) becomes even more interesting since it must contain all the essential elements of a learning mechanism. And very little else, since removal of the vertical lobe seems to have no effects other than on learning. It is fortunate in this respect that quite a lot is now known about the fine structure of the vertical lobe, thanks largely to the electron microscope studies of Gray (1970, Gray and Young, 1964) which have confirmed and extended Young's anatomical work at light-microscope level. What is known of the EM structure of the vertical lobe is summarized in Fig. 12.22. This gives a qualitative picture that must be considered in the light of quantitative information about the numbers of the different elements concerned. The important facts here are given in Table 12.4. The vertical lobe contains about 25 million neurones. All but about 65 000 of these are amacrines, with processes that do not run outside the lobe itself. The system has two inputs and one output. Numerically the most important input is from the median superior frontal lobe. A second comes from below, and is generally assumed to signal 'pain', though there is no proof of this and it might well carry 'pleasure' or any other signal. What is

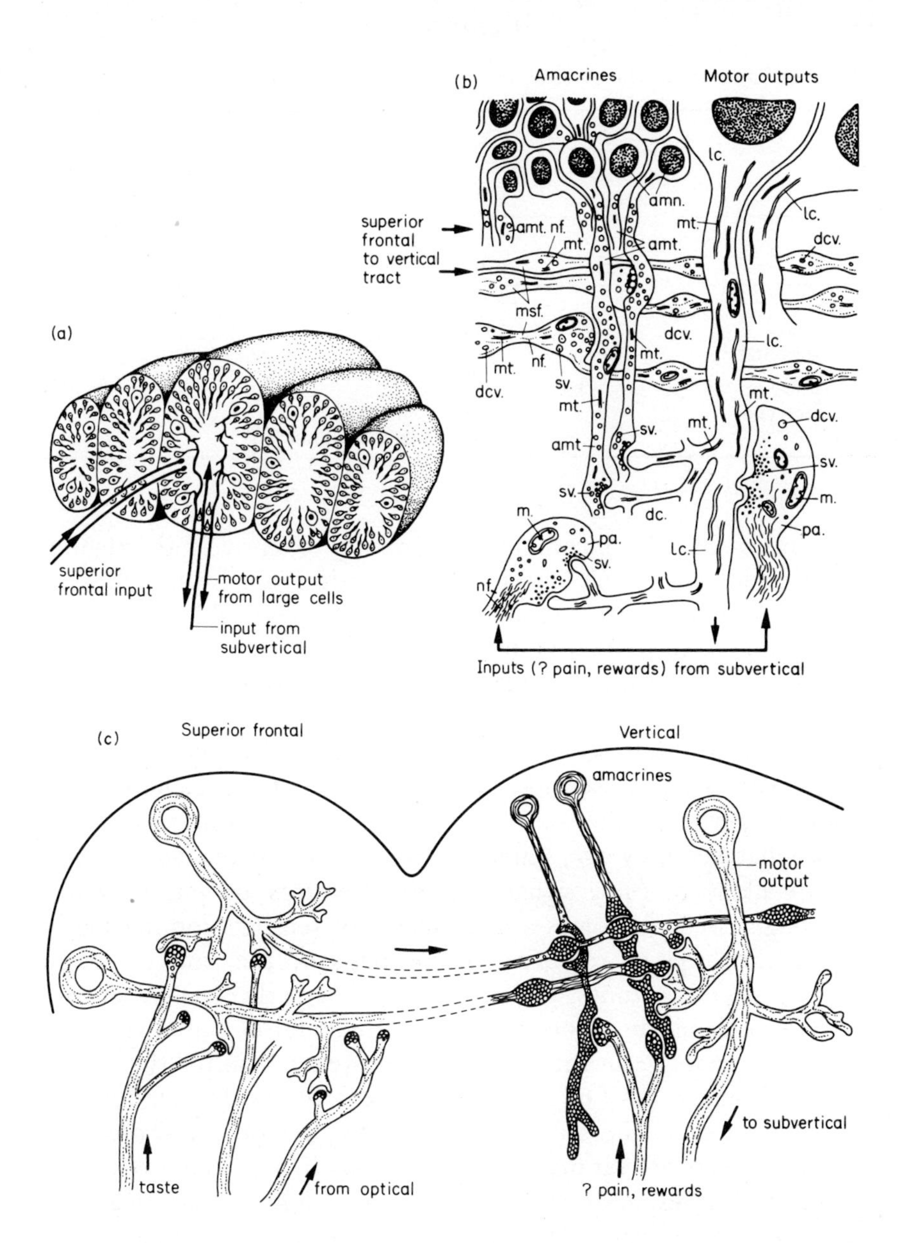

Fig. 12.22 Connectivity patterns in the vertical lobe of the brain of *Octopus.* (a) shows part of the vertical lobe. (b) is Gray's (1970) summary of his EM work on cellular connexions within the vertical lobe, amn, amt = amacrine nuclei and trunks; lc,dc = motor output cells and their dendritic collaterals; msf = axon from median superior frontal; pa = 'pain' or 'reward' cell process; m = mitochondrion, nf = neurofilament;

Table 12.4 Facts about the vertical lobe (after Young 1971).

Cells	Amacrines nuclei, 5 μm diameter, 25 × 10^6		Larger cells 65 × 10^3
Output		65 × 10^3 to subvertical	
Inputs			
from Median Superior frontal		17 × 10^5	
from below, origin uncertain		? × 10^4 (precise number uncertain, many branches)	

important is that the incoming axons ramify extensively throughout the vertical lobe. The only output for the vertical lobe is through some 65 000 large neurones that send axons to the subvertical lobe.

There appear to be two pathways through the system, a direct link from the superior frontal input to the vertical lobe output, and an indirect path, parallel to this, via the amacrines.

At this stage one passes (again) into theory. *Ex hypothesis,* the through path is not susceptible to long-term changes, though its transmission properties in the short term will clearly be influenced by the inputs, whatever they are, climbing from below. If one accepts that these are likely to carry signals denoting 'rewards' or 'punishments', the through circuit becomes a line capable of sensitization, raising or lowering the level of response on the basis of the sum of recent successes and misfortunes. *Ex hypothesis,* it is neither event-sequence sensitive nor labile in a long-term way. It should be noted that it is nevertheless adaptive in that it is tactically reasonable for the animal to respond on the basis of the sum of recent events. The randomized trial sequence is a laboratory artifact, in the sea potential prey and potential dangers are far more likely to appear in groups of events and any animal that responds on this assumption will stand a much greater than evens chance that its responses are 'correct' (Wells, 1968, 1975).

In parallel with the sensitizing pathway is an alternative route

---

mt = microtubule; sv = synaptic vesicle; dcv = dense cored vesicle. (c) is from Young (1971) and is based on his own and Gray's light microscope and EM investigations.

through the amacrines. This, it is suggested, is the part of the mechanism that changes when the animal learns to associate cause and effect, adding long-term cumulative changes to the short-term fluctuations in the manner already seen in Figs. 12.12 and 12.13. The amacrines synapse with the superior frontal input. Coming largely from the optic lobes, this probably represents the output from classifying cells, which have already abstracted elements from the patterns falling on the

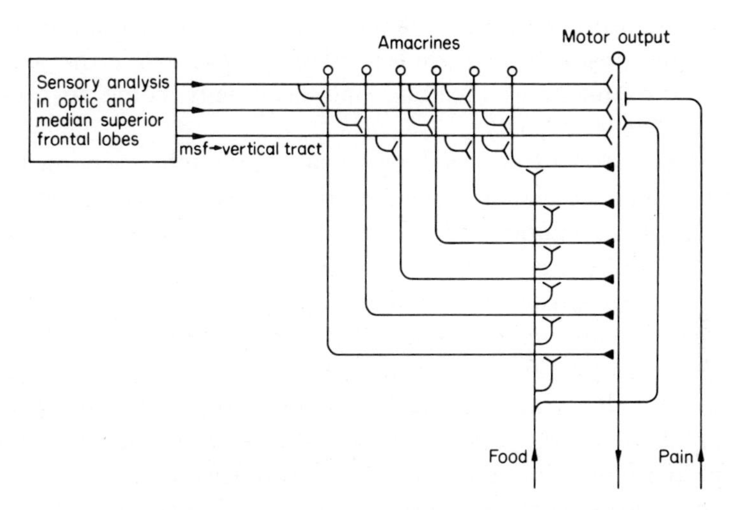

Fig. 12.23 A model of associative learning based on the pattern of connexions seen in the vertical lobe. The sensory input has already been analysed to some extent within the optic and superior frontal lobes (see Young, 1964a). This input can run through the vertical lobe directly (direct synapse with the 'motor output' neurones, see Fig. 12.22c), modulated by the sensitizing effects of the 'food' and 'pain' signals from below. An alternative route is through the amacrines. These are much more numerous than the msf input and are seen here as a further stage in analysis as well as being the 'memory' cells. They too receive the 'food' reinforcement signals. *Ex hypothesis,* the labile synapses linking amacrines to motor output (indicated ——◂) are activated and permanently facilitated only in the event that the reinforcement signals arrive in cells recently primed by msf inputs. The effects of the msf inputs persist for a matter of minutes only, giving sequence senstivity. Such a scheme would account for the relatively enormous numbers of amacrines and the capacity of the system to sensitize and habituate temporarily without obliterating long-term cumulative learning. Only one motor output is shown here, capable of triggering a positive response. A parallel set, facilitated by inputs from the 'pain' neurones, would activate a corresponding motor output triggering retreat or withdrawal (slightly modified from Wells, 1975).

retina. The amacrines, ten or twenty times as numerous, connect differing combinations of these elemental inputs; they represent a further stage in abstraction (Fig. 12.23). If one now supposes that the effect of stimulating an amacrine is to prime it so that it is for a while liable to generate an action potential if it receives a second input from the ascending ('pain' or 'pleasure') fibres, and further that the synapse through which it transmits alters whenever it is used, we have a learning system that will behave in the manner already observed. The labile synapses are protected, the system can continue to function producing adaptive responses through the sensitizing bypass while the pathways through the amacrines facilitate and eventually come to dominate responses (Wells, 1975).

A similar system is envisaged for the subfrontal lobe (Section 11.3.5). The subfrontal at least is a comparatively recent development in the evolution of cephalopods (decapods lack a subfrontal). Both this and the vertical lobe are, presumably, extensions of systems with similar properties in the posterior buccal and optic lobes.

The model outlined above has (the author feels) certain virtues. It could have evolved from a simpler, sensitizing, system, and there are grounds for believing that such systems have preceded associative learning (Wells, 1968, 1975). It fits the behavioural facts, and the known anatomy, qualitatively and quantitatively. It could explain why one always seems to find such enormous numbers of very small cells associated with parts of brains that are known or believed to be involved in learning. So it could, perhaps, be a model for learning in other animals as well as in *Octopus.*

References

Akimuskin, I.I. (1965), Cephalopods of the seas of the U.S.S.R. (Translated by A. Mercado). *Israel Program for Scientific Translations, Jerusalem,* 223 pp.

Alexandrowicz, J.S. (1927), Contribution á l'étude des muscles, des nerfs et du mécanisme de l'accommodation de l'oeil des céphalopodes. *Archives de zoologie expérimentale et générale,* **66**, 71–134.

Alexandrowicz, J.S. (1928a), Note sur l'innervation du tube digestif des céphalopodes. *Archives de Zoologie expérimentale et générale,* **67**, 69–90.

Alexandrowicz, J.S. (1928b), Sur la fonction des muscles intrinsiques de l'oeil des Céphalopodes. *Comptes rendus des séances la Société de Biologie,* **99**, 1161–1163.

Alexandrowicz, J.S. (1960a), Innervation of the hearts of *Sepia officinalis. Acta zoologica,* **41**, 65–100.

Alexandrowicz, J.S. (1960b), A muscle receptor organ in *Eledone cirrhosa. Journal of the Marine Biological Association of the United Kingdom,* **39**, 419–431.

Alexandrowicz, J.S. (1963), A pulsating ganglion in the Octopoda. *Proceedings of the Royal Society,* Series B, **157**, 562–573.

Alexandrowicz, J.S. (1964), The neurosecretory system of the vena cava in Cephalopoda, I. *Eledone cirrosa. Journal of the Marine Biological Association of the United Kingdom,* **44**, 111–132.

Alexandrowicz, J.S. (1965), The neurosecretory system of the vena cava in Cephalopoda, II. *Sepia officinalis* and *Octopus vulgaris. Journal of the Marine Biological Association of the United Kingdom,* **45**, 209–228.

Altman, J.S. (1967), The behaviour of *Octopus vulgaris* Lam. in its natural habitat; a pilot study. *Underwater Association Report,* 1966–67, (Lythgoe, J.N. and Woods, J.D., Eds.) T.G.W. Industrial & Research Promotions Ltd.

Altman, J.S. (1968), The nervous control of arm and buccal movements in *Octopus vulgaris* Lam. Ph.D. Thesis, University of London.

Altman, J.S. (1971). Control of accept and reject reflexes in the octopus. *Nature,* **229**, 204–206.

Altman, J.S. & Nixon, M. (1970), Use of the beaks and radula by *Octopus vulgaris* in feeding. *Journal of Zoology,* **161**, 25–38.

Amoore, J.E. Rodgers, K.W. & Young, J.Z. (1959), Sodium and potassium in the endolymph and perilymph of the statocyst and in the eye of *Octopus. Journal of Experimental Biology,* **36**, 709–714.

Arakawa, K.Y. (1962), An ecological account of the breeding behaviour of *Octopus luteus. Venus,* **22**, 176–180.

Arnold, J.M. (1971), Cephalopods. In: *Experimental embryology of marine and Freshwater Invertebrates.* (Reverberi, G., Ed.), pp. 265–311, North-Holland Publishing Company, Amsterdam.

Arnold, J.M. & Arnold, K.O. (1969), Some aspects of hole-boring predation by *Octopus vulgaris. American Zoologist,* **9**, 991–996.

Arnold, J.M. & Williams-Arnold, L.D. (1976), The egg cortex problem as seen through the eye of the squid. *American Zoologist,* **16**, 421–446.

Arvy, L. (1960). Histoenzymological data on the digestive tract of *Octopus vulgaris* Lamarck (Cephalopoda). *Annals of the New York Academy of Sciences,* **90**, 929–949.

Bacq, Z.M. & Ghiretti, F. (1953), Physiologie des glandes salivaires postérieures des Céphalopodes Octopodes isolées et perfusées *in vitro. Pubblicazioni della Stazione zoologica di Napoli,* **24**, 267–277.

Bacq, Z.M. & Mazza, F.P. (1935), Isolement de chloroaurate d'acétylcholine à partir d'un extrait de cellules nerveuses d'*Octopus vulgaris. Archives internnationales de physiologie,* **42**, 43–46.

Ballering, R.B. (1972), Octopus envenomation through a plastic bag via salivary proboscis. *Toxicon,* **10**, 245–248.

Barber, V.C. (1965), Preliminary observations on the fine structure of the *Octopus* statocyst. *Journal de Microscopie*, **4**, 547–550.

Barber, V.C. (1966a). The morphological polarisation of kinocilia in the *Octopus* statocyst. *Journal of Anatomy,* **100**, 685–686.

Barber, V.C. (1966b). The fine structure of the statocyst of *Octopus vulgaris. Zeitschrift für Zellforschung und mikroskopische Anatomie,* **70**, 91–107.

Barber, V.C. (1967), A neurosecretory tissue in *Octopus. Nature,* **213**, 1042–1043.

Barber, V.C. (1968), The structure of mollusc statocysts, with particular reference to cephalopods. *Symposia of the Zoological Society of London,* **23**, 37–62.

Barber, V.C. & Graziadei, P. (1965), The fine structure of cephalopod blood vessels. I. Some smaller peripheral vessels. *Zeitschrift für Zellforschung und mikroskopische Anatomie,* **66**, 765–781.

Barber, V.C. & Graziadei, P. (1967a), The fine structure of cephalopod blood vessels. II. The vessels of the nervous system. *Zeitschrift für Zellforschung und mikroskopische Anatomie,* **77**, 147–161.

Barber, V.C. & Graziadei, P. (1967b), The fine structure of cephalopod blood vessels. III. Vessel innervation. *Zeitschrift für Zellforschung und mikroskopische Anatomie,* **77**, 162–174.

Barlow, J.J., Juorio, A.V. & Martin, R. (1974), Monoamine transport in the *Octopus* posterior salivary gland nerves. *Journal of Comparative Physiology,* **89** (2), 105–122.

Bayne, C.J. (1973), Internal defence mechanisms of *Octopus dofleini.Malacological Review,* 6, 13–17.

Beer, T. (1897), Die Akkomodation des Kephalopodenauges. *Archiv fur die gesammte Physiologie des Menschen und der Thiere (Pflüger's Archiv),* **67**, 541–586.

Belonoschkin, B. (1929a). Die Geschlechtswege von *Octopus vulgaris* und ihre Bedeutung für die Bewegung der Spermatozoen. *Zeitschrift für Zellforschung und mikroskopische Anatomie,* **9**, 643–662.

Belonoschkin, B. (1929b), Das Verhalten der Spermatozoen zwischen Begattung und Befruchtung bei *Octopus vulgaris. Zeitschrift für Zellforschung und mikroskopische Anatomie,* **9**, 750–753.

Bern, H.A. (1967), On eyes that may not see and glands that may not secrete. *American Zoologist,* **7**, 815–821.

Berry, C.F. & Cottrell, G.A. (1970), Neurosecretion in the vena cava of the cephalopod *Eledone cirrosa. Zeitschrift für Zellforschung und mikroskopische Anatomie,* **104**, 107–115.

Bert, P. (1867), Mémoire sur la physiologie de la Seiche (*Sepia officinalis* Lam.). *Mémoires de la Société des Sciences physiques et naturelles de Bordeaux,* **5**, 115–138.

Bidder, A.M. (1950), The digestive mechanism of the European squids *Loligo vulgaris, Loligo forbesii, Alloteuthis media* and *Alloteuthis subulata. Quarterly Journal of Microscopical Science,* **91**, 1–43.

Bidder, A.M. (1957), Evidence for an absorptive function in the 'liver' of *Octopus vulgaris* Lam. *Pubblicazioni della Stazione zoologica di Napoli,* **29**, 139–150.

Bidder, A.M. (1966), Feeding and digestion in cephalopods. In: *Physiology of Mollusca.* Wilbur, K.M. & Yonge, C.M., Eds, Vol. 2, pp. 97–124, Academic Press, New York.

Bierens de Haan, J.A. (1926), Versuche über den Farbensinn und das psychische Leben von *Octopus vulgaris. Zeitschrift für vergleichende Physiologie,* **4**, 766–796.

Bierens de Haan, J.A. (1949), *Animal psychology.* Hutchinson, London.

Björkman, N. (1963), On the ultrastructure of the optic gland in *Octopus. Journal of Ultrastructure Research,* **8**, 195.

Blanchi, D. (1969), Esperimenti sulla funzione del sistema neurosecretorio della vena cava nei cefalopodi. *Bollettino della Società italiana di biologia sperimentale,* **45**, 1615–1619.

Blanchi, D. & De Prisco, R. (1971), Esperimenti sulla funzione del sistema neurosecretorio della vena cava nei cefalopodi. III. Purificazione e caratterizazione parziale del principio attivo. *Bollettino della Società italina di biologia sperimentale,* **47**, 477–480.

Blaschko, H. & Hawkins, J. (1952), Observations on amine oxidase in cephalopods. *Journal of Physiology,* **118**, 88–93.

Blaschko, H. & Philpot, F.J. (1953), Enzymic oxidation of tryptamine derivatives. *Journal of Physiology,* **122**, 403–408.

Bogoraze, D. & Cazal, P. (1944), Recherches histologiques sur le système nerveux du Poulpe. Les neurones, le tissue interstitiel et les éléments neuricrines. *Archives de zoologie experiméntale et générale,* **83**, 413–444.

Boletzky, S. von. (1968), Untersuchungen über die Organogenese des Kreislaufsystems von *Octopus vulgaris* Lam. *Revue suisse de Zoologie,* **75**, 765–812.

Boletzky, S. von. (1969), Zum Vergleich der Ontogenesen von *Octopus vulgaris, O. joubini* und *O. briareus. Revue suisse de Zoologie,* **76**, 716–726.

Boletzky, S. von.(1971), Rotation and first reversion in the *Octopus. Experientia,* **27** (5), 558–560.

Boletzky, S. von (1973), Structure et fonctionnement des organes de Kölliker chez les jeunes octopodes (Mollusca, Cephalopoda). *Zeitschrift für Morphologie und Ökologie der Tiere,* **75**, 315-327.

Boletzky, S. von. (1975), A contribution to the study of yolk absorption in the Cephalopoda. *Zeitschrift für Morphologie und Ökologie der Tiere,* **80**, 229–246.

Boletzky, S. von & Boletzky, M.V. (1969), First results in rearing *Octopus joubini* Robson. *Verhandlungen der Naturforschenden Gesellschaft in Basel,* **80**(1), 56-61.

Bolognari, A. (1951), Morfologia, struttura e funzione del 'corpo bianco' dei cefalopodi. Parte 2. Struttura e funzione. *Archivio zoologico italiano,* **36**, 253–287.

Bon, W.F., Dohrn, A. & Batink, H. (1967), The lens proteins of a marine invertebrate *Octopus vulgaris. Biochimica et biophysica acta,* **140**, 312–318.

Bonichon, A. (1967), Contribution à l'étude de la neurosécrétion et de l'endocrinologie chez les Céphalopodes. I. *Octopus vulgaris. Vie et Milieu,* **18**, 227–263.

Bonichon, A. (1968). Présence de cellules neurosécrétices dans le lobe buccal supérieur d'*Octopus vulgaris* Lam. *Comptes rendus hebdomadaire des séances de l'Académie des Sciences,* **D 266**, 1764–1766.

Borer, K.T. (1971), Control of food intake in *Octopus briareus* Robson. *Journal of Comparative and Physiological Psychology,* **75**, 171–185.

Borer, K.T. & Lane, C.E. (1971), Oxygen requirements of *Octopus briareus* Robson at different temperatures and oxygen concentrations. *Journal of experimental marine Biology and Ecology,* **7**, 263–269.

Bottazzi, F. & Enriques, P. (1901), Recherches physiologiques sur le système nerveux viscéral des Aplysies et de quelques Céphalopodes. *Archives italiennes de biologie,* **34**, 111–143.

Boucaud-Camou, E., Boucher-Rodoni, R. & Mangold, K. (1976), Digestive absorption in *Octopus vulgaris* (Cephalopoda: Octopoda). *Journal of Zoology, London,* **179**, 261–271.

Boucher-Rodoni, R. (1973), Vitesse de digestion d'*Octopus cyanea* (Céphalopoda, Octopoda). *Marine Biology,* **18**, 237–242.

Bouillon, J. (1960), Ultrastructure des cellules rénales des Mollusques: I. Gastéropodes pulmonés terrestres. *Annales des sciences naturalles (zoologie),* **12**, 719–749.

Boycott, B.B. (1953), The chromatophore system of cephalopods. *Proceedings of the Linnean Society of London,* **164**, 235–240.

Boycott, B.B. (1954), Learning in *Octopus vulgaris* and other cephalopods. *Pubblicazioni della Stazione zoologica de Napoli,* **25**, 67–93.

Boycott, B.B. (1960), The functioning of the statocysts of *Octopus vulgaris. Proceedings of the Royal Society,* Series B, **152**, 78–87.

Boycott, B.B. (1961), The functional organization of the brain of the cuttlefish, *Sepia officinalis. Proceedings of the Royal Society* Series B, **153**, 503–534.

Boycott, B.B. Lettvin, J.Y., Maturana, H.R. & Wall, P.D. (1965), Octopus optic responses. *Experimental Neurology,* **12**, 247–256.

Boycott, B.B. & Young, J.Z. (1950), The comparative study of learning. *Symposia of the Society for Experimental Biology,* **4**, 432–453.

Boycott, B.B. & Young, J.Z. (1955a), A memory system in *Octopus vulgaris* Lamarck. *Proceedings of the Royal Society,* Series B, **143**, 449–480.

Boycott, B.B. & Young, J.Z. (1955b), Memories controlling attacks on food objects by *Octopus vulgaris.* L. *Pubblicazioni della Stazione zoologica di Napoli,* **27**, 232–247.

Boycott, B.B. & Young J.Z. (1956a), The subpedunculate body and nerve and other organs associated with the optic tract of cephalopods. *Bertil Hanström: Zoological papers in honour of his sixty-fifth birthday.* K.G. Wingstrand, Ed. Zoological Institute, Lund., 76–165.

Boycott, B.B. & Young, J.Z. (1956b), Reactions to shape in *Octopus vulgaris* Lamarck. *Proceedings of the Zoological Society of London,* **126**, 491–547.

Boycott, B.B. & Young, J.Z. (1957), Effects of interference with the vertical lobe on visual discrimination in *Octopus vulgaris* Lamarck. *Proceedings of the Royal Society,* Series B, **146**, 439–459.

Brock, J. (1878), Ueber die Geschlechtsorgane der Cephalopoden. *Zeitschrift für wissenschaftliche Zoologie,* **32**, 1–116.

Brookes, V.J. (1969), The induction of yolk protein synthesis in the fat body of an insect, *Leucophaea maderae* by an analog of the juvenile hormone. *Developmental Biology,* **20**, 459–471.

Brown, R.K. & Brown, P.S. (1958), Visual pigments of the octopus and cuttlefish. *Nature,* **182**, 1288–1290.

Buckley, S.K.L. (1977), Oogenesis and its hormonal control in *Octopus vulgaris.* PhD. Thesis, University of Cambridge.

Budelmann, B.U. (1970), Die Arbeitsweise der Statolithenorgane von *Octopus vulgaris. Zeitschrift für vergleichende Physiologie,* **70**, 278–312.

Budelmann, B.U. (1976), Equilibrium receptor systems in Molluscs. In: *Structure and Function of Proprioceptors in the Invertebrates,* Mill, P.J., Ed., pp. 529–566, Chapman & Hall, London.

Budelmann, B.U. (1977), Structure and function of the angular acceleration receptor systems in the statocysts of cephalopods. In: *The Biology of Cephalopods,* Nixon, M. and Messenger, J.B., Eds., *Symposia of the Zoological Society of London,* **38**, 309–324.

Budelmann, B.U., Barber, V.C. & West, S. (1973), Scanning electron microscopical studies of the arrangements and numbers of hair cells in the statocysts of *Octopus vulgaris, Sepia officinalis* and *Loligo vulgaris. Brain Research,* **56,** 25–41.

Budelmann, B.U. & Wolff, H.G. (1973), Gravity response from angular acceleration receptors in *Octopus vulgaris. Journal of Comparative Physiology,* **85,** 283–290.

Burrows, T.M.O., Campbell, I.A., Howe, E.J. & Young, J.Z. (1965), Conduction velocity and diameter of nerve fibres of cephalopods. *Journal of Physiology,* **179,** 39–40.

Buytendijk, F.J.J. (1933), Das Verhalten von *Octopus* nach teilweisier Zerstörung des 'Gehirns'. *Archives néerlandaises de physiologie de l'homme et des animaux,* **18,** 24–70.

Cajal, S. Ramon (1917), Contribución al conocimiento de la retina y centros ópticos de los cefalópodos. *Trabajos del Laboratorio de investigaciones biologicas de la Universidad de Madrid,* **15,** 1–83.

Callan, H.G. (1940), The absence of a sex-hormone controlling regeneration of the hectocotylus in *Octopus vulgaris* Lam. *Pubblicazioni della Stazione zoologica di Napoli,* **18,** 15–19.

Cazal, P. & Bogoraze, D. (1943), Recherches sur les corps blancs du poulpe (*Octopus vulgaris* Lam.). Leur fonction globuligène et néphrocytaire. *Bulletin de l'Institut océanographique,* **842,** 1–12.

Chapman, D.D. & Martin, A.W. (1957), Application of weight specific oxygen consumption measurements to the determination of circulation rate in *Octopus hongkongensis. Anatomical Record,* **128,** 532.

Cloney R.A. & Florey, E. (1968), Ultrastructure of cephalopod chromatophore organs. *Zeitschrift für Zellforschung und mikroskipische Anatomie,* **89,** 250–280.

Coates, H.J., Hussey, R.E. & Nixon, M. (1965), An automatic food dispenser for *Octopus vulgaris. Journal of Physiology,* **183,** 51–52P.

Cousteau, J.Y. & Diolé, P. (1973), *Octopus and Squid: the soft intelligence.* Cassell, London.

Cowden, R.R. & Curtis, S.K. (1973), Observations on living cells dissociated from the leukopoietic organ of *Octopus briareus. Experimental and Molecular Pathology,* **19,** 178–185.

Crancher, P., King, M.G., Bennett, A. & Montgomery, R.B. (1972), Conditioning of a free operant in *Octopus cyaneus* Gray. *Journal of the Experimental Analysis of Behaviour,* **17,** 359–362.

Cuvier, G. (1829), Mémoire sur un ver parasite d'un nouveau genre (*Hectocotylus octopodis*). *Annales des sciences naturelles,* **18,** 127–156.

D'Aniello, A. & Scardi, V. (1971), Attività cellulasica nel polpo (*Octopus vulgaris*). *Bollettino della società italiana di biologica sperimentale,* **47,** 481–483.

Defretin, R. & Richard, A. (1967), Ultrastructure de la glande optique de *Sepia officinalis* L. (Mollusque, Céphalopode). Mise en évidence de la sécrétion et de son contrôle photopériodique. *Comptes rendus hebdomadaire des séances de l'Académié des Sciences,* **265,** 1415–1418.

Delaunay, H. (1931), L'excretion azotée des invertébrés. *Biological Reviews,* **6**, 265–302.

Delle Chiaje, S. (1828), Memoirie sulla storia a notomia degli animali senza vertebre. *Societa tipografica Napoli,* **4**, 154.

Denton, E.J. (1961), The buoyancy of fish and cephalopods. *Progress in Biophysics and biophysical Chemistry,* **11**, 177–234.

Denton, E.J. (1974), On buoyancy and the lives of modern and fossil cephalopods. *Proceedings of the Royal Society,* Series B, **185**, 273–299.

Denton, E.J. & Gilpin-Brown, J.B. (1961), The buoyancy of the cuttlefish. *Journal of the Marine Biological Association of the United Kingdom,* **41**, 319–342.

Denton, E.J. & Gilpin-Brown, J.B. (1966), On the buoyancy of the pearly *Nautilus. Journal of the Marine Biological Association of the United Kingdom,* **46**, 723–759.

Denton, E.J., Gilpin-Brown, J.B. & Howarth, J.V. (1967), On the buoyancy of *Spirula spirula. Journal of the Marine Biological Association of the United Kingdom,* **47**, 181–191.

Denton, E.J. & Land, M.F. (1971), Mechanism of reflexion in silvery layers of fish and cephalopods. *Proceedings of the Royal Society,* Series B, **178**, 43–61.

Deutsch, J.A. (1955), A theory of shape recognition. *British Journal of Psychology,* **46**, 30–37.

Deutsch, J.A. (1960), The plexiform zone and shape recognition in *Octopus. Nature,* **185**, 443–446.

Deutsch, J.A. & Sutherland, N.S. (1960), Theories of shape discrimination in *Octopus. Nature,* **188**, 1090–1094.

Dews, P.M. (1959), Some observations on an operant in the octopus, *Journal of the Experimental Analysis of Behaviour,* **8**, 57–63.

Dijkgraaf, S. (1961), The statocyst of *Octopus vulgaris* as a rotation receptor. *Pubblicazioni della Stazione zoologica di Napoli,* **32**, 64–87.

Dilly, P.N. (1963), Delayed responses in *Octopus. Journal of Experimental Biology,* **40**, 393–401.

Dilly P.N. & Messenger, J.B. (1972), The branchial gland: A site of haemocyanin synthesis in *Octopus. Zeitschrift für Zellforschung und mikroskopische Anatomie,* **132**, 193–201.

Dodwell, P.C. (1957a), Shape recognition in rats. *British Journal of Psychology,* **48**, 221–229.

Dodwell, P.C. (1957b), Shape discrimination in the octopus and the rat. *Nature,* **179**, 1088.

Dodwell, P.C. & Sutherland, N.S. (1961), Facts and theories of shape discrimination. *Nature,* **191**, 578–583.

Donovan, D.T. (1964), Cephalopod phylogeny and classification. *Biological Reviews,* **39**, 259–287.

Drew, G.A. (1910), Some points on the physiology of Lamellibranch blood corpuscles. *Quarterly Journal of Microscopical Science,* **54**, 605–621.

Drew, G.A. (1911), Sexual activities of the squid *Loligo pealii* (Les). I. Copulation. egg-laying and fertilization. *Journal of Morphology,* **22**, 327–359.

Drew, G.A. (1919), Sexual activities of the squid, *Loligo pealii* (Les). II. The spermatophore; its structure, ejaculation and formation. *Journal of Morphology,* **32**, 379–436.

Drew, G.A. & De Morgan, W. (1910), The origin and formation of fibrous tissue as a reaction to injury in *Pecten maximus. Quarterly Journal of Microscopical Science,* **55**, 595–610.

Ducros, C. (1971a), L'innervation des glandes salivaires postérieures chez *Octopus vulgaris. Archives d'anatomie microscopique et de morphologie expérimentale,* **60**, 27–36.

Ducros, C. (1971b), Étude histochemique de la distribution des amines biogènes dans les glandes salivaires postérieuses des céphalopodes. *Archives d'anatomie microscopique et de morphologie expérimentale,* **60**, 407–420.

Ducros, C. (1972a), Étude ultrastructurale de l'innervation des glandes salivaires postérieures chez *Octopus vulgaris.* I. Les troncs nerveux du canal salivaire. *Zeitschrift für Zellforschung und mikroskopische Anatomie,* **132**, 35–49.

Ducros, C. (1972b), Étude ultrastructurale de l'innervation des glandes salivaires postérieures chez *Octopus vulgaris.* II. Innervation de la musculature du canal et des glandes. *Zeitschrift für Zellforschung und mikroskopische Anatomie,* **132**, 51–65.

Ducros, C. (1972c), Étude ultrastructurale de l'innervation des glandes salivaires postérieures chez *Octopus vulgaris.* III. L'innervation des tubules de la glande. *Zeitschrift für Zellforschung und mikroskopische Anatomie,* **132**, 67–78.

Dulhunty, A. & Gage, P.W. (1970), The cellular basis of paralysis caused by the blue-ringed octopus. *Proceedings of the Australian physiological and pharmalogical Society,* **1**, 46–47.

Dulhunty, A. & Gage, P.W. (1971), Selective effects of an octopus toxin on action potentials. *Journal of Physiology,* **218**, 433–445.

Durchon, M. & Richard, A. (1967), Étude, en culture organotypique, du rôle endocrine de la glande optique dans la maturation ovarienne chez *Sepia officinalis* L. (Mollusque, Céphalopode). *Comptes rendus hebdomadaire des séances de l'Académie des Sciences,* D, **264**, 1497–1500.

Emmanuel, C.F. & Martin, A.W. (1956), The composition of octopus renal fluid. I. Inorganic constitutents. *Zeitschrift für vergleichende Physiologie,* **39**, 226–234.

Emmanuel C.F. (1957), The composition of octopus renal fluid. II. A chromatographic examination of the constituents. *Zeitschrift für vergleichende Physiologie,* **39**, 477–482.

Enriques, P. (1902), Il fegato dei molluschi e le sue funzione. *Mitteilungen aus der Zoologischen Station zu Neapel,* **15**, 281–406.

Falloise, A. (1906), Contribution à la physiologie comparée de la digestion. La digestion chez les céphalopodes. *Archives internationales de Physiologie,* **3**, 282–296.

Fänge, R. & Østlund, E. (1954), The effects of adrenaline, noradrenaline, tyramine and other drugs on the isolated heart from marine vertebrates and a cephalopod. (*Eledone cirrhosa*). *Acta zoologica,* **35**, 289–305.

Fioroni, P. (1962a), Die embryonale Entwicklung der Hautdrüsene und des Trichterorganes von *Octopus vulgaris* Lam. *Acta anatomica,* **50**, 264–295.

Fioroni, P. (1962b), Die embryonale Entwicklung der Köllikerischen Organe von *Octopus vulgaris* Lam. *Revue suisse de Zoologie,* **69**, 497–511.

Fioroni, P. (1970), Die embryonale Genese der Chromatophoren bei *Octopus vulgaris* Lam. *Acta anatomica,* **75**, 199–224.

Flecker, H. & Cotton, B.C. (1955), Fatal bite from octopus. *Medical Journal of Australia,* **2**, 329–331.

Flock, Å. (1965), Transducing mechanisms in the lateral line canal organ receptor. *Cold Spring Harbor Symposia on Quantitative Biology,* **30**, 133–145.

Florey, E. (1969). Ultrastructure and function of Cephalopod chromatophores. *American Zoologist,* **9**, 429–442.

Florey, E. & Florey, E. (1954), Über die mögliche Bedeutung von Enteramine (5 oxy-tryptamin) als nervöse Aktionssubstanz bei Cephalopoden und dekapoden Crustacean. *Zeitschrift für Naturforschung,* **9**, 58–68.

Florey, E. & Kriebel, M.E. (1969), Electrical and mechanical responses of chromatophore muscle fibres of the squid, *Loligo opalescens,* to nerve stimulation and drugs. *Zeitschrift für vergleichende Physiologie,* **65**, 98–130.

Fort, G. (1937), Le spermatophore des céphalopodes. Étude du spermatophore d'*Eledone cirrhosa* (Lamarck). *Bulletin biologique de la France et de la Belgique,* **71**, 357–373.

Fort, G. (1941), Le spermatophore des céphalopodes. Étude du spermatophore d'*Eledone moschata* (Lamarck, 1799). *Bulletin biologique de la France et de la Belgique,* **75**, 249–256.

Fredericq, H. (1914), Recherches expérimentales sur la physiologie cardiaque d'*Octopus vulgaris. Archives internationales de Physiologie,* **14**, 126–151.

Fredericq, H. & Bacq, Z.M. (1939), Analyse quantitative des effects cardiaques de la stimulation du nerf viscéral des céphalopodes. *Archives internationales de physiologie,* **49**, 490–496.

Fredericq, L. (1878), Recherches sur la physiologie du poulpe commun (*Octopus vulgaris*). *Archives de zoologie expérimentale et générale,* **7**, 525–483.

Froesch, D. (1973a), Projection of chromatophore nerves on the body surface of *Octopus vulgaris. Marine Biology,* **19**, 153–155.

Froesch, D. (1973b), On the fine structure of the *Octopus* iris. *Zeitschrift für Zellforschung und mikroskopische Anatomie,* **145**, 119–129.

Froesch, D. (1974), The subpedunculate lobe of the octopus brain: evidence for dual function. *Brain Research,* **75**, 277–285.

Froesch, D. & Mangold, K. (1976), On the structure and function of a neurohemal organ in the eye cavity of *Eledone cirrosa* (Cephalopoda). *Brain research*, **111**, 287–293.

Froesch, D. & Marthy, H.-J. (1975), The structure and function of the oviducal gland in octopods (Cephalopoda). *Proceedings of the Royal Society,* Series B, **188,** 95-101.

Fröhlich, F.W. (1914), Beiträge sur allgemeinen Physiologie der Sinnesorgane. *Zeitschrift für Sinnesphysiologie,* **48,** 354–438.

Fugii, T. (1960), Comparative biochemical studies on the egg-yolk proteins of various animal species. *Acta Embryologiae et Morphologiae Experimentalis,* **3,** 260–285.

Fuchs, S. (1895), Beiträge zur Physiologie des kreislaufs bei dem Cephalopoden. *Pflügers Archiv fur die gesamte Physiologie des Menschen und der Tiere,* **60,** 173.

Furia, M., Gianfreda, L. & Scardi, V. (1972). Presenza di attivita cellulasica in alcuni cephalopodi. *Bolletina della Società italiana di biologia sperminental,* **48** (23), 1127–1130.

Geiger, W. (1956), Quantitative Untersuchungen über das Gehirn der Knochenfische, mit besonderer Berücksichtigung seines relativen Wachstums, Part 1. *Acta anatomica,* **26,** 121–163.

Gennaro, J.F.J., Lorincz, A.E. & Brewster, H.B. (1965), The anterior salivary gland of the octopus (*Octopus vulgaris*) and its mucous secretion. *Annals of the New York Academy of Sciences,* **118,** 1021–1025.

Ghiretti, A., Guiditta, A. & Ghiretti, F. (1958), Pathways of terminal respiration in marine invertebrates. *Journal of Cellular and Comparative Physiology,* **52,** 389–429.

Ghiretti, F. (1950), Enzimi delle ghiandole salivari posteriori dei cefalopodi. *Bollettino della Società italiana di biologia sperimentale,* **26,** 776–780.

Ghiretti, F. (1953), Les excitants chimiques de la sécrétion salivaire chez les Céphalopodes octopodes. *Archives internationales de physiologie,* **61,** 10–21.

Ghiretti, F. (1959), Cephalotoxin: the crab-paralysing agent of the posterior salivary glands of cephalopods. *Nature,* **183,** 1192–1193.

Ghiretti, F. (1960), Toxicity of octopus saliva against crustacea. *Annals of the New York Academy of Sciences,* **90,** 726–741.

Ghiretti, F. (1966), Molluscan Hemocyanins. In: *Physiology of Mollusca,* Wilbur, K.M. & Yonge, C.M., Eds., Vol. 2, Academic Press, New York.

Ghiretti, F. & Violante, U. (1964), Richerche sul metabolismo del rame in *Octopus vulgaris. Bollettino di zoologia, pubblicato dall' Unione zoologica italiana,* **31,** 1081–1092.

Giersberg, M. (1926), Ueber den chemischen Sinn von *Octopus vulgaris. Zeitschrift für vergleichende Physiologie,* **3,** 827–838.

Goddard, C.K. (1968), Studies on the blood sugar of *Octopus dofleini.* Comparative *Biochemistry and Physiology,* **27,** 275–285.

Golding, D.W. (1974), A survey of neuroendocrine phenomena in non-arthropod invertebrates. *Biological Reviews,* **49,** 161–224.

Goldsmith, M. (1917a), Psychologie animale. Quelques réactions sensorielles chez le Poulpe. *Comptes rendus hebdomadaire des séances de l'Académie des Sciences,* **164**, 448–450.

Goldsmith, M. (1917b), Acquisition d'une habitude chez le Poulpe. *Comptes rendus hebdomadaire des séances de l'Académie des Sciences,* **164**, 737–738.

Goldsmith, M. (1917c), Quelques réactions du Poulpe; contribution à la psychologie des invertébres. *Bulletin de l'Institut général psychologique,* **17**, 24-44.

Gonzalez-Santander, R. & Garcia-Blanco, E. Socastro. (1972), Ultrastructure of the obliquely striated or pseudostriated muscle fibres of the cephalopods: *Sepia, Octopus* and *Eledone. Journal of Submicroscopic Cytology,* **4**, 233–245.

Gray, E.G. (1969), Electron microscopy of the glio-vascular organization of the brain of *Octopus. Philosophical Transactions of the Royal Society,* Series B, **255**, 13–32.

Gray, E.G. (1970), The fine structure of the vertical lobe of *Octopus. Philosophical Transactions of the Royal Society,* Series B, **258**, 379–395.

Gray, E.G. & Young, J.Z. ((1964), Electron microscopy of synaptic structure of *Octopus* brain. *Journal of Cell Biology,* **21**, 87–103.

Gray, J.A.B. (1960), Mechanically excitable receptor units in the mantle of the octopus and their connexions. *Journal of Physiology,* **153**, 573–582.

Graziadei, P. (1964), Electron microscopy of some primary receptors in the sucker of *Octopus vulgaris. Zeitschrift für Zellforschung und mikroskopische Anatomie,* **64**, 510–522.

Graziadei, P. (1965a), Sensory receptor cells and related neurones in cephalopods. *Cold Spring Harbour Symposia on Quantitative Biology,* **30**, 45–47.

Graziadei, P. (1965b), Muscle receptors in cephalopods. *Proceedings of the Royal Society,* Series B, **161**, 392–402.

Graziadei, P. (1965c) Electron microscope observations of some peripheral synapses in the sensory pathway of the sucker of *Octopus vulgaris. Zeitschrift für Zellforschung und mikroskopische Anatomie,* **65**, 363–379.

Graziadei, P. (1971), The nervous system of the arms. In: *The Anatomy of the Nervous System of Octopus vulgaris,* Young, J.Z., pp. 44–61, Clarendon Press, Oxford.

Graziadei, P.P.C. & Gagne, H.T. (1973), Neural components in the octopus sucker. *Journal of Cell Biology,* **59**, 121a.

Graziadei, P.P.C. & Gagne, H.T. (1976), An unusual receptor in the octopus. *Tissue and Cell,* **8**, 229–240.

Grimpe, G. (1913), Das Blutgefässystem der dibranchiaten Cephalopoden I. Octopoda. *Zeitschrift für Wissenschaftliche Zoologie,* **104**, 531–621.

Guérin, J. (1908), Contribution à l'étude des systèmes cutané, musculaire et nerveux de l'appareil tentaculaire des céphalopodes. *Archives de zoologie expérimentale et générale,* **8**, 1–178.

Hagins, W.A., Zonana, H.V. & Adams, R.G. (1962), Local membrane current in the outer segments of squid photoreceptors. *Nature,* **194**, 844–847.

Hamasaki, D.I. (1968a), The electroretinogram of the intact anaesthetised octopus. *Vision Research,* **8**, 247–258.

Hamasaki, D.I. (1968b), The ERG-determined spectral sensitivity of the octopus. *Vision Research,* **8**, 1013–1024.

Hancock, A. (1852), On the nervous system of *Ommastrephes todarus. Annals and Magazine of Natural History,* **10**, 1–13.

Hanlon, R.T. (1975), A study of growth in *Octopus briareus*, with notes on its laboratory rearing, mating and field behaviour. Masters thesis, University of Miami.

Hanson, J. & Lowy, J. (1960), Structure and function of the contracile apparatus in the muscles of invertebrate animals. In: *Structure and Function of Muscles,* Bourne, G.H., Ed., pp. 265–335, Academic Press, New York.

Hara, T., Hara, R. & Takeuchi, J. (1967), Rhodopsin and retinochrome in the octopus retina. *Nature,* **214**, 573–575.

Harrison, F.M. & Martin, A.W. (1965), Excretion in the Cephalopod, *Octopus dofleini. Journal of Experimental Biology,* **42**, 71–92.

Hartman, W.J., Clarke, W.G., Cyr, S.D., Jordan A.C. & Leibhold, R.A. (1960), Pharmacologically active amines and their biogenesis in the octopus. *Annals of the New York Academy of Sciences,* **90**, 637–666.

Hazelhof, E.H. (1939). Über die Ausnützung des Sauerstoffs bei verschiedenen Wassertieren. *Zeitschrift für vergleichende Physiologie,* **26**, 306–327.

Heidermans, C. (1928), Messende Untersuchungen über das Formensehn der Cephalopoden und ihre optische Orientierung im Raume. *Zoologische Jahrbücher Abteilungen 3.Allgemeine Zoologie und Physiologie der Tiere,* **45**, 609–650.

Heine, L. (1907), Über die Verhältnisse der Refraktion, Akkomodation und des Augenbinnendruckes in dem Tierreiche. *Medizinisch-naturwissenschaftliches Archiv,* **1**, 323–344.

Heldt, J.H. (1948), Observations sur une ponte d'*Octopus vulgaris.* Lmk. *Bulletin de la Société des sciences naturelles de Tunisie,* **1**, 87–90.

Hemmingsen, A.M. (1960), Energy metabolism as related to body size and respiratory surfaces, and its evoltuion. *Reports of the Steno Memorial Hospital and Insulinlaboratorium,* **9**, 1–110.

Hess, C. (1910), Neue Untersuchungen über den Lichtsinn bei wirbellosen Tieren. *Archiv für die gesammte Physiologie,* **136**, 282–367.

Hill R.B. & Welsh, J.H. (1966), Heart, circulation and blood cells. Chapter 4. In: *Physiology of Mollusca*, Vol. 2. Wilbur, K.M. and Yonge, C.M., Eds. Academic Press, New York.

Hoar, W.S. (1966), *General and comparative physiology.* Prentice Hall, New Jersey.

Hobbs, M.J. & Young, J.Z. (1973), A cephalopod cerebellum. *Brain Research,* **55**, 424–430.

Horridge, G.A. (1968), *Interneurones.* Freeman, New York and London.

Hubel, D.H. & Wiesel, T.H. (1959), Receptive fields of single neurones in the cat's striate cortex. *Journal of Physiology,* **148,** 574–591.

Isgrove, A. (1909), *Eledone.* Liverpool Marine Biological Committee Memoirs No. 18.

Itami, K. (1964), (Mark and release study in the *Octopus* – in Japanese.) (Quoted in van Heukelem 1973). *Aquiculture,* **12**(2), 119–125.

Itami, K., Izawa, Y., Maeda, S. & Nakai, K. (1963), Notes on the laboratory culture of the *Octopus* larvae. *Bulletin of the Japanese Society of Scientific Fisheries,* **29,** 514–519.

Ito, S., Karita, K., Tsukahara, Y. & Tasaki, K. (1973), Electrical activity of perfused and freely swimming squids as compared with *in vitro* responses. *Tohoka Journal of Experimental Medicine,* **109,** 223–233.

Ivanoff, A. & Waterman, T.H. (1958), Factors, mainly depth and wavelength, affecting the degree of underwater light polarisation. *Journal of Marine Research,* **16,** 283–307.

Johansen, K. & Huston, M.J. (1962), Effects of some drugs on the circulatory system in the intact, non-anaesthetized cephalopod *Octopus dofleini. Comparative Biochemistry and Physiology,* **5,** 177–184.

Johansen, K. & Martin, A.W. (1962), Circulation in the cephalopod, *Octopus dofleini. Comparative Biochemistry and Physiology,* **5,** 161–176.

Jolyet, F. & Regnard, P. (1877), Recherches physiologiques sur la respiration des animaux aquatiques. *Archives de physiologie normale et pathologique* Series 2, **4,** 584–633.

Juorio, A.V. (1971), Catecholamines and 5-hydroxytryptamine in nervous tissue of cephalopods. *Journal of Physiology,* **216,** 213–226.

Juorio, A.V. & Molinoff, P.B. (1974), The normal occurrence of octopamine in neural tissues of the octopus and other cephalopods. *Journal of Neurochemistry,* **22,** (2), 271–280.

Kawaguti, S. (1970), Electron microscopy on muscles fibres in blood vessels and capillaries of cephalopods. *Biological Journal of Okayama University,* **16,** 19–28.

Kawaguti, S. & Ikemoto, N. (1957), Electron microscopy of the smooth muscle of a cuttlefish, *Sepia esculenta. Biological Journal of Okayama University,* **3,** 196–208.

Kayes, R.J. (1974), The daily activity pattern of *Octopus vulgaris* in a natural habitat. *Marine Behaviour & Physiology,* **2,** 337–343.

Kellicott, W.E. (1908), Growth of the brain and viscera in the smooth dogfish (*Mustelus canis*). *American Journal of Anatomy,* **8,** 319–353.

Krijgsman, B.J. & Divaris, G.A. (1955), Contractile and pacemaker mechanisms of the heart of molluscs. *Biological Reviews,* **30,** 1–39.

Kühn, A. (1950), Über Farbwechsel und Farbensinn von Cephalopoden. *Zeitschrift für vergleichende Physiologie,* **32,** 572–598.

Lane, F.W. (1957), *Kingdom of the Octopus.* Jarrolds, London.

Lankester, E.R. (1875), Contributions to the developmental history of the Mollusca. *Philosophical Transactions of the Royal Society,* **165,** 2–48.

Laubier-Bonichon, A. (1971), Sur la prèsence de cellules neurosécrétrices dans le lobe subpédonculé d'un Céphalopode. Relation avec la maturation sexuelle. *Comptes rendus hebdomadaire des séances de l'Académie des Sciences*, **D, 272,** 2086–2088.

Laubier-Bonichon, A. (1973), Arguments expérimentaux sur l'activité des cellules sécrétrices du lobe viscéral d'un mollusque Céphalopode *Octopus vulgaris. Comptes rendus hebdomadaire des séances de l'Academie des Sciences,* D, **276,** 1593–1596.

Ledrut, J. & **Ungar, G.** (1937), Action de la sécrétine chez l'*Octopus vulgaris. Archives internationales de physiologie,* **44,** 205–211.

Lettvin, J.Y., Maturana, H.R., McCulloch, W.S. & **Pitts, W.H.** (1959), What the frog's eye tells the frog's brain. *Proceedings of the Institute of Radio Engineers, New York,* **47,** 1940–1951.

Lettvin, J.Y. & Maturana, H.R. (1965), Octopus vision. *Quarterly Progress Report of the Research Laboratory of Electronics at the Massachusetts Institute of Technology,* **27,** 194–209.

Lo Bianco, S. (1909), Notizie biologiche riguardanti specialmente il periodo di maturità sessuale degli animali del Golfo di Napoli *Mitteilungen aus der zoologischen Station zu Neapel,* **19,** 513–761.

Lund, R.D. (1971), Stellate ganglion. In: *The Anatomy of the Nervous system of Octopus vulgaris.* J.Z. Young., pp. 621–640, Clarendon Press, Oxford.

Mackintosh, J. (1962), An investigation of reversal learning in *Octopus vulgaris* Lamarck. *Quarterly Journal of Experimental Psychology,* **14,** 15–22.

Mackintosh, N.J. (1965), Discrimination learning in the octopus. *Animal Behaviour,* Supplement 1, 129–134.

Mackintosh, N.J. & **Mackintosh, J.** (1963), Reversal learning in *Octopus vulgaris* Lamarck with and without irrelevant cues. *Quarterly Journal of Experimental Pyschology,* **15,** 236–242.

Mackintosh, N.J. & **Mackintosh, J.** (1964a), Performance of *Octopus* over a series of reversals of a simultaneous discrimination. *Animal Behaviour,* **12,** 321–324.

Mackintosh, N.J. & **Mackintosh, J.** (1964b), The effect of overtraining on a non-reversal shift in *Octopus. Journal of Genetic Psychology* **106,** 373–377.

Mackintosh, N.J., Mackintosh, J. & **Sutherland, N.S.** (1963), The relative importance of horizontal and vertical extents in shape discrimination by *Octopus. Animal Behaviour,* **11,** 355–358.

Maginniss, L.A. & **Wells, M.J.** (1969), The oxygen consumption of *Octopus cyanea. Journal of Experimental Biology,* **51,** 607–613.

Maldonado, H. (1963a), The visual attack system in *Octopus vulgaris. Journal of Theoretical Biology,* **5,** 470–488.

Maldonado, H. (1963b), The positive learning process in *Octopus vulgaris. Zeitschrift für vergleichende Physiologie,* **47,** 191–214.

Maldonado, H. (1963c), The general amplification function of the vertical lobe in *Octopus vulgaris. Zeitschrift für vergleichende Physiologie,* **47,** 215–229.

Maldonado, H. (1964), The control of attack by *Octopus. Zeitschrift für vergleichende Physiologie,* **47**, 656–674.

Maldonado, H. (1968), Effect of electroconvulsive shock on memory in *Octopus vulgaris. Zeitschrift für vergleichende Physiologie,* **59**, 25–37.

Maldonado, H. (1969), Further investigations on the effect of electroconvulsive shock (ECS) on memory in *Octopus vulgaris. Zeitschrift für vergleichende Physiologie,* **63**, 113–118.

Mangold, K. & Boletzky, S.V. (1973), New data on reproductive biology and growth of *Octopus vulgaris. Marine Biology,* **19**, 7–12.

Mangold-Wirz, K. (1963), Biologie des Céphalopodes benthiques et nectoniques de la Mer Catalane. *Vie et Milieu,* **13** (suppl.), 1–285.

Mangold, K. & Boucher-Rodoni, R. (1973), Role de jeûne dans l'induction de la maturation génitale chez les femelles d'*Eledone cirrhosa* (Cephalopoda: Octopoda). *Comptes rendus hebdomadaire des séances de l'Académie des Sciences,* D, **276**, 2007–2010.

Mann, T., Martin, A.W. & Thiersch, J.B. (1970), Male reproductive tract, spermatophores and spermatophoric reaction in the giant octopus of the North Pacific, *Octopus dofleini martini. Proceedings of the Royal Society,* Series B, **175**, 31–61.

Marchand, W. (1907), Studien über Cephalopoden I. Der männliche Leitung-support der Dibranchiaten. *Zeitschrift für wissenschaftliche Zoologie,* **86**, 311–615.

Marchand, W. (1913), Studien über Cephalopoden II. Ueber die Spermatophoren. *Zoologica,* **67**, 171–200.

Marshall, N.B. (1965), *The Life of Fishes.* Weidenfeld and Nicolson, London.

Marthy, H.-J. (1968), Die Organogenese des Coelomsystems von *Octopus vulgaris* Lam. *Revue suisse de Zoologie,* **75**, 723–763.

Marthy, H.-J. (1970), Beobachtungen beim Transplantieren von Organanlagen am Embryo von *Loligo vulgaris* (Cephalopoda, Decapode). *Experientia,* **26**, 160–161.

Marthy, H.-J. (1972), Sur la localisation et la stabilité du plan d'ébauches d'organes chez l'embryon de *Loligo vulgaris* (Mollusque, Céphalopode). *Comptes rendus hebdomadaire des séances de l'Académie des Sciences* D, **275**, 1291–1293.

Marthy, H.-J. (1973), An experimental study of eye development in the cephalopod *Loligo vulgaris*: determination and regulation during formation of the primary optic vesicle. *Journal of Embryology and experimental Morphology,* **29**, 347–361.

Marthy, H.-J. (1975), Organogenesis in the Cephalopoda: further evidence of blastodisc-bound developmental information. *Journal of Embryology and experimental Morphology*, **33**, 75–83.

Martin, A.W. (1965), The renopericardial canal as the reabsorptive structure of an octopus urinary tract. *American Zoologist,* **5**, 207.

Martin, A.W. (1975), Physiology of the excretory organs of cephalopods. *Fortschritte der Zoologie*, **23**, 112–123.

Martin, A.W. & Harrison, F.M. (1966), Excretion, Chapter 11. In: *Physiology of Mollusca*, Vol. 2. Wilbur, K.M. and Yonge, C.M., Eds. Academic Press, New York.

Martin, A.W., Harrison, F.M., Huston, M.G. & Stewart, D.M. (1958), The blood volume of some representative molluscs. *Journal of Experimental Biology*, **35**, 260–279.

Martin, R. (1968), Fine structure of the neurosecretory system of the vena cava in *Octopus. Brain Research*, **8**, 201–205.

Martin, R. & Barlow, J. (1971), Localisation of monoamines in nerves of the posterior salivary gland and salivary centre in the brain of *Octopus. Zeitschrift für Zellforschung und mikroskopische Anatomie*, **125**, 16–30.

Martoja, R. & May, R.M. (1956), Comparison de l'innervation brachiale des Céphalopodes *Octopus vulgaris* Lamarck et *Sepiola rondeleti* Leach. *Archives de zoologie expérimentate et générale*, **94**, 1–60.

Maturana, H.R. & Sperling, S. (1963), Unidirectional response to angular acceleration recorded from the middle cristal nerve in the statocyst of *Octopus vulgaris. Nature*, **197**, 815–816.

Matus, A.I. (1971), Fine structure of the posterior salivary gland of *Eledone cirrosa* and *Octopus vulgaris. Zeitschrift für Zellforschung und mikroskopische Anatomie*, **122**, 111–121.

Matus, A.I. (1973), Histochemical localization of biogenic monoamines in the cephalic ganglia of *Octopus vulgaris. Tissue and Cell*, **5** (4), 591–601.

Mauro, A. & Baumann, F. (1968), Electrophysiological evidence of photoreceptors in the epistellar body of *Eledone moschata. Nature*, **220**, 1332–1334.

Messenger, J.B. (1965), The peduncle lobe and associated structures in cephalopods. Ph.D. Thesis, University of London.

Messenger, J.B. (1967a), The peduncle lobe: a visuomotor centre in *Octopus. Proceedings of the Royal Society*, Series B, **167**, 225–251.

Messenger, J.B. (1967b), The effects on locomotion of lesions to the visuo-motor system of *Octopus. Proceedings of the Royal Society*, Series B, **167**, 252–281.

Messenger, J.B. (1968), Monocular discrimination of mirror images in *Octopus. Pubblicazionie della Stazione zoologica di Napoli*, **36**, 103–111.

Messenger, J.B. (1971), The optic tract lobes. In: *The anatomy of the nervous system of Octopus vulgaris*, pp. 481–506, Young, J.Z., Clarendon Press, Oxford.

Messenger, J.B. (1973a), Learning in the cuttlefish, *Sepia. Animal Behaviour*, **21**, 801–826.

Messenger, J.B. (1973b), Learning performance and brain structure: a study in development. *Brain Research*, **58**, 519–523.

Messenger, J.B. (1974), Reflecting elements in cephalopod skin and their importance for camouflage. *Journal of Zoology*, **174**, 387–395.

Messenger, J.B., Muzii, E.O., Nardi, G. & Steinberg, H. (1974), Haemocyanin synthesis and the branchial gland of *Octopus. Nature*, **250**, 154–155.

Messenger, J.B. & Sanders, G.D. (1971), The inability of *Octopus vulgaris* to discriminate monocularly between oblique rectangles. *International Journal of Neuroscience,* **1**, 171–173.

Messenger, J.B. & Sanders, G.D. (1972), Visual preference and two cue discrimination learning in octopus. *Animal Behaviour,* **20**, 580–585.

Messenger, J.B., Wilson, A.P. & Hedge, A. (1973), Some evidence for colour blindness in *Octopus. Journal of Experimental Biology,* **59**, 77–94.

Meyer, W.T. (1911), Die Spermatophore von *Polypus (Octopus) vulgaris. Zoologischer Anzeiger,* **37**, 404–5.

Mikhailoff, S. (1921), Expériences réflexologiques. Expériences nouvelles sur *Eledone moschata. Bulletin de l'Institut océanographique,* **398**, 1–11.

Mirow, S. (1972a), Skin colour in the squids *Loligo pealii* and *Loligo opalescens* I, Chromatophores. *Zeitschrift für Zellforschung und mikroskopische Anatomie,* **125**, 143–175.

Mirow, S. (1972b), Skin colour in the squids *Loligo pealii* and *Loligo opalescens* II, Iridiophores. *Zeitschrift für Zellforschung und mikroskopische Anatomie,* **125**, 176–190.

Mislin, H. (1950), Nachweis einer reflektorischen Regulation des peripheren Kreislaufs bei den Cephalopoden. *Experientia,* **6**, 467–468.

Mislin, H. (1955), Die rhythmischer Spontanentladungen in Zentralnervensystem der Tintenfische. *Acta medica scandinavica,* Suppl 307, **152**, 58.

Mislin, H. & Kauffmann, M. (1948), Der aktive Gefässpuls in der Arm-Schirmhaut der Cephalopoden. *Revue suisse de zoologie,* **55**, 267–271.

Montalenti, G. & Vitagliano, G. (1946), Richerche sul differenziamento dei sessi negli embryoni di *Sepia officinalis. Pubblicazioni della Stazione zoologica di Napoli,* **20**, 1–18.

Montuori, A. (1913), Les processus oxydatifs chez les animaux marins en rapport avec la loi de superficie. *Archives italienne de biologie,* **59**, 213–214.

Moody, M.F. (1962), Evidence for the intraocular discrimination of vertically and horizontally polarized light by *Octopus. Journal of Experimental Biology,* **39**, 21–30.

Moody, M.F. & Parriss, J.R. (1960), The discrimination of polarized light by *Octopus. Nature,* **186**, 839–840.

Moody, M.F. & Parriss, J.R. (1961), The discrimination of polarised light by *Octopus*; a behavioural and morphological study. *Zeitschrift für vergleichende Physiologie,* **44**, 268–291.

Muntz, W.R.A. (1961a), Interocular transfer in *Octopus vulgaris. Journal of Comparative and Physiological Psychology,* **54**, 49–55.

Muntz, W.R.A. (1961b), The function of the vertical lobe system of *Octopus* in interocular transfer. *Journal of Comparative and Physiological Psychology,* **54**, 186–191.

Muntz, W.R.A. (1961c), Interocular transfer in *Octopus vulgaris;* bilaterality of the engram. *Journal of Comparative and Physiological Psychology,* **54**, 192–195.

Muntz, W.R.A. (1963), Intraocular transfer and the function of the optic lobes in *Octopus. Quarterly Journal of Experimental Pyschology,* **15**, 116–124.

Muntz, W.R.A (1970), An experiment on shape discrimination and signal detection in *Octopus. Quarterly Journal of Experimental Pyschology,* **22**, 82–90.

Muntz, W.R.A., Sutherland, N.S. & Young, J.Z. (1962), Simultaneous shape discrimination in *Octopus* after removal of the vertical lobe. *Journal of Experimental Biology,* **39**, 557–566.

Naef, A. (1928). Die Cephalopoden (Embryologie). *Fauna und Flora des Golfes von Neapel und der angrenzenden Meeres-Abschnitte. Monographia,* **35**, 1–357, Friedländer und Sohn, Berlin.

Necco, A. & Martin, R. (1963), Behaviour and estimation of the mitotic activity of the white body cells in *Octopus vulgaris,* cultured *in vitro. Experimental Cell Research,* **30**, 588–590.

Nishioka, R.S., Bern, H.A. & Golding, D.W. (1970), Innervation of the Cephalopod optic gland. In: *Aspects of Neuroendocrinology,* W. Bargman & B. Scharer, Eds. Springer-Verlag, Berlin.

Nishioka, R.S., Hagadorn, I.R. & Bern, H.A. (1962), Ultrastructure of the epistellar body of the octopus. *Zeitschrift für Zellforschung und mikroskopische Anatomie,* **57**, 406–421.

Nixon, M. (1966), Changes in body weight and intake of food by *Octopus vulgaris. Journal of Zoology,* **150**, 1–9.

Nixon, M. (1969a), The time and frequency of responses by *Octopus vulgaris* to an automatic food dispenser. *Journal of Zoology,* **158**, 475–483.

Nixon, M. (1969b), Growth of the beak and radula of *Octopus vulgaris. Journal of Zoology,* **159**, 363–379.

Nixon, M. (1969c), The lifespan of *Octopus vulgaris* Lamarck. *Proceedings of the Malacological Society of London.* **38**, 529–540.

Nixon, M. (1973), Beak and radula growth in *Octopus vulgaris. Journal of Zoology,* **170**, 451–462.

O'Dor, R.K. & Wells, M.J. (1973), Yolk protein synthesis in the ovary of *Octopus vulgaris* and its control by the optic gland gonadotropin. *Journal of Experimental Biology,* **59**, 665–674.

O'Dor, R.K. & Wells, M.J. (1975), Control of yolk protein synthesis by octopus gonadotropin *in vivo* and *in vitro. General and Comparative Endocrinology,* **27**, 129–135.

Orelli, M. von (1962), Die Übertragung der Spermatophore von *Octopus vulgaris* und *Eledone. Revue suisse de zoologie,* **69**, 193–202.

Orlov, O. Yu & Byzov, A.M. (1961), Colorimetric research on the vision of molluscs (Cephalopoda) (in Russian). *Doklady Akademii nauk SSSR,* **139**, 723–725.

Orlov, O. Yu & Byzov, A.L. (1962), Vision in cephalopod molluscs (in Russian). *Priroda (Moscow),* **3**, 115–118.

Owen, R. (1832), Memoir on the pearly Nautilus (*Nautilus pompilius,* Linn.). Royal College of Surgeons, London.

Packard, A. (1961), Sucker display of *Octopus. Nature,* **190**, 736–737.

Packard, A. (1963), The behaviour of *Octopus vulgaris. Bulletin de l'Institut océanographique,* numéro spécial, **1D**, 35–49.

Packard, A. (1969), Visual acuity and eye growth in *Octopus vulgaris* (Lamarck). *Monitore zoologico italiano (N.S.),* **3**, 19–32.

Packard, A. (1972), Cephalopods and fish: the limits of convergence. *Biological Reviews,* **47**, 241–307.

Packard, A. & Albergoni, V. (1970), Relative growth, nucleic acid content and cell numbers of the brain in *Octopus vulgaris. Journal of Experimental Biology,* **52**, 539–553.

Packard, A. & Hochberg, F.G. (1977), Skin patterning in *Octopus* and other genera. *Symposia of the Zoological Society of London*, **38**, 191–231.

Packard, A. & Sanders, G.D. (1969), What the octopus shows to the world. *Endeavour,* **28**, 92–99.

Packard, A. & Sanders, G.D. (1971), Body patterns of *Octopus vulgaris* and maturation of the response to disturbance. *Animal Behaviour,* **19**, 780–790.

Packard, A. & Trueman, E.R. (1974), Muscular activity of the mantle of *Sepia* and *Loligo* (Cephalopoda) during respiratory movements and jetting, and its physiological interpretation. *Journal of Experimental Biology,* **61**, 411–419.

Parriss, J.R. (1963), Interference in learning and lesions in the visual system of *Octopus vulgaris. Behaviour,* **21**, 233–245.

Parsons, T.R. & Parsons, W. (1923), Observations on the transport of carbon dioxide in the blood of some marine invertebrates. *Journal of General Physiology,* **6**, 153–166.

Peterson, R.P. (1959), The anatomy and histology of the reproductive systems of *Octopus bimaculoides. Journal of Morphology,* **104**, 61–82.

Pickford, G.E. (1955), A revision of the Octopodinae in the collections of the British Museum. *Bulletin of the British Museum (Natural History) Zoology,* **3**, 151–167.

Pickford, G.E. & McConnaughy, B.H. (1949), The *Octopus bimaculatus* problem: a study in sibling species. *Bulletin of the Bingham Oceanographic Collection,* **12**, Article 4, 1–66.

Piéron, H. (1911), Contribution à la pyschologie du poulpe. *Bulletin de l'Institut général pyschologique,* **11**, 111–119.

Piéron, H. (1914), Contribution à la psychologie du poulpe; la mémoire sensorielle. *Année Psychologique,* **20**, 182–185.

Polimanti, O. (1913), Sui rapporti fra peso del corpo e ritmo respiratorio in *Octopus vulgaris* Lamarck. *Zeitschrift für allgemeine Physiologie,* **15**, 449–455.

Potts, W.T.W. (1965), Ammonia excretion in *Octopus dofleini. Comparative Biochemistry and Physiology,* **14**, 339–355.

Potts, W.T.W. (1967), Excretion in the Molluscs. *Biological Reviews,* **42**, 1–41.

Potts, W.T.W. & Todd, M. (1965), Kidney function in octopus. *Comparative Biochemistry and Physiology,* **16**, 479–489.

Pringle, J.W.S. (1963), The proprioceptive background to mechanisms of orientation. *Ergebnisse der Biologie,* **26**, 1–11.

Pumphrey, R.J. (1961), Concerning vision. In: *The Cell and the Organism,* Ramsay, J.A. & Wigglesworth, V.B., Eds. pp. 193–208, Cambridge University Press, Cambridge.

Ransom, W.B. (1884), On the cardiac rhythm of invertebrates. *Journal of Physiology,* **5**, 261–341.

Rees, W.J. (1950), The distribution of *Octopus vulgaris* Lamarck in British waters. *Journal of the the Marine Biological Association of the United Kingdom,* **29**, 361–378.

Rees, W.J. & **Lumby, J.R.** (1954), The abundance of *Octopus* in the English Channel. *Journal of the Marine Biological Association of the United Kingdom,* **33**, 515–536.

Rhodes, J.M. (1963), Simultaneous discrimination in *Octopus. Pubblicazioni della Stazione zoologica di Napoli,* **33**, 83–91.

Richard, A. (1967), Rôle de la photopériode dans le déterminisme de la maturation génitale femelle du Céphalopode *Sepia officinalis* L. *Comptes rendus hebdomadaire des séances de l'Académie des Sciences,* D, **265**, 1998–2001.

Richard, A. (1970a), Différentiation sexuelle des céphalopodes en culture *in vitro. Année Biologique,* **9**, 409-415.

Richard, A. (1970b), Analyse du cycle sexuel chez les céphalopodes: mise en évidence expérimentale d'une rhythme conditionné par les variations des facteurs externes et internes. *Bulletin de la Société zoologique de France,* **95**, 461–469.

Richard, A. (1971), Action qualitative de la lumière dans le déterminisme du cycle sexuel chez le Céphalopode *Sepia officinalis* L. *Comptes rendus hebdomadaire des séances de l'Académie des Sciences,* D, **272**, 106–109.

Robertson, J.D. (1949), Ionic regulation in some marine invertebrates. *Journal of Experimental Biology,* **26**, 182–200.

Robertson, J.D. (1953), Further studies on ionic regulation in marine invertebrates. *Journal of Experimental Biology,* **30**, 377–296.

Robertson, J.D. (1964), Osmotic and ionic regulation. In: *The Physiology of Molluscs,* Wilbur, K.M. & Yonge, C.M., Eds., Vol. 1, pp. 283–311, Academic Press, New York.

Robson, G.C. (1929), A monograph of the recent Cephalopoda. Part I. Octopodinae. *British Museum (Natural History) of London.*

Rocca, E. & **Ghiretti, F.** (1963), Richerche sull Emocianine, VII: Sulla capacità dell' emocianina di *Octopus vulgaris* di legare l'ossido di carboni. *Bollettino della Società italiana di biologia sperimentale,* **39**, 2075–2077.

Romanini, M.G. (1952), Osservazioni sulla ialuronidasi delle ghiandole salivari anteriori e posteriori degli octopodi. *Pubblicazioni della Stazione zoologica di Napoli,* **23**, 251–270.

Romanini, M.G. (1954), Attività mucinolitiche nel regno animal. *Revue de Biologie,* **46**, 29–93.

Rossi, F. & Graziadei, P. (1958), Nouvelles contributions à la connaissance du système nerveux du tentacule des Céphalopodes. *Acta anatomica*, **34**, Suppl. 32, 1–79.

Rowell, C.H.F. (1963), Excitatory and inhibitory pathways in the arm of *Octopus. Journal of Experimental Biology*, **40**, 257–270.

Rowell, C.H.F. (1966), Activity of interneurones in the arm of *Octopus* in response to tactile stimulation. *Journal of Experimental Biology*, **44**, 589–605.

Rowell, C.H.F. & Wells, M.J. (1961), Retinal orientation and the discrimination of polarized light by octopuses. *Journal of Experimental Biology*, **38**, 827–831.

Russell, F.E. (1965). Marine toxins and venemous and poisonous marine animals. *Advances in Marine Biology*, **3**, 255–384.

Sacarrão, G.F. (1945), Études embryologiques sur les Céphalopodes. *Arquivos do Museu Bocage*, **16**, 33–70.

Sacarrão, G.F. (1956), Contribition à l'étude du développement embryonnaire du ganglion stellaire et de la glande épistellaire endocrine des Céphalopodes. *Arquivos do Museu Bocage*, **27**, 137–152.

Sacarrão, G.F. (1961), Sur quelques aspects des rapports entre l'ontogénie et l'évolution chez les Céphalopodes (Dibranchiata). *Revista Faculdade de Ciências Universidade de Lisboa*, (C), **8**, 167–202.

Sacarrão, G.F. (1965), On the origin and development of the epistellar body of the octopus (*Octopus vulgaris* Lamarck). *Revista Faculdade de Ciências Universidade de Lisboa* (C), **13**, 215–224.

Sakaguchi, H. (1968), Studies on digestive enzymes of devilfish. *Bulletin of the Japanese Society of Scientific Fisheries*, **34**, 716–721.

Sanders, F.K. & Young, J.Z. (1940), Learning and other functions of the higher nervous centres of *Sepia. Journal of Neurophysiology*, **3**, 501–526.

Sanders, G.D. (1970a), Long-term memory of a tactile discrimination in *Octopus vulgaris* and the effect of vertical lobe removal. *Brain Research*, **20**, 59–73.

Sanders, G.D. (1970b), Long-term tactile memory in *Octopus*: further experiments on the effect of vertical lobe removal. *Brain research*, **24**, 169–178.

Sanders, G.D. (1970c), The retention of visual and tactile discriminations by *Octopus vulgaris*. Ph.D. Thesis, University of London.

Sanders, G.D. (1975), The Cephalopods. In: *Invertebrate learning*, Corning, W.C., Dyal, J.A. & Willows, A.O.D., Eds. Vol. 3, pp. 1–101, Plenum Press, New York.

Sanders, G.D. & Young, J.Z. (1974), Reappearance of specific colour patterns after nerve regeneration in *Octopus. Proceedings of the Royal Society of London*, Series B, **186**, 1–11.

Sawano, E. (1935), Contributions to the knowledge of the digestive enzymes in Marine Invertebrates. I. Proteolytic enzymes in *Polypus vulgaris* (Lamarck). Preliminary reports. *Science Reports of the Tokyo Bunrika Daigaku*, **2**, (34), 101–126.

Schiller, P.H. (1948), Studies on learning in the octopus. *Report of the Committee on Research for the National Academy of Science*, 158–160.

Schiller, P.H. (1949a), Delayed detour response in the octopus. *Journal of Comparative Physiology and Psychology,* **42**, 220–225.

Schiller, P.H. (1949b), Studies on learning in the octopus. *Yearbook of the American Philosophical Society,* **12**, 158–160.

Schipp, R. & Boletzky S. von, (1975), Morphology and function of the excretory organs in dibranchiate Cephalopods. *Fortschritte der Zoologie,* **23**, 89–111.

Schipp, R., Boletzky S. von, & Doell, G. (1975), Ultrastructural and cytochemical investigations on the renal appendages and their concrements in Dibranchiate Cephalopods (Mollusca, Cephalopoda). *Zeitschrift für Morphologie der Tiere,* **81**, 279–304.

Schipp, R., Höhn, P., & Ginkel, G. (1973), Elektronmikroskopische und histochemische Untersuchungen zur Funktion den Branchialdrüse (Parabranchialdrüse) der Cephalopoda. *Zeitschrift für Zellforschung und mikroskopische Anatomie,* **139**, 252–269.

Schöne, H. & Budelmann, B.U. (1970), Function of the gravity receptor of *Octopus vulgaris. Nature,* **226**, 864–865.

Sereni, E. (1930), The chromatophores of the cephalopods. *Biological Bulletin, Marine Biological Laboratory, Woods Hole, Mass.* **59**, 247–268.

Sereni, E. & Young, J.Z. (1932), Nervous degeneration and regeneration in Cephalopods. *Pubblicazioni della Stazione zoologica di Napoli,* **12**, 173–208.

Sheppard, C.W. & Beyl, G.E. (1951), Cation exchange in mammalian erythrocytes. III. The prolytic effect of X-rays on human cells. *Journal of General Physiology,* **34**, 691–704.

Skramlik, E. von (1941), Über den Kreislauf bei den Weichtieren. *Ergebnisse der Biologie,* **18**, 88–286.

Smith, L.S. (1962), The role of venous peristalsis in the arm circulation of *Octopus dofleini. Comparative Biochemistry and Physiology,* **7**, 269–275.

Songdahl, J.H. & Shapiro, B.I. (1974), Purification and composition of a toxin from the posterior salivary gland of *Octopus dofleini. Toxicon,* **12**(2), 109–115.

Spiess, P.E. (1972), Organogenese des Schalendrüsenkomplexes bei einigen coleoiden Cephalopoden des Mittelmeeres. *Revue suisse de zoologie,* **79**, 167–226.

Stephens, R.O. (1974). Electrophysiological studies of the brain of *Octopus vulgaris. Journal of Physiology,* **240**, 19–20P.

Stuart, A.E. (1968), The reticulo-endothelial apparatus of the lesser octopus, *Eledone cirrosa. Journal of Pathology and Bacteriology,* **96**, 401–412.

Sutherland, N.S. (1957a), Visual discrimination of orientation by *Octopus. British Journal of Psychology,* **48**, 55–71.

Sutherland, N.S. (1957b), Visual discrimination of orientation and shape by the octopus. *Nature,* **179**, 11–13.

Sutherland, N.S. (1958a), Visual discrimination of the orientation of rectangles by *Octopus vulgaris* Lamarck. *Journal of Comparative and Physiological Psychology,* **51**, 452–458.

Sutherland, N.S. (1958b), Visual discrimination of shape by *Octopus.* Squares and triangles. *Quarterly Journal of Experimental Psychology,* **10,** 40–47.

Sutherland, N.S. (1959a), A test of theory of shape discriminations in *Octopus vulgaris* Lamarck. *Journal of Comparative and Physiological Psychology,* **52,** 135–141.

Sutherland, N.S. (1959b), Visual discrimination of shape by *Octopus.* Circles and squares, and circles and triangles. *Quarterly Journal of Experimental Psychology.* **11,** 24–32.

Sutherland, N.S. (1960a), Visual discrimination of orientation by *Octopus:* mirror images. *British Journal of Psychology,* **51,** 9–18.

Sutherland, N.S. (1960b), The visual discrimination of shape by *Octopus:* squares and rectangles. *Journal of Comparative and Physiological Psychology,* **53,** 95–103.

Sutherland, N.S. (1960c), Visual discrimination of shapes by *Octopus:* open and closed forms. *Journal of Comparative and Physiological Psychology,* **53,** 104–112.

Sutherland, N.S. (1960d), Theories of shape discrimination in *Octopus. Nature,* **186,** 840–844.

Sutherland, N.S. (1961), Discrimination of horizontal and vertical extents by *Octopus. Journal of Comparative and Physiological Psychology,* **54,** 43–48.

Sutherland, N.S. (1962a), Shape discrimination by animals. *Experimental Psychology Society Monograph,* **1.**

Sutherland, N.S. (1962b), Visual discrimination of shape by *Octopus*: squares and crosses. *Journal of Comparative and Physiological Psychology,* **55,** 939–943.

Sutherland, N.S. (1963a), Shape discrimination and receptive fields. *Nature,* **197,** 118–122.

Sutherland, N.S. (1963b), Visual acuity and discrimination of stripe widths in *Octopus vulgaris* Lamarck. *Pubblicazioni della Stazione zoologica di Napoli,* **33,** 92–109.

Sutherland, N.S. (1963c), The shape discrimination of stationary shapes by octopuses. *American Journal of Psychology,* **76,** 177–190.

Sutherland, N.S. (1964), Visual discrimination in animals. *British Medical Bulletin,* **20,** 54–59.

Sutherland, N.S. (1968), Outlines of a theory of visual pattern recognition in animals and men. *Proceedings of the Royal Society,* Series B, **171,** 297–317.

Sutherland, N.S. (1969), Shape discrimination in rat, octopus and goldfish: a comparative study. *Journal of Comparative and Physiological Psychology,* **67,** 160–176.

Sutherland, N.S. & **Carr, A.E.** (1963), The visual discrimination of shape by *Octupus*: the effects of stimulus size. *Quarterly Journal of Experimental Psychology,* **15,** 225–235.

Sutherland, N.S., Mackintosh, J. & **Mackintosh, N.J.** (1963), The visual discrimination of reduplicated patterns by *Octopus. Animal Behaviour,* **11,** 106–110.

Sutherland, N.S., Mackintosh, N.J. & Mackintosh, J. (1963), Simultaneous discrimination training of octopus and transfer of discrimination along a continuum. *Journal of Comparative and Physiological Psychology,* **56**, 150–156.

Sutherland, N.S., Mackintosh, N.J. & Mackintosh, J. (1965), Shape and size discrimination in *Octopus*: the effects of pretraining along different dimensions. *Journal of Genetic Psychology,* **107**, 1–10.

Sutherland, N.S. & Muntz, W.R.A. (1959), Simultaneous discrimination training and preferred direction of motion in visual discrimination of shape in *Octopus vulgaris* Lamarck. *Pubblicazioni della Stazione zoologica di Napoli,* **31**, 109–126.

Taki, I. (1944), Studies on *Octopus* (2) Sex and the genital organ (in Japanese). *Japanese Journal of Malacology (Venus),* **13**, 267–310.

Taki, I. (1964), On the morphology and physiology of the branchial gland in Cephalopoda. *Journal of the Faculty of Fisheries and Animal Husbandry, Hiroshima University,* **5**, 345–417.

Tasaki, K. & Karita, K. (1966), Discrimination of horizontal and vertical planes of polarised light by the cephalopod retina. *Japanese Journal of Physiology,* **16**, 205–216.

Tasaki, K. & Norton, A.C. (1963), Regional and directional differences in the lateral spread of retinal potentials in the octopus. *Nature,* **198**, 1207–1208.

Tasaki, K., Oikawa, T. & Norton, A.C. (1963), The dual nature of the octopus electroretinogram. *Vision Research,* **3**, 61–73.

Ten Cate, J. (1928), Contribution à l'innervation des ventouses chez *Octopus vulgaris. Archives neérlandaises de physiologie de l'homme et des animaux,* **13**, 407–422.

Ten Cate, J. & Ten Cate, B. (1938), Les *Octopus vulgaris,* peuvent-ils discerner les formes? *Archives neérlandaises des sciences exactes et naturelles,* **23**, 541–551.

Thore, S. (1939), Beiträge zur Kenntnis der vergleichenden Anatomie des zentralen Nervensystems der dibranchiaten Cephalopoden. *Pubblicazioni della Stazione zoologica di Napoli,* **17**, 313–506.

Tinbergen, L. (1939), Zur Fortpflanzungethologie von *Sepia officinalis* L. *Archives neérlandaises de zoologie,* **3**, 323–364.

Tranter, D.J. & Augustine, O. (1973), Observations on the life history of the blue-ringed octopus, *Hapalochlaena maculosa. Marine Biology,* **18**, 115–128.

Tsukahara, Y., Tamai, M. & Tasaki, K. (1973), Oscillatory potentials of the octopus retina. *Proceedings of the Japanese Academy* **49**, 57.

Turchini, J. (1923), Contribution à l'étude de l'histologie de la cellule rénale. L'excrétion urinaire chez les Mollusques. *Archives de morphologie générale et expérimentale,* **18**, 7–253.

Twarog, B.M. (1954), Effects of acetylcholine and 5-hydroxytryptamine on the contraction of a molluscan smooth muscle. *Journal of Physiology,* **152**, 236–242.

Uexküll, J. von (1894), Physiologische Untersuchungen an *Eledone moschata.* 2. Die Reflexe des Armes. *Zeitschrift für Biologie,* **30**, 179–184.

Uexküll, J. von (1895), Physiologische Untersuchungen an *Eledone moschata*. 4. Zur Analyse der Funktionen des Centralnervensystems. *Zeitschrift für Biologie,* **31**, 584–609.

Van Heukelem, W.F., (1966), Ecology and ethology of *Octopus cyanea* Gray. Master's thesis, University of Hawaii.

Van Heukelem, W.F., (1973), Growth and lifespan of *Octopus cyanea* (Mollusca: Cephalopoda). *Journal of Zoology,* **169**, 299–315.

Vernon, H.M. (1896), The respiratory exchange of lower marine invertebrates. *Journal of Physiology,* **19**, 18–70.

Vevers, H.G. (1961), Observations on the laying and hatching of octopus eggs in the society's aquarium. *Proceedings of the Zoological Society of London,* **137**, 311–315.

Voss, G.L. (1977), Present status and new trends in cephalopod systematics. The Biology of Cephalopods, eds. Nixon, M. & Messenger, J.B. *Symposia of the Zoological Society of London,* **38**, 49–60.

Walker, J.J., Longo N. & Bitterman, M.E. (1970), The octopus in the laboratory; handling, maintenance and training. *Behaviour Research Methods & Instrumentation,* **2**, 15–18.

Wallace, R.A. & Jared, D.W. (1969), Studies of amphibian yolk, VIII. Estrogen-induced hepatic synthesis of a serum lipophosphoprotein and its selective uptake by ovary and transformation into yolk platelet proteins by *Xenopus laevis. Developmental Biology,* **19**, 498–526.

Ward, D.V. & Wainwright, S.A. (1972), Locomotory aspects of squid mantle structure. *Journal of Zoology,* **167**, 437–449.

Watkinson, G.D. (1909), Untersuchungen über die sogennanten Geruchsorgane der Cephalopoden. *Jenaische Zeitschrift für Naturwissenschaft,* **37**, 353–414.

Weel, P.B. van & Thore, S. (1936), Ueber die pupillarreaktion von *Octopus vulgaris. Zeitschrift für vergleichende Physiologie,* **23**, 26–33.

Wells, M.J. (1958), Factors affecting reactions to *Mysis* by newly hatched *Sepia. Behaviour,* **13**, 96–111.

Wells, M.J. (1959a), Functional evidence for neurone fields representing the individual arms within the central nervous system of *Octopus. Journal of Experimental Biology,* **36**, 501–511.

Wells, M.J. (1959b), A touch learning centre in *Octopus. Journal of Experimental Biology,* **36**, 590–612.

Wells, M.J. (1960a), Proprioception and visual discrimination of orientation in *Octopus. Journal of Experimental Biology,* **37**, 489–499.

Wells, M.J. (1960b), Optic glands and the ovary of *Octopus. Symposia of the Zoological Society of London,* **2**, 87–101.

Wells, M.J. (1961a), Weight discrimination by *Octopus. Journal of Experimental Biology,* **38**, 127–133.

Wells, M.J. (1961b), Centres for tactile and visual learning in the brain of *Octopus. Journal of Experimental Biology,* **38**, 811–826.

Wells, M.J. (1962), Early learning in *Sepia. Symposia of the Zoological Society of London.* **8**, 149–169.

Wells, M.J. (1963a), The orientation of *Octopus. Ergebnisse der Biologie,* **26**, 40–54.

Wells, M.J. (1963b), Taste by touch: some experiments with *Octopus. Journal of Experimental Biology,* **40**, 187–193.

Wells, M.J. (1964a), Tactile discrimination of surface curvature and shape by octopuses. *Journal of Experimental Biology,* **41**, 435–445.

Wells, M.J. (1964b), Detour experiments with octopuses. *Journal of Experimental Biology,* **41**, 621–642.

Wells, M.J. (1964c), Tactile discrimination of shape by *Octopus. Quarterly Journal of Experimental Psychology,* **16**, 156–162.

Wells, M.J. (1965), The vertical lobe and touch learning in the octopus. *Journal of Experimental Biology,* **42**, 233–255.

Wells, M.J. (1966a), Learning in the octopus. *Symposia of the Society for Experimental Biology,* **20**, 477–507.

Wells, M.J. (1966b), Cephalopod sense organs. In: *The Physiology of Mollusca,* Wilbur, K.M. & Yonge, C.M., Eds., Vol. 2, pp. 523–545, Academic Press, New York.

Wells, M.J. (1967), Short-term learning and interocular transfer in detour experiments with octopuses. *Journal of Experimental Biology,* **47**, 393–408.

Wells, M.J. (1968), Sensitization and the evolution of associative learning. In: *Neurobiology of Invertebrates,* Salanki, J., Ed., pp. 391–411, Plenum Press, New York.

Wells, M.J. (1970), Detour experiments with split-brain octopuses. *Journal of Experimental Biology,* **53**, 375–389.

Wells, M.J. (1974), A location for learning. In: *Essays on the Nervous System,* A Festschrift for Professor J.Z. Young, Bellairs, R. & Gray, E.G., Eds., pp. 407–430. Clarendon Press, Oxford.

Wells, M.J. (1975), Evolution and associative learning. In *'Simple' Nervous Systems,* Usherwood, P.N.R. & Newth, D.R., Eds., pp. 446–473, Edward Arnold, London.

Wells, M.J. (1976a), *In vitro* assays for octopus gonadotropin. In: *Actualités sur les hormones d'invertébrés.* Durchon, M.M., Ed., Colloques internationaux du centre national de la recherche scientifique, No. 251, 149–159, Editions CNRS, Paris.

Wells, M.J. (1976b), Proprioception and learning. In: *Structure and Function of the Proprioceptors of Invertebrates.* P.J. Mill, Ed., pp. 567–603, Chapman and Hall, London.

Wells, M.J., Freeman, N.H. & Ashburner, M. (1965), Some experiments on the chemotactile sense of octopuses. *Journal of Experimental Biology,* **43**, 553–563.

Wells, M.J., O'Dor, R.K. & Buckley, S.K.L. (1975), An *in vitro* bioassay for a molluscan gonadotropin. *Journal of Experimental Biology,* **62**, 433–446.

Wells, M.J. & **Wells, J.** (1956). Tactile discrimination and the behaviour of blind *Octopus. Pubblicazioni della Stazione zoologica di Napoli,* **28**, 94–126.

Wells, M.J. & **Wells, J.** (1957a), The function of the brain of *Octopus* in tactile discrimination. *Journal of Experimental Biology,* **34**, 131–142.

Wells, M.J. & **Wells, J.** (1957b), The effect of lesions to the vertical and optic lobes on tactile discrimination in *Octopus. Journal of Experimental Biology,* **34**, 378–393.

Wells, M.J. & **Wells, J.** (1957c), Repeated presentation experiments and the function of the vertical lobe in *Octopus. Journal of Experimental Biology,* **34**, 469–477.

Wells, M.J. & **Wells, J.** (1958a), The influence of preoperational training on the performance of octopuses following vertical lobe removal. *Journal of Experimental Biology,* **35**, 324–336.

Wells, M.J. & **Wells, J.** (1958b), The effect of vertical lobe removal on the performance of octopuses in retention tests. *Journal of Experimental Biology,* **35**, 337–348.

Wells, M.J. & **Wells, J.** (1959), Hormonal control of sexual maturity in *Octopus. Journal of Experimental Biology,* **36**, 1–33.

Wells, M.J. & **Wells, J.** (1969), Pituitary analogue in the octopus. *Nature,* **222**, 293–294.

Wells, M.J. & **Wells, J.** (1970), Observations on the feeding, growth rate and habits of newly settled *Octopus cyanea. Journal of Zoology,* **161**, 65–74.

Wells, M.J. & **Wells, J.** (1972a), Sexual displays and mating of *Octopus vulgaris* Cuvier and *Octopus cyanea* Gray and attempts to alter performance by manipulating the glandular condition of the animals. *Animal Behaviour,* **20**, 293–308.

Wells, M.J. & **Wells, J.** (1972b), Optic glands and the state of the testis in *Octopus. Marine Behaviour and Physiology,* **1**, 71–83.

Wells, M.J. & **Wells, J.** (1975), Optic gland inplants and their effects on the gonads of *Octopus. Journal of Experimental Biology,* **62**, 579–588.

Wells, M.J. & **Wells, J.** (1977), Cephalopoda: Octopoda. In: *Reproduction in Marine Invertebrates*, Giese, A.C. & Pearse, J.S., Eds., Vol. 4, Academic Press, New York, (in press).

Wells, M.J. & **Young, J.Z.** (1965), Split-brain preparations and touch learning in the octopus. *Journal of Experimental Biology,* **43**, 565–579.

Wells, M.J. & **Young, J.Z.**(1966), Lateral interaction and transfer in the tactile memory of the octopus. *Journal of Experimental Biology,* **45**, 383–400.

Wells, M.J. & **Young, J.Z.** (1968a), Changes in textural preferences in *Octopus* after lesions. *Journal of Experimental Biology,* **49**, 401–412.

Wells, M.J. & **Young, J.Z.** (1968b), Learning with delayed rewards in *Octopus. Zeitschrift für vergleichende Physiologie.* **61**, 103–128.

Wells, M.J. & **Young, J.Z.** 1969a), Learning at different rates of training in the octopus. *Animal Behaviour,* **17**, 406–415.

Wells, M.J. & Young, J.Z. (1969b). The effect of splitting part of the brain or removal of the median inferior frontal lobe on touch learning in *Octopus. Journal of Experimental Biology,* **50**, 515–526.

Wells, M.J. & Young, J.Z. (1970a), Single session learning by octopuses. *Journal of Experimental Biology,* **53**, 779–788.

Wells, M.J. & Young, J.Z. (1970b), Stimulus generalisation in the tactile system of *Octopus. Journal of Neurobiology,* **2**, 31–46.

Wells, M.J. & Young, J.Z. (1973), The median inferior frontal lobe and touch learning in the octopus. *Journal of Experimental Biology,* **56**, 381–402.

Wells, M.J. & Young, J.Z. (1975), The subfrontal lobe and touch learning in the octopus. *Brain research,* **92**, 103–121.

Wersäll, J., Flock, A. & Lundquist, P.G. (1965), Structural basis for directional sensitivity in cochlear and vestibular sensory receptors. *Cold Spring Harbour Symposia on Quantitative Biology,* **30**, 115–132.

Westerman, G.E.G. (1971), Form, structure and function of shell and siphuncle in coiled Mesozoic ammonoids. *Life Sciences Contribution, Royal Ontario Museum,* **78**.

Wilson, D.M. (1960), Nervous control of movement in cephalopods. *Journal of Experimental Biology,* **37**, 57–72.

Winterstein, H. (1909), Zur Kenntnis der Blutgase wirbelloser Seetiere. *Biochemische Zeitschrift,* **19**, 384–424.

Winterstein, H. (1925), Über die chemische Regulierung der Atmung bei den Cephalopoden. *Zeitschrift für vergleichende Physiologie,* **2**, 315–328.

Wirz, K. (1954), Études quantatives sur le système nerveux des Céphalopodes. *Comptes Rendus de l'Académie des sciences,* **238**, 1353–1355.

Witmer, A. & Martin, A.W. (1973), The fine structure of the branchial heart appendage of the cephalopod *Octopus dofleini martini. Zeitschrift für Zellforschung,* **136**, 545–568.

Wodinsky, J. (1969), Penetration of the shell and feeding on gastropods by *Octopus. American Zoologist,* **9**, 997–1010.

Wolff, H.G. & Budelmann, B.U. (1977), Properties of angular acceleration receptors in *Octopus vulgaris* (in preparation).

Wolvekamp, H.P. (1938), Über den Sauerstofftransport durch Hämocyanin von *Octopus vulgaris* L. und *Sepia officinalis* Lam. *Zeitschrift für vergleichende Physiologie,* **25**, 541–547.

Wolvekamp, H.P., Baerends, G.P., Kok, B. & Mommaerts, W.F.H.M. (1942). O_2- and CO_2-binding properties of the blood of the cuttlefish (*Sepia officinalis*) and the common squid (*Loligo vulgaris* Lam.). *Archives neérlandaises de physiologie de l'homme et des animaux,* **26**, 203–218.

Wood, F.G. (1963), Observations on the behaviour of *Octopus. International Congress of Zoology,* **16**, (1), 73.

Woodhams, P.L. (1977), The ultrastructure of a cerebellar analogue in *Octopus. Journal of Comparative Neurology,* **174**, 329–346.

Woodhams, P.L. & Messenger, J.B. (1974), A note on the ultrastructure of the *Octopus* olfactory organ. *Cell and Tissue Research,* **152**, 253–258.

Yamamoto, T., Tasaki, K., Sugawara, Y. & Tonosaki, A. (1965), Fine structure of the octopus retina. *Journal of Cell Biology,* **25**(2), 345–359.

Yarnall, J.L. (1969), Aspects of the behaviour of *Octopus cyanea* Gray. *Animal Behaviour,* **17**, 747–754.

Young, J.Z. (1936), The giant nerve fibres and epistellar body of cephalopods. *Quarterly Journal of Microscopical Science,* **78**, 367–386.

Young, J.Z. (1951), Growth and plasticity in the nervous system. *Proceedings of the Royal Society,* Series B, **139**, 18–37.

Young, J.Z. (1958a), Effect of removal of various amounts of the vertical lobes on visual discrimination by *Octopus. Proceedings of the Royal Society,* Series B, **149**, 441–462.

Young, J.Z. (1958b), Responses of untrained octopuses to various figures and the effect of removal of the vertical lobe. *Proceedings of the Royal Society,* Series B, **149**, 463–483.

Young, J.Z. (1960a), Regularities in the retina and optic lobes of *Octopus* in relation to form discrimination. *Nature,* **186**, 936–845.

Young, J.Z. (1960b), The statocysts of *Octopus vulgaris. Proceedings of the Royal Society*, Series B, **152**, 3–29.

Young, J.Z. (1960c), Unit processes in the formation of representations in the memory of *Octopus. Proceedings of the Royal Society,* Series B, **153**, 1–17.

Young, J.Z. (1960d), The failure of discrimination learning following the removal of the vertical lobes in *Octopus. Proceedings of the Royal Society,* Series B, **153**, 18–46.

Young, J.Z. (1961a), Learning and discrimination in the octopus. *Biological Reviews,* **36**, 32–96.

Young, J.Z. (1961b), Rates of establishment of representations in the memory of octopuses with and without vertical lobes. *Journal of Experimental Biology,* **38**, 43–60.

Young, J.Z. (1962a), The retina of Cephalopods and its degeneration after optic nerve section. *Philosophical Transactions of the Royal Society,* Series B, **245**, 1–18.

Young, J.Z. (1962b), The optic lobes of *Octopus vulgaris. Philosphical Transactions of the Royal Society,* Series B, **245**, 19–58.

Young, J.Z. (1962c), Repeated reversal of training in *Octopus. Quarterly Journal of Experimental Psychology,* **14**, 206–222.

Young, J.Z. (1963a), The number and sizes of nerve cells in *Octopus. Proceedings of the Zoological Society of London,* **140**, 229–254.

Young, J.Z. (1963b), Light and dark adaptation in the eyes of some cephalopods. *Proceedings of the Zoological Society of London,* **140**, 255–271.

Young, J.Z. (1963c), Some essentials of neural memory systems. Paired centres that regulate and address the signals of results of actions. *Nature,* **198**, 626–630.

Young, J.Z. (1964a), *A Model of the Brain.* Clarendon Press, Oxford.

Young, J.Z. (1964b), Paired centres for the control of attack by *Octopus. Proceedings of the Royal Society,* Series B, **159**, 565–588.

Young, J.Z. (1965a), Two memory stores in one brain. *Endeavour,* **24**, 13–20.

Young, J.Z. (1965b), The nervous pathways for poisoning, eating, and learning in *Octopus. Journal of Experimental Biology,* **43**, 581–593.

Young, J.Z. (1965c), The central nervous system of *Nautilus. Philosophical Transactions of the Royal Society,* Series B, **249**, 1–25.

Young, J.Z. (1965d), The diameters of the fibres of the peripheral nerves of *Octopus. Proceedings of the Royal Society,* Series B, **162**, 47–79.

Young, J.Z. (1965e), The organization of a memory system. *Proceedings of the Royal Society,* Series B, **163**, 285–320.

Young, J.Z. (1967), The visceral nerves of *Octopus. Proceedings of the Royal Society,* Series B, **253**, 1–22.

Young, J.Z. (1968), Reversal of a visual preference in *Octopus* after removal of the vertical lobe. *Journal of Experimental Biology,* **49**, 413–419.

Young, J.Z. (1970a), Short and long memories in *Octopus* and the influence of the vertical lobe system. *Journal of Experimental Biology,* **52**, 385–393.

Young, J.Z. (1970b), Neurovenous tissues in cephalopods. *Philosophical Transactions of the Royal Society,* Series B, **257**, 309–321.

Young, J.Z. (1971), *The Anatomy of the Nervous System of Octopus vulgaris.* Clarendon Press, Oxford.

Young, J.Z. (1972), The organisation of a cephalopod ganglion. *Philosophical Transactions of the Royal Society,* Series B, **263**, 409–429.

Yung, Ko Ching, M. (1930), Contribution à l'étude cytologique de l'ovogénèse, du développement et de quelques organes chez les Céphalopodes. *Annales de l'Institut Océanographique Monaco,* **7** (8), 300–364.

Author index

Subject index